THE RESURRECTION OF THE SHROUD

Sandra A. French

MARK ANTONACCI

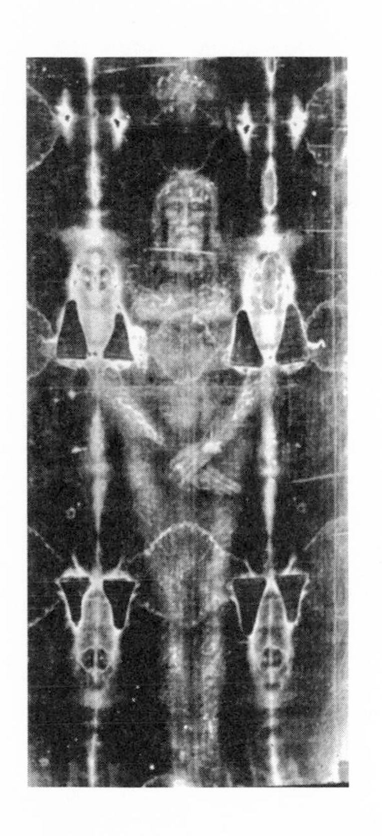

M. EVANS AND COMPANY, INC.
NEW YORK

M. Evans and Company, Inc.
216 East 49th Street
New York, New York 10017

Library of Congress Cataloging-in-Publication Data

Antonacci, Mark, 1949–
The resurrection of the Shroud / Mark Antonacci.
p. cm.
Includes bibliographical references and index.
ISBN 0-87131-890-3
1. Holy shroud. I. Title.
BT587.S4 A58 1999
232.96'6—dc21 99-046809

Book design by Rik Lain Schell

Printed in the United States of America

9 8 7 6 5 4 3 2 1

CONTENTS

To the memories of
Nick J. Antonacci and Louise DeLong

To Mary Rose,
the love and joy of my life.

To Joe Marino,
who helped me in every way possible.

To Art Lind,
the smartest and nicest scientist I ever met.

To Debbie Bauer and Esther Noble,
the best lawyer, landlady, business partners and
friends I could have.

To Dick Nieman, John Schulte, Chuck Hampton and
Ed Cherbonnier, whose concern for others
invariably exceeds their own.

To Mom, Dad, Mike and Steve,
from whom I benefited far more than they know.

To Dave and Kay,
who are like family.

To Nick, Barb, Kelly, Ryan, Brandon and future generations.

PREFACE

Many people have asked me how I first became interested in the Shroud of Turin. As anyone who knew me before would tell you, I was an unlikely candidate to show such an interest. Not only that, I never had an inkling or desire to write a book, let alone a book on a subject like this.

It actually began out of nowhere on a Saturday afternoon. I had just had a serious argument with a girlfriend. The problem was that she was religious, and I was not. She was very committed to her religious views, and I was definitely a committed agnostic. Because it was over a matter that threatened our relationship, it caused me great anxiety. Not wanting to hang around the apartment and continue to think about it all day, I decided to go to work at my law office to take my mind off the argument.

When I got to my office, I couldn't concentrate on my work. Hoping to get my mind off the disagreement, I decided to go across the street and catch some lunch at Hummel's, my favorite restaurant, and read the paper. On the way, I happened to buy a particular newspaper for the only time in my life.

As I was moving down the serving line in Hummel's with the newspaper on my tray, I couldn't help but notice the words across the top of the paper, advertising what was inside the weekend edition. In the upper right-hand corner was a facial picture of the man in the Shroud; next to it, a caption asked if this was a picture of the historical Jesus Christ. The paper was featuring an article on a recently published book about the Shroud of Turin. The picture was staring up right at me, and as I went down the line, I muttered to myself, "Great, this is just what I need."

I sat down and began eating. As I was scanning the sports page, out of the corner of my eye I noticed the same picture on the front page, looking at me. I did not like seeing this picture and being reminded of the subject and the argument, so I pushed the paper a little bit away from me. As I continued reading, I noticed the picture again and pushed the paper farther away still. I did the same thing another time until, noticing the picture yet again, I finally said, "All right, I'll read the !*$#?% article," and angrily threw down the sports page and picked up the article on the Shroud.

The article told what a group of scientists from various professional and personal backgrounds had found after examining the entire Shroud—the cloth and its full-length front and back images. These scientists had the most sophisticated equipment available and were from some of the most prestigious scientific institutions in the world. With such an array of sophisticated equipment, many of them thought they'd be able to prove in twenty minutes that the Shroud was the work of an artist. However, that is not what they found at all.

The article summarized some of the main findings from the book and the scientists' comprehensive examination. As I began contemplating some of the possible implications of these findings, I found myself back at my apartment, pacing anxiously back and forth across two rooms. These findings and their possible implications threatened my entire philosophical foundation. While I was thinking and pacing for what seemed like a couple of hours, it suddenly hit me in midstep: if all of the possible implications from the scientific examination were true, it would not be bad news—it would be good news.

I immediately stopped pacing and sat down. I looked at the subject without feeling threatened or uncomfortable. I purchased the book featured in the article, as well as other books and articles on the subject. In the following years, I acquired access to practically everything that has ever been printed in the English language on the Shroud, as well as translations of major foreign publications. I have always been interested in history and am an attorney by training, and as I seriously considered the potential implications of all of the evidence, I realized this one matter could be more important than all of the cases that had ever been argued before the Supreme Court.

As luck would have it, the paper that I purchased soon ceased publication, and it turned out to be the only one in the region that featured an article on the Shroud for the next few years. This incident, which occurred almost twenty years ago, is how I literally stumbled onto the subject. What, you may ask, ever happened to the girlfriend and me? We had numerous other disagreements about other subjects and soon concluded neither one was good for the other. Yet, if it weren't for her and the argument, I never would have stumbled onto the article or, if I had, would probably never have read it.

Mark Antonacci

CHAPTER ONE

RESURRECTING THE SHROUD

The Shroud of Turin is either the most awesome and instructive relic of Jesus Christ in existence . . . or it is one of the most ingenious, most unbelievably clever products of the human mind and hand on record. It is one or the other; there is no middle ground.

—John Walsh, *The Shroud*

The Shroud of Turin, a large linen cloth 14 feet 3 inches long and 3 feet 7 inches wide (4.34 m x 1.10 m), takes its present name from the city where it has been kept for the last 422 years—Turin, Italy. In ancient times, burial shrouds were wrapped lengthwise around a body as shown below. Jesus would have been wrapped and buried in such a linen shroud.

The Shroud of Turin purports to contain evidence of some of the most startling events in all of history: that a man who was beaten and scourged, his head pierced repeatedly about the crown; a man who was crucified, pierced in the side, from which blood and water flowed, a man who was dead and buried—that this man was resurrected; and, further, that he was Jesus Christ.

Unlike any other burial garment, this cloth contains the front and back images of the body of a man, which can be seen in the photographs on the following pages.

Fig. 1. Part of a sixteenth-century painting by della Rovere of the Shroud wrapped over the body of Christ.

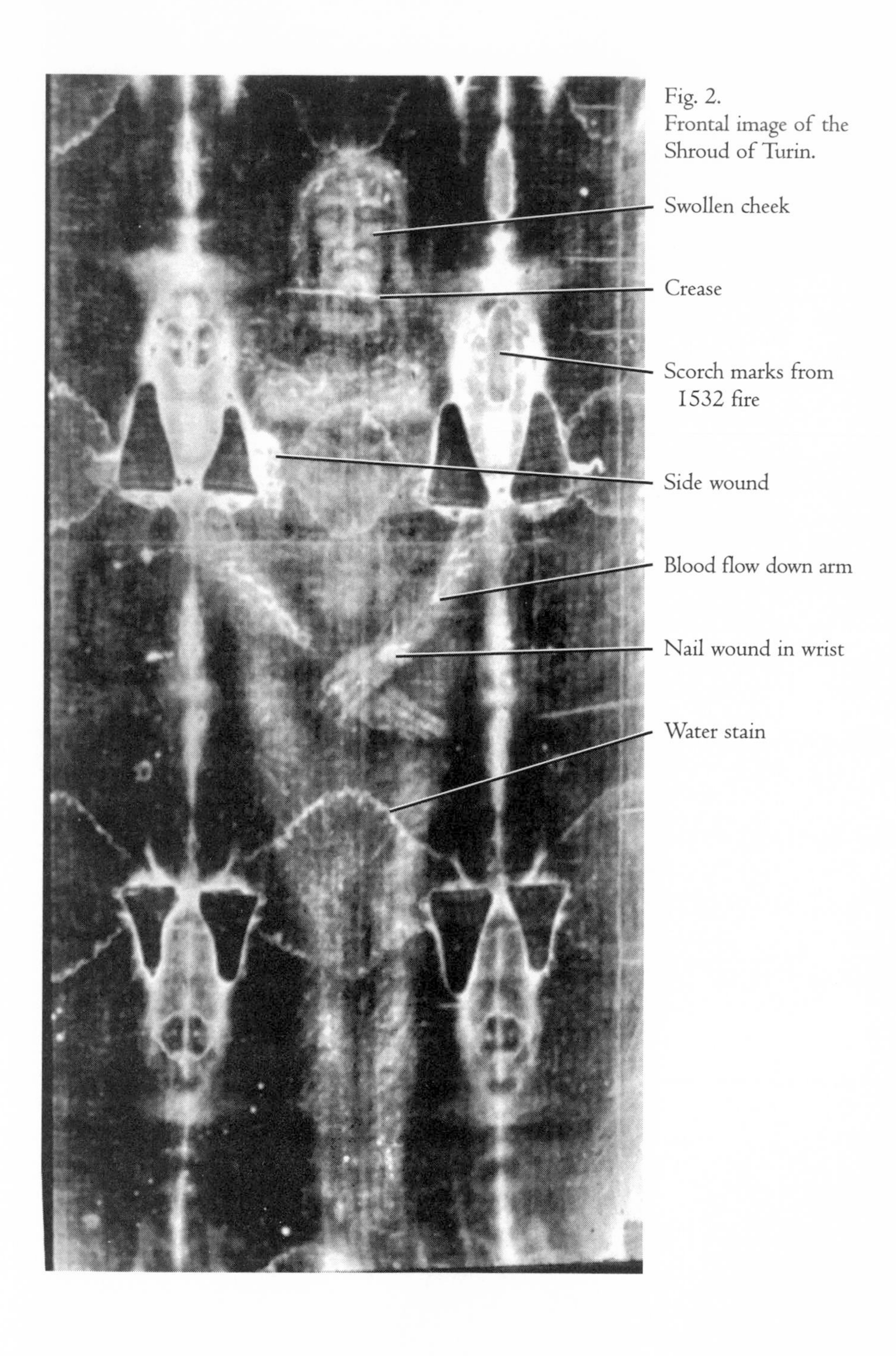

Fig. 2.
Frontal image of the Shroud of Turin.

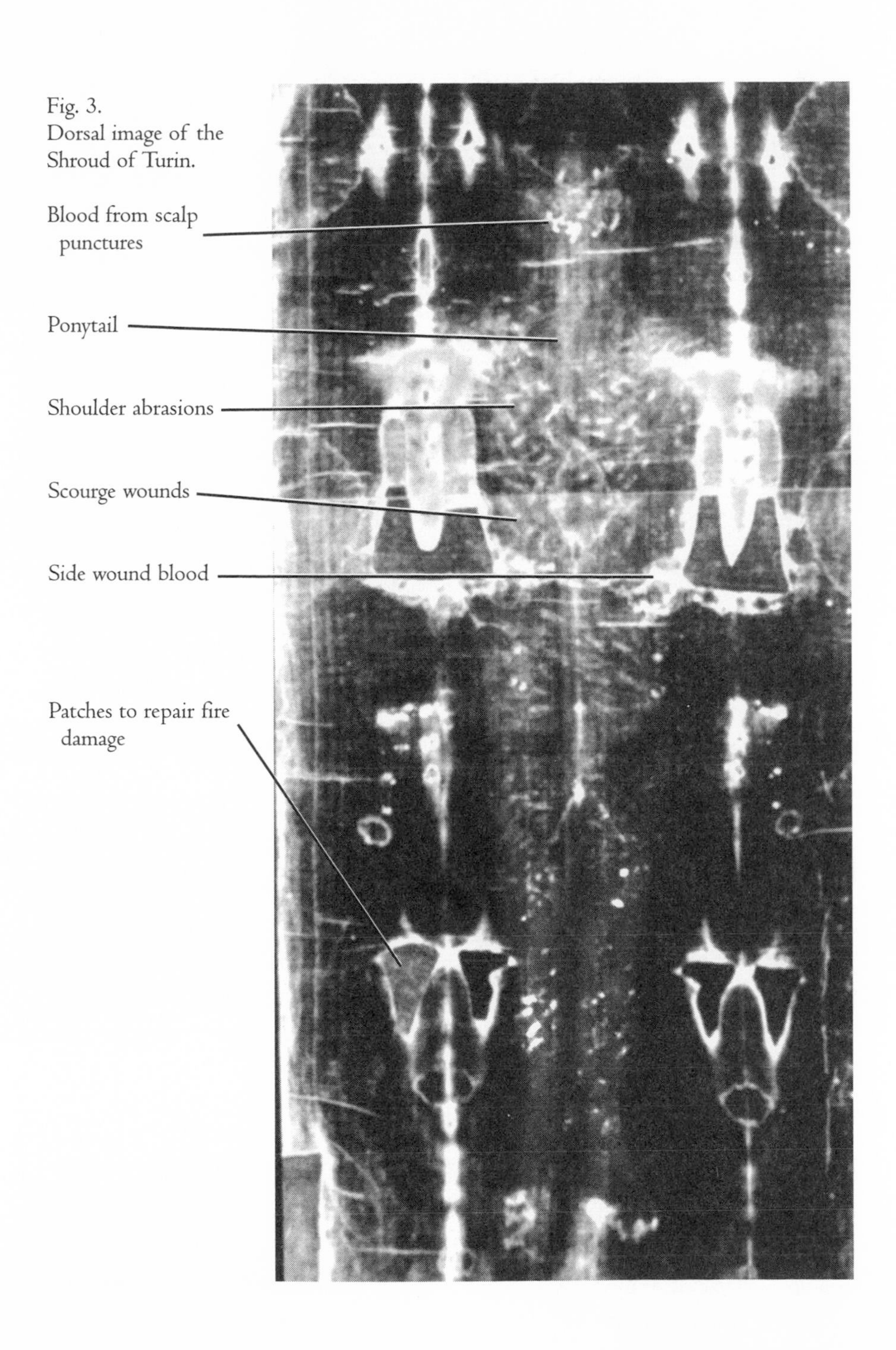

Fig. 3.
Dorsal image of the Shroud of Turin.

Of course, we might be tempted to dismiss such claims as colorful legends from the far reaches of the past, assumptions carried forth by our own philosophical desires. But the passage of time and the expansion of our knowledge about the world around us have, surprisingly, borne out these claims to greater and greater degrees. As the greatest advances in our knowledge have come about with increasing intensity during the twentieth century, so, too, has our knowledge about the enigmatic Shroud of Turin.

PREVIOUS SCIENTIFIC TREATMENT OF THE SHROUD

It was near the turn of the twentieth century that scientists first began to take an interest in the Shroud of Turin. This resulted from the first photographs ever taken of the Shroud, by an Italian photographer named Secondo Pia, in 1898.[1] In 1900, Yves Delage, a professor of anatomy at the Sorbonne and a director of the Museum of Natural History, showed his assistant, Paul Vignon, the Pia photographs and encouraged him to begin an investigation of the Shroud. Delage was an agnostic whose range of scientific interest and expertise included many fields. The photos disturbed Vignon so much that he agreed to undertake the investigation. This would be the beginning of a lifetime of investigation of the Shroud for Vignon, whose knowledge of biology and art would lend itself perfectly to this study. Delage lent Vignon his supervision and assistance for this investigation and put his laboratory at Vignon's disposal.

From 1900 to 1902, Vignon and Delage, assisted by three other scientists, undertook their investigation, conferring at every stage and agreeing on their conclusions. Though the investigation did not have access to the Shroud itself, it yielded some startling results, and Delage was emboldened to give a full-scale report to no less an audience than the French Academy of Science, the foremost scientific body in the world at the time. Aware of the subject of the report, a capacity crowd of academics and members of the general public gathered on April 21, 1902, for the regular meeting of the academy to hear Delage speak. Not since Louis Pasteur had reported on the vaccine for the cure of rabies had the long narrow hall of the academy accommodated so large a crowd.[2]

Delage's talk would last about half an hour as he reviewed the findings of the investigation. Among the findings, the scientists determined that the images were those of a dead human male. They found that the images could not have resulted from painting but involved direct and indirect contact between the cloth and the body. The investigators went so far as to identify the body as being that of the historical Jesus and to declare the Shroud his burial garment. Delage concluded his presentation by proposing that the academy itself appoint a special committee to petition the Italian authorities for a more thorough examination of this amazing object.[3]

In contrast to the customary conversation during presentations at the academy, the audience gave Delage their rapt attention. Upon his conclusion, however, a mur-

mur of excited discussion broke out among the spectators on the public benches. After examining the glass photographic plates, members of the academy admitted that the images on the Shroud could not have been the work of an artist.[4]

When the meeting adjourned, the president of the academy announced that a secret committee would go into session immediately. At that time, the academy was composed, to a great extent, of religious skeptics whose work and interests were in completely different areas from the subject of the presentation. In fifteen minutes, the committee returned to decline Delage's invitation to request a more extensive examination of the Shroud on the grounds that the actual owner of the Shroud, the royal house of Italy, had not made the request. They further stated that Delage's proposal was beyond the scope of the academy.[5]

The permanent secretary of the academy, Pierre Berthelot, was a leader of the "free thought" school in France. Knowing beforehand the results of Delage and Vignon's investigation, he attempted to prevent Delage from even presenting his report to the academy; however, he was overruled. Thereafter, he abused his authority as secretary by eliminating from the journal of the academy any part of Delage's lecture that even mentioned the Shroud or Delage's reasons for holding that the images were of the historical Jesus Christ.[6]

From Berthelot and others who behaved like him, Vignon and Delage received a great deal of criticism and abuse, particularly Delage, because he was a prominent scientist and professor. He was derided as a traitor to his agnosticism and to science. In a response to his critics, Delage published an open letter to the editors of the *Revue Scientifique.* It contained a statement of the facts from the investigation and included the portions of his lecture that Berthelot had excluded from the journal of the academy. Thereafter, he addressed the accusations against his integrity as a scientist: "I recognize Christ as a historical personage, and I see no reason why anyone should be scandalized by the fact that there still exists material traces of his earthly life."[7]

He then pointed out that the real reasons for the criticisms actually had nothing at all to do with science. "If our proofs have not been received by certain persons as they deserve to be, it is only because a religious question has been needlessly injected into a problem which in itself is purely scientific, with the result that feelings run high and reason has been led astray. If, instead of Christ, there were questions of some person like a Sargon or Achilles, or one of the Pharaohs, no one would have thought of making any objection."[8]

The numerous awards, achievements, and honors that Delage achieved during his lifetime have now been forgotten, but his part in the scientific investigation and his dedication as a scientist remain as an example for the world. "I have been faithful to the true spirit of Science in treating this question, intent only on the truth, not concerned in the least whether the truth would affect the interests of any religious party. There are those, however, who have let themselves be swayed by this consideration and have betrayed the scientific method."[9]

Those would be the last words that Delage ever published on the subject, about which he never recanted. Immediately following Delage's lecture at the Academy, newspaper and magazine accounts of his presentation and articles on the Shroud appeared all over Paris. Numerous books were soon published throughout France, the most notable of which was by Vignon, containing a fuller explanation of the investigation into the Shroud. About half the publications favored the Shroud's authenticity as declared by the investigation. Similar reactions came from the rest of the world, though the response was less vociferous.

Regrettably, the debate remained unsettled. All efforts to conduct a thorough scientific examination of the cloth itself failed and would not occur for another seventy-five years. Gradually, the debate and the relic became forgotten outside of continental Europe. The reasons behind both sides of the debate, along with a much larger set of evidence than was ever present at that time, will be developed in the chapters that follow. The important thing to note about this debate, however, was that it was based on very little information. This was the first scientific investigation ever undertaken concerning the Shroud of Turin, and it was based on the first photographs ever taken of the Shroud. Three fourths of the twentieth century would pass, years in which horrors of an unprecedented scale would result in the deaths of more than 100 million people throughout the world. Two world wars, the Holocaust, massive killings by nation's rulers of their own residents, and numerous wars and conflicts, some of which are continuing today, in which religion is an underlying element or contributing cause, would all take place before a comprehensive, scientific examination of the cloth would finally occur in 1978.

The overwhelming majority of evidence supporting the claims that the man in the Shroud is Jesus Christ has only recently come to light. In fact, some of the most astonishing aspects of it were discovered only within the past few years. Most people are still completely unaware of this evidence.

Only in the last twenty years has the Shroud gained widespread publicity, almost exclusively in the western world. Throughout its history, it has rarely been displayed to the public. Prior to 1978, it had not been publicly displayed in the twentieth century except in the early 1930s. And it was not until 1978, when the first comprehensive scientific examination of the entire cloth was performed, that the most intriguing and, in some cases, supernatural qualities of the Shroud image surfaced. This first examination was conducted by a team of esteemed scientists who called themselves the Shroud of Turin Research Project (STURP).

STURP was formed after some of its members discovered that an actual three-dimensional image could be derived from the Shroud image contained on the two-dimensional cloth, a quality impossible to achieve with traditional painting and photographic methods. In 1902, scientist Paul Vignon first observed that a correlation existed between light and dark on the Shroud image and those parts of the cloth that would have been closest to and farthest from the body. Not until 1976 had space-age computer technology advanced to the point that this phenomenon could be tested.

When images of planets and moons in outer space are produced, they are not recorded with a traditional camera. The photons of light from a three-dimensional object in outer space are measured electronically, and based on the distance that each of these light impulses has traveled from the object to the space probe, a true, three-dimensional, computer-generated image of the object can be interpreted and displayed. When a photograph of the man in the Shroud is put in a similar device, a miraculous three-dimensional image results.

The three-dimensional image is produced because the lightness or darkness on the image of the man in the Shroud is directly correlated with how close the body was to the cloth at the time the image was generated. A normal photograph is created by light reflecting off many surfaces, generated from many different light sources. The human mind can imagine the contours of a face from a flat photograph because we see faces so often from so many angles, in so many lighting situations. A computer does not have this advantage. To a computer, a normal photo is impossible to interpret correctly (see fig. 5). The image on the shroud, however, is not like a photograph, but is a perfect contour map of the body of the man in the shroud.

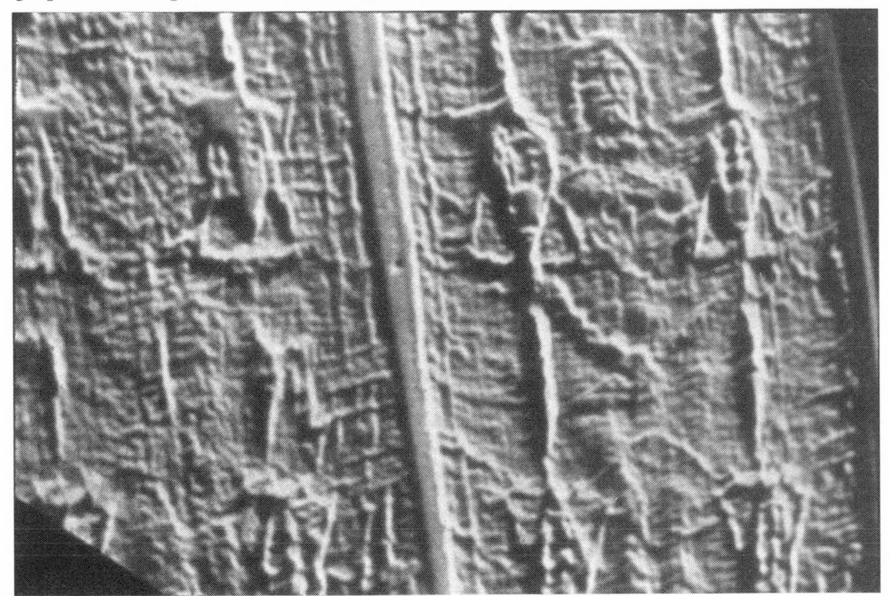

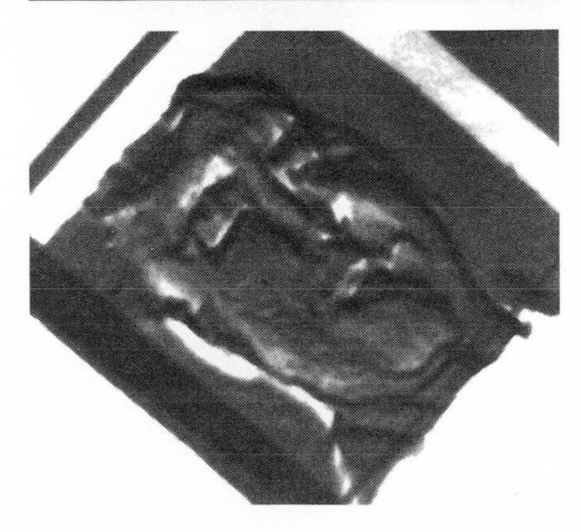

(above) Fig. 4. Three-dimensional image of the Shroud derived from the VP-8 Image Analyzer.

(left) Fig. 5. The distorted image from a normal photograph analyzed by the VP-8.

In 1976, computer technology had evolved to the point that it could confirm the three-dimensionality of the Shroud image. This discovery and its dissemination among other world-class scientists provided the impetus for the organization of STURP. Coming from some of the most advanced institutions in the world, the scientists who founded and joined STURP had access to the most highly developed technology available. When scientists viewed the three-dimensionality of the Shroud image for the first time, they were stunned.

Having learned of a few of the Shroud's amazing characteristics, the STURP team wrote the cloth's custodians in Turin in 1978 for permission to scientifically examine it. The Shroud was scheduled to be exhibited that year in celebration of the four-hundredth anniversary of its arrival in Turin, where it had been brought and kept by the Savoys, a family that would ultimately become the first royal family of Italy. The STURP team negotiated with the custodians in Turin for a hands-on examination to be undertaken at the conclusion of the 1978 Shroud exposition in Turin.

In October 1978, STURP assembled in Turin, armed with about seventy-two crates of scientific equipment weighing approximately eight tons and costing nearly $3 million. Physicists, chemists, pathologists, engineers, and photographers, they came from some of the most prestigious institutions in the world, including NASA, Sandia Laboratory, Jet Propulsion Laboratory, the U.S. Air Force, and Los Alamos Scientific Laboratory. For 120 consecutive hours, the STURP team members were given an unprecedented degree of access to the cloth and permission to conduct a battery of tests and examinations on the ancient relic. The researchers applied an extensive array of sophisticated, nondestructive scientific testing to the Shroud of Turin for the first time in its history. In addition, the researchers were allowed to bring samples taken from the Shroud back to their laboratories for further examination and study.

The STURP examination yielded a wealth of information that began to appear in scientific journals during the 1980s. Many of the STURP scientists had assumed before the landmark examination was undertaken that the Shroud would not live up to its traditional claims as the burial garment of Jesus Christ. After all, there had never been a scientific examination conducted on the cloth; therefore, there was very little substantial evidence to support its authenticity. But in the years to follow, as the scientists' conclusions were released in scientific journals, the Shroud's mystique quickly deepened, as did the scientists' respect for it.

The scientists realized that this was no ordinary relic they were studying. The Shroud of Turin is perhaps the most extraordinary relic in existence. There is nothing else we know of that even comes close to it. The characteristics present in the Shroud are so unprecedented and difficult to reproduce that the notion is preposterous that an artist used any media or technique to create it. Such an artist would have had to have a knowledge of light negativity, light spectrometry, microscopy, radiology, human physiology, pathology, hematology, endocrinology, forensics, and archaeology. In fact, even with all the technology available to us today at the dawn of the twenty-first century, the Shroud's unique characteristics still cannot be duplicated.

As a result of the first thorough scientific examination in 1978, the Shroud received a physical such as no other relic has ever received before or since. It was magnified; illuminated; photographed; thermographed; bombarded with visible, infrared, and ultraviolet light as well as radiation and X rays, and vacuumed. The cloth was examined in microscopic detail. Fiber composition and discoloration were analyzed, as well as chemical content. Pollen and other minute particles were identified. And those first 120 hours the team spent with the cloth were just the beginning.

As the team members returned to their laboratories, mounds of data awaited organization and analysis. From it would come a plethora of theories and hypotheses as to what the image was. Little by little, the team found out what the Shroud was not.

- It was not a painting.
- It was not a vapograph.
- It was not an imprint due to body heat or funeral anointing.
- It was not made by any natural means of draping a cloth over a human body.
- It was not a block print.
- It was not a scorch.
- It was not formed by draping the cloth over a statue or bas relief.
- It was not made from rubbing dry compounds on the cloth.

As for what it was, the team still could not say. What they did discover was that every bit of scientific and medical data that had been gathered from the testing and examination of the cloth—its samples, the blood, the particles removed from between or attached to the cloth's threads, and the thousands of photographs taken of the Shroud in every wavelength possible—was entirely consistent with what the cloth was reputed to be: the burial garment of the historical Jesus Christ.

At the same time the scientific and medical examination was occurring, an extensive history on the Shroud was published by historian Ian Wilson, titled *The Shroud of Turin.* This book provided the first plausible explanation of the Shroud's entire history going all the way back to the first century. Archaeologists, numismatics, polynologists, ethnologists, art historians, exegetes, textile experts, limestone analysts, botanists, and various other experts have added their own critiques and analyses to this body of knowledge over the past two decades. The results of their studies have been published in various archaeological and historical journals and other publications, and have been consistent with the findings from the scientific and medical examinations.

This book discusses this comprehensive evidence derived from or pertaining to the Shroud and its history. In the process of obtaining this evidence, I acquired unlimited access to the most up-to-date and, perhaps, extensive research sources on the Shroud in the world.[10] This material was supplemented by evidence derived from years of personal interviews, documentaries, conferences, and seminars conducted around the world. While the majority of the important evidence pertaining to the Shroud has been acquired since 1978, this book examines evidence from various sources, some of which are many centuries old. In addition, this book will reveal a good deal of new evidence being published for the first time.

At times, this book criticizes various scientists and other people. Such criticisms are mentioned only so that truth and knowledge critical to the future testing of the cloth, and the world's understanding of the Shroud and its evidence, are revealed. At times, including recent ones, crucial evidence about the shroud has been held back or obscured. This book will examine those setbacks as it reveals the evidence.

The goal of science has always been to ascertain the truth. In the coming chapters you will witness actions and attitudes of scientists involved in the carbon dating of the Shroud who intentionally helped prevent the acquisition of facts about the cloth by another group of scientists, who were the most knowledgeable in the world on the subject. Some of this additional evidence would also have related to the questions of the age, origin, and history of this burial garment.

This book will challenge and discredit the results of the 1988 radiocarbon dating of the Shroud. Due to the unprofessional, unscientific, and even incompetent behavior of certain scientists, many otherwise open-minded people are under the misapprehension that the Shroud is a fraud. This book will not hold back in exposing the real frauds. Moreover, further evidence concerning the age of the Shroud, could be much more accurate than any radiocarbon dating of the cloth, or any ancient object, could ever be.

Many other incredible aspects of the image on the Shroud will be discussed in the forthcoming chapters. This incredible image has never been explained or accounted for, despite centuries of efforts from people in various fields of study throughout the world. Instead of revealing how the image was encoded, science has been able to reveal only how it was *not* encoded. The more science has learned about the image, the more it appears to transcend the laws of physics, requiring something very extraordinary to account for all of the Shroud's unique features.

This book lays before you all the facts available, and the one theory that is consistent with the various scientific, medical, archaeological, and historical evidence.

In 1978, the scientific perception of the Shroud of Turin began to change forever. The catalyst for this change, the world-class innovative research study by STURP, allowed the Shroud cloth to receive intensive attention from a group of scientists representing a variety of disciplines. This inquiry, conducted by a team of thirty-two investigators, had a scope and dimension of unprecedented magnitude. The comprehensive sixty-three-page "Operations Test Plan" that the team launched on the Shroud remains the only thorough scientific examination of the cloth thus far conducted. It was never the scientists' intention or purpose to draw conclusions about the Shroud's authenticity as Jesus' burial cloth. Devising their own plan completely independent of any restrictions from the cloth's custodians (except that the tests be nondestructive), the STURP scientists traveled to Turin to collect the physical and chemical evidence contained on the cloth and test different theories related to how the image might have been formed. As highly trained and disciplined scientists guided by facts, objectivity, and logic, STURP members were attracted to the project because no serious scientific explanation of the image on the Shroud had ever been proposed. As engineer Eric Jumper and physicist Robert Mottern put it, "Ordinarily science remains detached from [religious relics], but in this case, the unusual quality of the image intrigues the scientific mind."[11]

The idea for forming a scientific group to investigate the Shroud came from John Jackson and Eric Jumper, both of whom worked at the U.S. Air Force Academy in the mid-1970s. They soon acquired the interest of Robert Dinegar of Los Alamos National Scientific Laboratory, who helped recruit several other scientists. Although many were employed by the government at the time, STURP members volunteered their personal time and money to study the Shroud. Neither the government nor any other employer of STURP members has ever funded or endorsed the project. When the Shroud's three-dimensional features became known, other scientists took up the challenge of deciphering the body image and joined the STURP effort. With religious orientations ranging from Catholic to Jewish, Protestant to agnostic, many were highly skeptical at first and even believed the Shroud would be declared a fraud within minutes after they began the examination.

After enduring prolonged delays with customs officials, the scientists were finally allowed to unload numerous boxes of fragile high-tech equipment in seven rooms that were part of the Savoy royal palace. The palace itself is part of a complex that also includes St. John the Baptist Cathedral and the Chapel of the Holy Shroud, where the cloth was kept high above the altar, rolled on a large spool, inside a long, rectangular reliquary. (The part of the splendid palace where the examination was conducted was completely destroyed in a fire in 1997. Had the Shroud been kept at its normal location above the chapel, it, too, would have been destroyed. Fortunately, the Shroud had been moved to temporary facilities within the cathedral several years earlier due to extensive remodeling at the chapel. Several firemen were able to carry the reliquary containing the Shroud to safety before any flames could even reach the several layers of thick, bulletproof glass that completely surrounded it. We will see later, however, that the Shroud had a much narrower escape from a fire in 1532.)

STURP began its examination of the Shroud on the night of October 8, 1978, and by the time the scientists finished at 2:00 A.M. on October 14, the Shroud and its image had been probed in every nondestructive manner the team could devise. Knowing their time with the cloth would be limited, STURP project leaders had apportioned every available minute of the 120 hours for testing. Because they had to work around the clock, STURP members ate carry-out food and slept on cots whenever they got breaks; outside, five different police forces guarded the palace that temporarily housed the scientists and the object of their intense investigation.

The STURP team's first act was to transfer the Shroud to a specially designed aluminum mounting frame that could be rotated through 360° and would thereby make testing easier. After the cloth was secured in the frame with small, polyethylene-coated magnets, a wire grid marked with coordinates was placed over the Shroud so the exact location of all tests and each thread or particle removed could be documented. Every action by every person was photographed, logged, and recorded on audiotape to ensure that all data could be accounted for and verified.[12]

Once the fabric was mounted, STURP members studied the cloth itself, the image on the cloth, the threads and their fibrils, even the foreign particles lodged within and between the threads. Imaging specialists took between five thousand and seven thousand photographs at various wavelengths of the light spectrum (including X ray, gamma ray, infrared, and ultraviolet). The cloth was scanned with a thermograph, which measures differences in the emission of infrared heat energy. Areas of the body image, the bloodstains, the background cloth, the thread fibrils, and the marks left by fire and water in 1532 were photographed extensively through microscopes.

Since all tests were nondestructive, samples of particles and thread fibrils were removed from the Shroud with sticky tape. A specially designed pressure-sensitive roller applied the tape to selected areas of the cloth. When removed, the tapes were stuck, adhesive side up, to microscopic slides and carefully identified. The back of the cloth was vacuumed to collect some of the loose debris, such as dust and pollen, that has collected on the linen over the centuries. Through visual and chemical tests, many substances contained on and in the cloth were identified, and the absence of other materials was confirmed. By the time STURP left Turin, the scientists had collected samples needed for the more than one thousand chemical experiments later carried out to determine "the nature of the image and blood marks, as well as the history of the linen, water stains, miscellaneous fibrils, particles and debris, and the presence of organic and inorganic pigments and vehicles, oxidants, reductants, and all known human means for creating the image on the Shroud."[13] The STURP team used every scientific testing method it could think of to inspect the image on the Shroud and gain a better understanding of its composition and characteristics. The studies revealed information never before suspected by allowing observation of features not visible to the human eye. This was the first necessary step in investigating how the image might have been formed.

All data collected in Turin were taken back to laboratories throughout the world for more detailed analysis. Interpretation of the results continues today as numerous scientific disciplines apply their expertise in an effort to explore and explain the Shroud image. Professions involved in evaluating the STURP data include nuclear and molecular physics, along with the fields of optics, spectroscopy, radiography, volcanology, and meteorology. In biology, the specialties of entomology, microscopy, botany, mycology, physiology, bacteriology, pathology, endocrinology, anatomy, immunology, and hematology have participated in the study. Investigation has also incorporated the chemical science disciplines of analytical, inorganic, organic, biological, physiological, geological, pharmaceutical, and textile chemistry.[14]

It is not necessary for lay readers to understand in detail the many scientific tests conducted on the Shroud. Instead, people need to grasp the broad scope of the scientific assault that has been directed at this one, very old piece of linen. In addition to tests already mentioned, investigation of the Shroud (both in Turin and in laboratories) has included three-dimensional analysis with the VP-8 Image Analyzer; computer-image enhancement and mathematical-image analysis; mapping or directional-function analysis; topographic imaging; reflectance spectroscopy at ultraviolet, visible, and infrared wavelengths; microdensitometry; macroscopy; fluorescent, phase contrast, and electron microscopic study; biostereometry; laser microprobe Raman spectroscopy; electron energy dispersive spectroscopy; microspectrophotometric transmission spectra; wet chemistry; cyanmethemoglobin and hemochromagen testing; protease lysis; generation of porphyrin fluorescence; and immunofluorescence.[15] These words may not mean much to us, but they give no doubt that science has used every means available to shed light on the mystery of the Shroud.

As a result of the STURP investigation and the continual scientific study during the last twenty years—based on the voluminous data acquired from this examination, we can now shed an enormous amount of light on this debate, an overwhelmingly greater amount than those at the turn of the century had available to them. Moreover, the light that will be shed will contain not only an enormous amount of scientific information, but it will contain comparable amounts of medical, archaeological, and historical information.

There is so much about the Shroud of Turin that the general public is not aware of. A small portion of this has been introduced in the preceding pages. Much more is to come. As a result of the number and length of the worldwide exhibits of the Shroud of Turin occurring at the end of the second millennium, the discrediting of the radiocarbon dating of the Shroud, and the bringing to light of the voluminous amount of information learned about the Shroud since its examination in 1978, the debate and entire subject of the Shroud is today experiencing what one would call a resurrection. It has been not only resurrected but placed prodigiously on the world stage where everyone can learn of it. Despite the desires of its many skeptics, the Shroud image will not fade from view. He continues to challenge and enlighten us, the man of the Shroud, to agitate us into reconsidering some of the most important questions humankind has ever faced. These are all questions worth knowing. There is much at stake. With an open mind and a keen eye, let us begin our inquiry into the mysterious Shroud of Turin.

CHAPTER TWO

EXAMINATION OF THE MAN IN THE SHROUD

Numerous pathologists, anatomists, and doctors have studied the images on the Shroud of Turin, beginning with Drs. Paul Vignon and Yves Delage in 1900–1902. The study of the Shroud was advanced further during the 1931 and 1933 Shroud expositions, which allowed modern physicians to see for themselves the accurate wounds and blood flows displayed on the cloth. But the real scrutiny of science began in 1978 when thousands of photographs of the images and wounds were taken in all wavelengths, and samples from the Shroud were removed for further examination.

Based on all the various examinations through the present day, experts agreed that the Shroud depicts the frontal and dorsal imprints of an adult human male who was well proportioned and of average height and weight. This man suffered serious wounds, died during his crucifixion, then was covered with a linen burial cloth. According to these medical authorities, the anatomically flawless wounds and body images visible on the Shroud resulted from direct or indirect contact between the linen and the man's corpse.

This chapter relates the specific findings of the most notable medical scientists (listed in table 1) who have studied the detailed physiological evidence contained in the Shroud image.[1] To understand the many features discussed below, readers should refer to figures 2 and 3 and the individual photos cited throughout the following pages.

Table I: Medical Examiners of the Shroud

Name	Position and Affiliation	Dates of Study
Dr. Paul Vignon	Professor of Biology—Institute Catholique, Paris	1900–1943
Dr. Pierre Barbet	Chief Surgeon, Autopsy Surgeon, Professor of Anatomy—St. Joseph's Hospital, Paris	1932–1961
Dr. Robert Bucklin	Chief of Forensic Medical Division—Los Angeles County Coroner–Medical Examiners Office; Clinical Professor of Pathology—University of Southern California	1940s–present
Dr. Yves Delage	Professor of Comparative Anatomy—Sorbonne, Paris	1900-1902
Dr. Alan Adler	Physical Chemist—Western Connecticut State University	1978–present
Dr. John Heller	Professor of Internal Medicine and Medical Physics—Yale; Biophysicist—New England Institute	1978–1995
Dr. Pierluigi Baima Bollone	Chair of Forensic Medicine in the Faculty of Law Institute of Forensic Medicine—University of Turin	1978–present
Dr. Giuseppe Caselli	Physician and Radiologist—Fano, Italy	1939–present
Dr. Sebastiano Rodante	Pediatrician, Medical Consultant for International Center of Sindonology—Turin, Italy	1940–present
Dr. Pietro Scotti	Priest, Professor, and Doctor of Medicine, Surgery and the Natural Sciences—University of Geneva, Switzerland	1930s–1982

Dr. David Willis	General Practitioner—England	1960s–1976
Dr. Rudolf Hynek	Battlefield and ship's surgeon—Prague, Czechoslovakia	1936–1954
Dr. Anthony Sava	Surgeon—Brooklyn, New York	1940–1993
Dr. Hermann Moedder	St. Francis Hospital—Cologne, Germany	1949–present
Dr. Luigi Gedda	Director of Institute of Medical Genetics—Rome	1939–present
Dr. Giovanni Judica-Cordiglia	Professor of Legal Medicine—University of Milan	1930s–1980
Dr. Maurizio Masera	Professor of Chemistry and Natural Sciences—University of Genoa	1939–present
Dr. Frederick Zugibe	Chief Medical Examiner—Rockland County, New York	1953–present
Dr. Leopoldo L. Gomez	Professor of Legal Medicine, Royal Academy of Medicine—Valencia, Spain	1950–present
Dr. Francesco La Cava	Pathologist—Naples, Italy	1953–present
Dr. Joseph Gambescia	Chairman of Medicine—Halinemann University Professor— St. Agnes Medicine Center—Philadelphia	1955–present
Dr. Edward A. Brucker	Deputy Medical Examiner—Pima County and Southern Arizona	1960–present
Dr. J. Malcolm Cameron	Pathologist—British Home Office	1978–present
Dr. Gilbert R. Lavoie	Chief of Epidemiology—U.S. Army Headquarters Medical Command (Europe); Clinical Instructor—Tufts Medical School; Staff Physician—Faulkner and New England Diaconess Hospitals	1978–present

THE MAN'S WOUNDS

Facial and Head Wounds

The man on the Shroud has a mustache, beard, and hair that falls to his shoulders from a central part. His cheeks appear swollen, and the area below the right cheek contains a triangular-shaped wound.[2] On close examination, his nose, which is bruised and swollen, shows a slight deviation that indicates the cartilage may be separated from the bone.[3] Microscopic study also reveals that scratches and dirt are on the nose.[4] The areas above and below each eye, especially the right eye, look swollen, and the face appears to have been beaten with a hard object (such as a fist or stick) and/or injured in a fall. The man's eyes are closed, and small round objects seem to lie on top of them. (Chapters 3, 6, 9 and 10 discuss in greater detail the three-dimensional studies of the face and eye area; see fig. 6.)

A number of wounds are visible on the top, middle, and sides of the man's forehead. Altogether, more than a dozen blood flows have been counted on the front of the head alone (see fig. 7). The blood marks associated with the frontal head wounds seem to run in different directions from their points of origin, which suggests the head was in different positions as the blood was flowing.

Circling the top and middle of the back of the head is another series of blood marks (see fig. 8). Since these scalp wounds are covered by hair, the exact number of blood rivulets is difficult to determine, but Dr. Sebastiano Rodante has identified as many as twenty separate blood flows on the back of the head (fig. 9), which brings the total number of head wounds to more than thirty.[5] Like the frontal image head wounds, the dorsal blood flows also run in different directions, until they seem to stop along a concave line just below the middle of the head. Above this line, wounds can be seen in the middle and near the top of the back of the head. When considered with the wounds evident on the top, middle, and sides of the forehead, the head wounds give the impression that the man was wearing something like a cap made of sharp, pointed objects. Several physicians have noted that a cap made of thorns would produce head wounds identical to those on the man in the Shroud.[6]

Hand and Arm Wounds

The man's left wrist has been wounded or pierced, and blood flows from the wrist area toward the left elbow. Although the right wrist is covered by the left hand, similar blood flows are also visible extending along the right forearm toward the elbow. As shown in figure 10, both forearm blood flows run in two nearly parallel streams, with one stream measuring approximately 65° from the horizontal axis of the arm and the other stream measuring about 55° from the horizontal axis.[7] These unusual blood marks flowing from the wrists toward the elbows proved to be an important piece of evidence, helping physicians determine that the man on the Shroud had been crucified. During crucifixion, a victim's hands would be higher than his head. Since the blood flows from the pierced wrists toward the elbows of the man in the Shroud, we know that his arms were elevated, not hanging at his sides, while his wrists were bleeding.

The two parallel streams running at slightly different angles from the horizontal are also significant. Because of the hanging position, a person being crucified can exhale only if he pushes himself up with his feet to raise his shoulders and expand the ribcage. This movement alters the horizontal axis of the arms by approximately 10° (see fig. 11). Pushing upward in this fashion would temporarily lessen some of the constant pain in the victim's wrists and arms, but it would increase the pressure and pain in his feet. While this up-and-down motion was arduous, it did allow a crucifixion victim to breathe and forestall inevitable death—at least until he was too exhausted or in too much pain to push himself up anymore. Often, the executioners would break the legs of the crucified to stop this movement and hasten death. However, the legs of the man in the Shroud were not broken.

Scourge Marks

The man's front and back, from shoulders to lower legs, are covered with an estimated one hundred or more scourge marks.[8] These dumbbell-shaped patterns, which are most noticeable on the dorsal image, generally run parallel and diagonal across the body in groups of two or three (see fig. 12). Although all are approximately the same size, these scourge marks vary in intensity from light contusions to deep punctures, and close examination reveals the presence of blood in many of them.[9]

Because the scourge marks have two lobes (see fig. 13), these wounds must have been inflicted with a bifid instrument. The form and distribution of these marks led medical examiners of the Shroud to believe they were caused by a whip or cord-like device containing metal or some other sharp object at the end capable of tearing flesh. In particular, these wounds match, in size and shape, the Roman *flagrum* (shown in fig. 14).[10] The flagrum, a whip used for flagellation, had pellets of lead (or sometimes bone) at the end of a pair of leather thongs. These unusual marks and their similarity to the flagrum led medical experts to conclude that the man in the Shroud must have been whipped or scourged.

Since the scourge marks are more visible on the dorsal view, physicians further believe the man was whipped from behind. The central point from which the blows radiate is a little higher on the victim's right side than on his left, so two men probably carried out the beating; further, it seems the "scourger" standing to the right was taller and tended to lash more at the man's legs in addition to his back. The scourge marks decrease in number and depth toward the ankle, where some fade into lines visible only under ultraviolet light.[11] Since the man's arms, head, and feet seem to be the only areas that escaped scourging, we can assume that either his arms

Fig. 6. Small round objects are clearly visible over the man's eyes in the three-dimensional image.

Fig. 7. Number of forehead clots in reduced size.

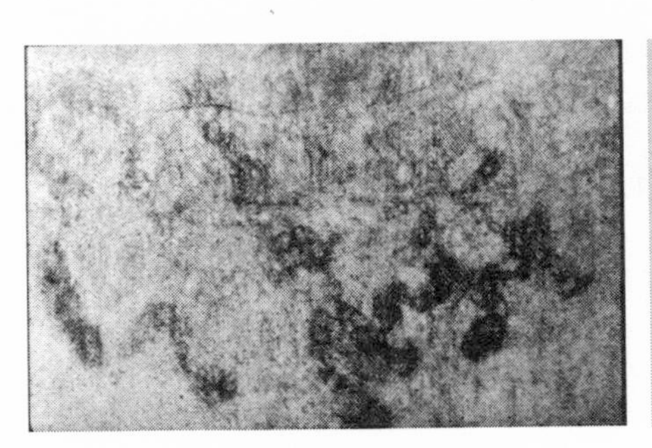

Fig. 8. Blood marks on the back of the head.

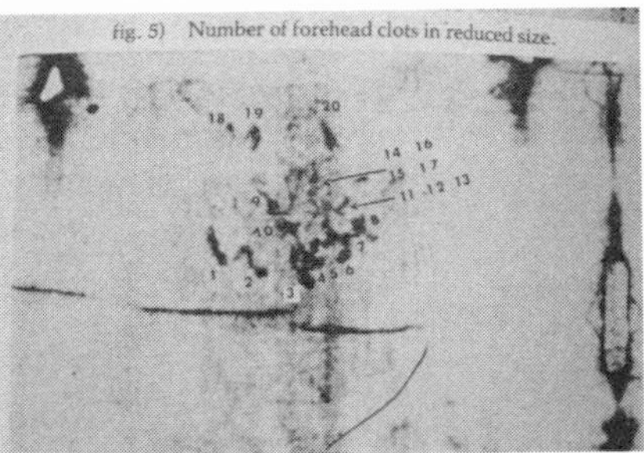

Fig. 9. Number of blood marks on the back of the head in reduced size.

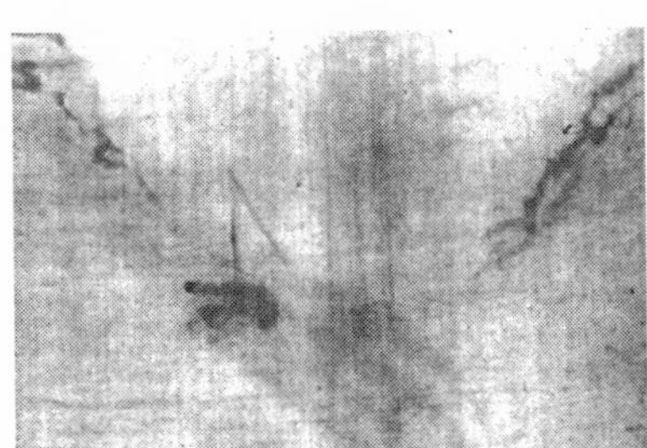

Fig. 10. Blood flows on arms and wrist.

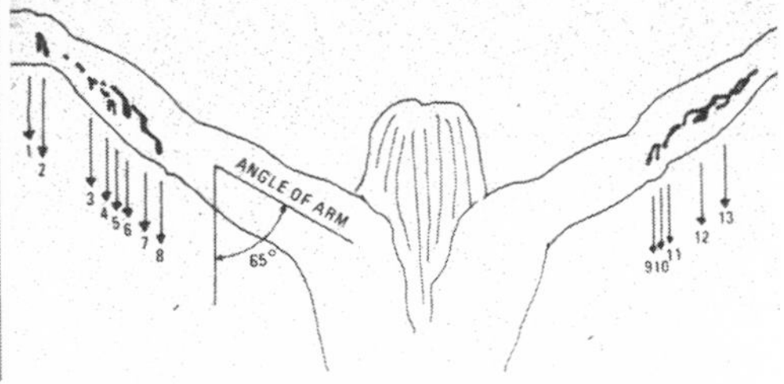

Fig. 11. The main angle appears to be 65°, but there is evidence that at some stages the forearms were at 55°, indicating that the man of the Shroud sought to raise himself, probably continually, during crucifixion.

were elevated above his body during the scourging, or that his hands were tied to a post or pillar in front of him while he was whipped from behind. There are so many scourge marks on the back of the man that they are easily the dominant feature of the dorsal image.

Chest Wound

On the right side of the man's chest, a large side wound is apparent. This wound accounts for the most massive concentration of blood on the Shroud (see fig. 15), and all medical authorities cited agree that this large blood flow resulted from a postmortem wound. The blood from this wound appears to have oozed out and flowed due to the force of gravity, not driven by a pumping heart. Also, no swelling surrounds the side wound. Although this wound's bloodstain is partly hidden by one of the patches sewn on the cloth to repair damage from the fire of 1532, so much fluid poured from this wound that it collected in a puddle along the small of the man's back when he was placed in a horizontal position after death (fig. 16). Curiously, not all of this stained area is composed of blood; instead, this stain "very clearly shows separation of blood from a clear watery material."[12] While serum separated from blood may account for a portion of this stain, there is far too much watery fluid to be explained by the process of serum release from a blood clot. Forensic examination reveals that the blood and fluid from the side wound flows from an elliptical-shaped lesion approximately 4.4 cm long and 1.1 cm wide. The size and shape of this wound match excavated examples of the Roman leaf-shaped *lancea,* an instrument used by foot soldiers of the Roman militia.[13]

Shoulder Injuries

Two broad excoriated areas are present across the victim's shoulder blades (fig. 12). These scrapes are consistent with surface abrasions caused by contact between skin and a heavy rough object.[14] Because some of the scourge marks within this area are slightly different when compared to the clearly defined marks elsewhere on the body,[15] the scourging must have preceded the shoulder abrasions. We know that many crucifixion victims were forced to carry their own crossbars to the execution site. Carrying such a large chunk of wood could easily have caused some of the man's shoulder wounds, especially if he fell under the weight of the beam and was struck by the wood falling on top of him. The Shroud contains evidence consistent with such a fall or falls. Scratches, lesions, and abrasions on the front of the man's knees have been revealed by white light photos (see fig. 17)[16] and by ultraviolet fluorescent lighting.[17] Microscopic examination of the Shroud image also discloses particles of dirt on the front of the knees, nose, and bottom of the feet.[18] Being struck by the crossbeam during a fall may also explain some of the wounds on the back of the man's head.

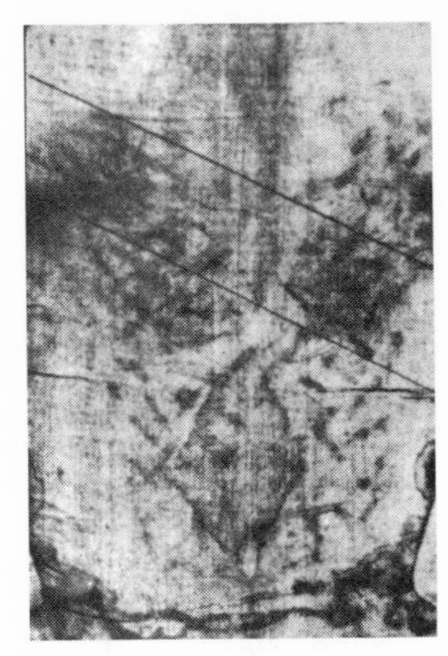

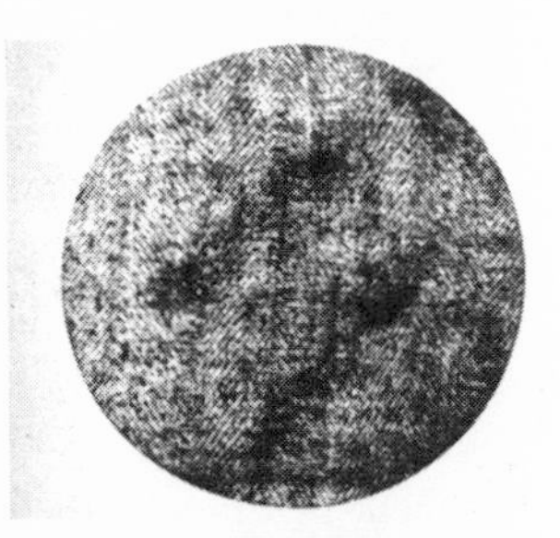

(above) Fig. 13. Scourge marks.

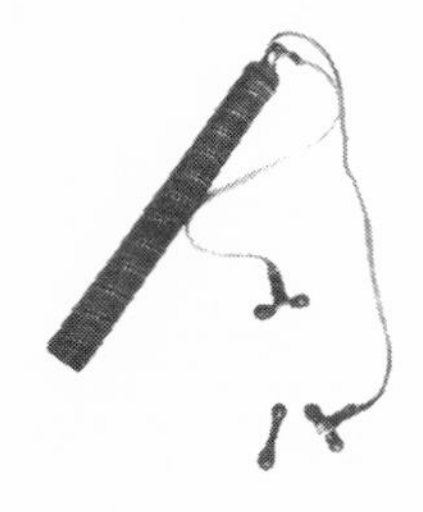

Fig. 14. Roman flagrum with pellets of lead.

(left) Fig. 12. Back of the man covered with scourge marks, along with abrasions over the backs of his shoulders.

Some pathologists have identified another injury on the man that may or may not be postmortem. As he appears on the Shroud, the man's arms have been forcibly bent so his hands cover the groin. To accomplish this, the shoulder girdle would have had to be broken or dislocated, a practice common to morticians when positioning a body for burial.[19] In this procedure, the muscles between the neck and shoulder are massaged to release rigor mortis so the arms can be moved. Some medical investigators have noted that the man's right shoulder is about five degrees lower than the left, a feature most apparent on the dorsal view.[20] Dr. Pierre Barbet believes this indicates a dislocated shoulder, which may have occurred either during the hand-positioning at burial, when the victim fell, or when he was raising and lowering himself on the cross.[21] If the man's shoulder had been dislocated while he was still alive, that injury would have been another source of intense pain.

Leg and Foot Wounds

Detailed study of the lower extremities reveals two large blood marks on the front of the feet, the larger of which has a surrounding border that fluoresces under ultraviolet light.[22]

On the dorsal image, two bloodstained imprints of the feet are evident, with the right foot impression being more complete and showing the outline from heel to toes (fig 18). Some blood has flowed off the right heel area and onto the cloth. Medical experts agree that this large amount of blood resulted from a piercing wound to the foot[23] and Dr. Robert Bucklin has identified the source of the blood flow: "a square image surrounded by a pale hole" in the metatarsal zone.[24] From this wound, some blood runs vertically toward the toes, but most flows toward the heels and horizontally onto the cloth.[25] This tells us the man was bleeding while in different positions—vertically while on the cross and horizontally when being carried after he was dead. The blood flow running toward the heels and onto the cloth, deeper in color, has been identified as postmortem.[26] The most likely explanation for this postmortem bloodstain is that most of the blood that accumulated in the front and lower part of the foot while in the vertical position flowed from the wound after the piercing instrument was removed and the body was laid flat. The Shroud's medical examiners have concluded that this piercing instrument must have been the nail or spike typically used for crucifixion. Since this foot wound (shown in fig. 19) is surrounded by the metatarsal bones, a large nail would provide the support necessary to prop up the victim's weight.

When viewing the back of the man's legs and feet, we see that the left foot and leg images are less defined than the right ones. In addition, the left heel is elevated above the right. These facts indicate that the left knee was flexed to some degree. While this is most apparent on the dorsal view, the left leg visible on the frontal image also appears slightly raised. In light of these findings, most pathologists contend that the right foot was placed directly against a flat surface, while the left leg was bent at the knee and the left foot rotated to rest on top of the right foot. With a body in this position, a single nail driven between the metatarsal bones could affix both feet in a stationary position.

WOUNDS DEPICTED ON THE SHROUD ARE THOSE OF A HUMAN BEING

After pathologists and other physicians who examined the man in the Shroud considered the body's position and condition, wound locations, and instruments used to inflict the injuries, they concluded that his many wounds reacted exactly as a living human being's would. When people are similarly injured, their bodies respond in specific and identical ways; for example, puncture wounds always bleed in the same manner, and abrasions undergo the same physical processes on all people. In the case of the man in the Shroud, the wounds are so anatomically accurate that the details are startling in their precision.

As an example, the wounds in the man's wrists clearly demonstrate the natural physiological reactions of an actual human body. For centuries, people believed that crucifixion victims were nailed to crosses through the palms of their hands. No doubt this perception stemmed from the many artistic representations of Jesus' crucifixion in such a manner. In fact, those who have doubted the Shroud's authenticity pointed to the absence of hand wounds as proof that the man in the Shroud could not be Jesus. Artists who depicted the crucifixion were inspired not by an understanding of physiology but by the words in the Bible: the Gospels state that Jesus' "hands" were pierced.[27] In actuality, the original Greek word used in the Gospels is *cheir*, which also means wrist and forearm.[28]

Medical experiments first conducted in the 1930s and repeated since then have proved that it is anatomically impossible for a person to be nailed to a cross through the center of the palms. The upper extremities cannot be anchored because the body's weight causes the nail to tear through the flesh of the palm. In 1968, when

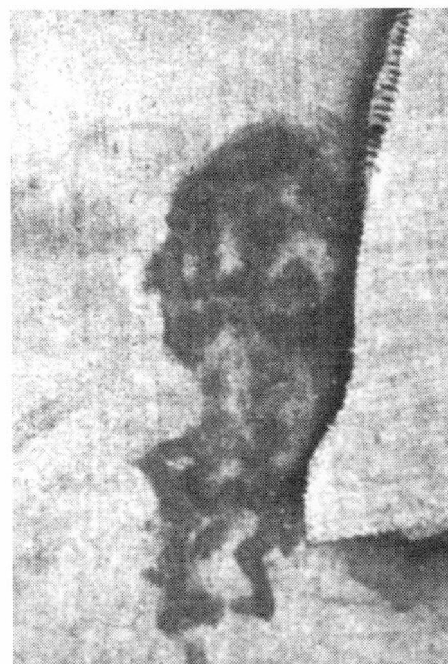

Fig. 15. The side wound.

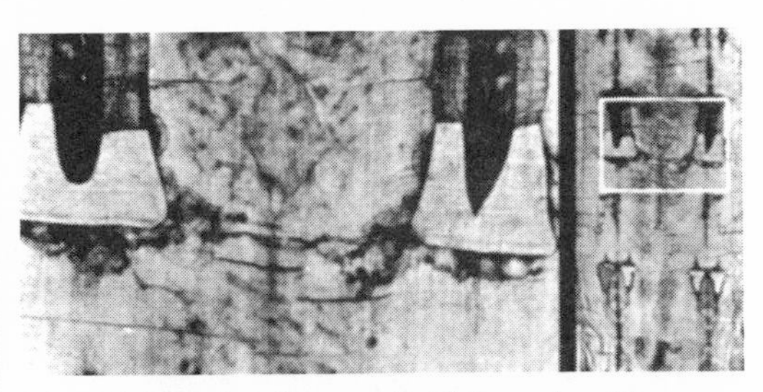

Fig. 16. Blood and watery fluid from the side wound puddled along small of back.

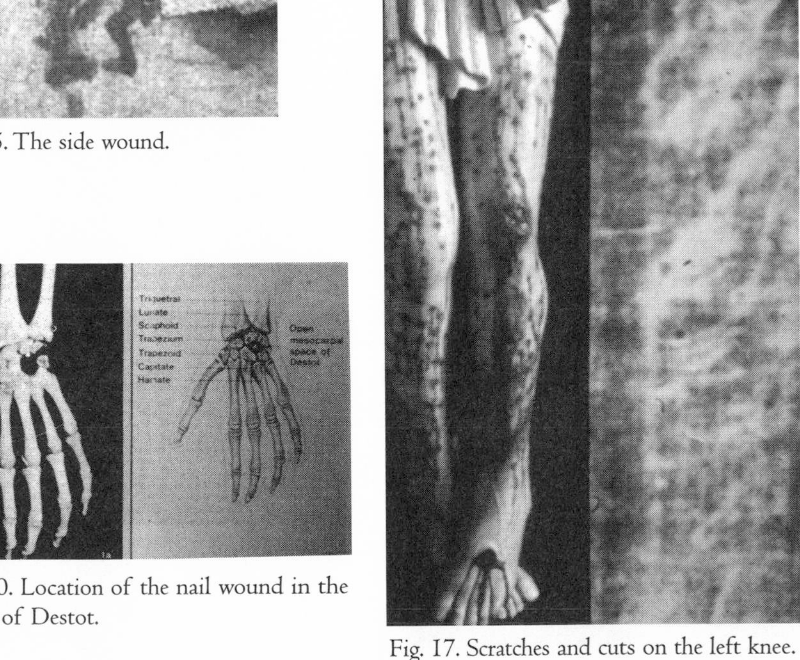

Fig. 20. Location of the nail wound in the Space of Destot.

Fig. 17. Scratches and cuts on the left knee.

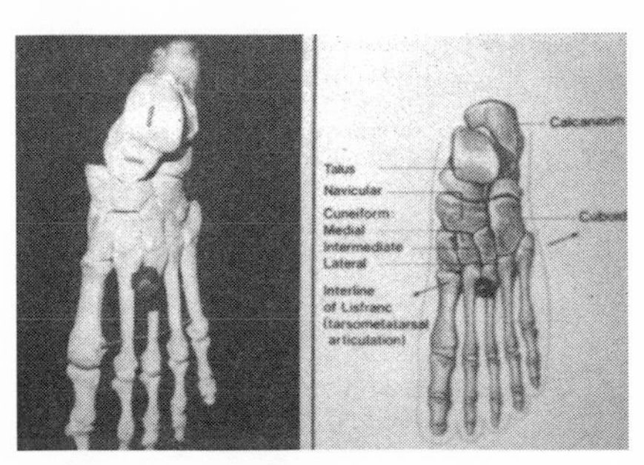

Fig. 19. Location of the nail wound in the foot.

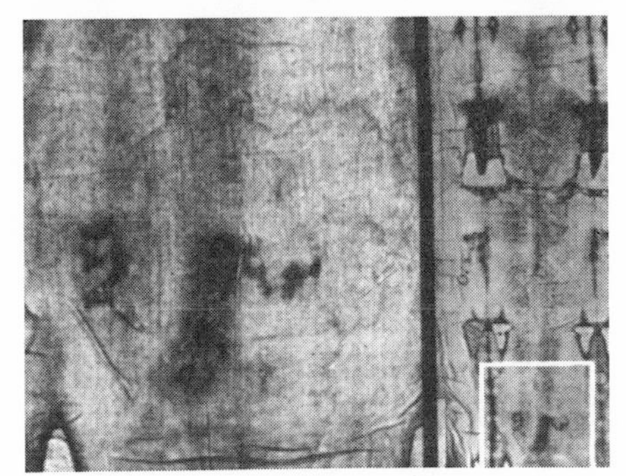

Fig. 18. Dorsal image of the feet.

excavating a site at Giv'at ha-Mivtar northeast of Jerusalem in a Jewish cemetery of the Second Temple period (the time of Jesus), archaeologists unexpectedly discovered the only known remains of a crucifixion victim. Professor Vassilios Tzaferis, the archaeologist with Israel's Department of Antiquities who excavated the site, determined that the man, called Yehohanan, had been nailed to the crossbeam through the flesh near the wrists, as evidenced by a scratch found on the wrist end of the man's right forearm radius bone. Tzaferis concluded: "The scratch was produced by the compression, friction, and gliding of an object on the fresh bone. This scratch is the osteological evidence of the penetration of the nail between the two bones of the forearm, the radius and ulna."[29] The friction and gliding to which Dr. Tzaferis refers resulted from the up-and-down motion discussed earlier: In order to breathe, crucifixion victims pushed themselves up and down until the executioners broke their legs. The remains found in the Jerusalem cemetery showed that both of this man's legs had been broken by a single strong blow.[30]

What does all of this mean? The locations of the wrist wounds on the man in the Shroud are anatomically accurate, yet only in this century did scientists learn through experiments and previously unknown remains that crucifixion victims were not nailed through the palms, as is commonly depicted in art. The locations actually constitute points of authenticity for they match those of another first-century crucifixion victim.

Some critics also doubted the Shroud's authenticity due to an Old Testament passage that says, "A bone of him shall not be broken,"[31] and the wrist area contains many small bones that could easily be broken. However, medical experiments would also explain this concern as well as answer, perhaps, the most puzzling feature of all concerning the hands. This visible feature is the apparent absence of the man's thumbs. When Dr. Barbet conducted experiments with cadavers to see what would happen when he drove a nail into the exact wrist area indicated on the Shroud, he discovered that the nail diverted upward into what is called the Space of Destot. When struck with a few more blows, the nail pushed aside the four bones surrounding the Space of Destot so the space widened and allowed the nail to pass freely through the flesh without breaking any wrist bones (fig. 20). Much to Barbet's surprise, driving the nail still further caused the thumb to contract spontaneously inward toward the palm. He found a simple explanation for this previously unknown physiological phenomenon: When a nail is driven into the Space of Destot, the median nerve controlling the thumb is injured and stimulated, automatically causing the thumb to contract inward and lie across the palm. Injury to the median nerve, then, would account for the anatomical reaction visible on the Shroud.[32]

When searching through centuries of artistic tradition, you might find one or two portrayals of Jesus' crucifixion that show nails in the wrist area, but no known work of art both depicts nails through the wrists and the absence of thumbs. The image on the Shroud is unique because the wrists and hands are anatomically precise in their illustration of crucifixion wounds, even though knowledge of where crucifixion victims were nailed and what would happen to their thumbs was not known until this century. At minimum, these anatomical facts are recognized by medical examiners to be the spontaneous and natural reactions of a real human being who has been crucified. Moreover, these anatomical characteristics could be unique points of authenticity as the image and burial garment of the historical Jesus Christ, for there is no depiction or reference in all of history like this of Jesus, or anyone else.

That the Shroud covered an actual person is further substantiated by the fact that actual blood is present on the cloth. This determination was made throughout this century by physicians and other scientists who studied the characteristics, forms, and configurations of the Shroud's wounds and blood flows.[33] During the 1978 testing conducted by STURP, fibril samples were removed from various bloodstained locations on the cloth. The following list summarizes the tests that led STURP to the conclusion that real blood is present on the Shroud:

- High iron content (in heme-bound form) in bloodstained areas detected by X ray fluorescence tests;
- The spectral fingerprint of blood revealed by reflection spectra;
- Indications of blood from microspectrophotometric transmission spectra;
- Characteristic generation of porphyrin fluorescence disclosed by ultraviolet imaging (porphyrin is a derivative of hemoglobin);
- Positive hemochromagen tests;
- Positive cyanmethemoglobin tests;
- Positive detection of bile pigments;
- Positive demonstration of protein;
- Positive indication of human albumin in immunological tests;
- Positive results from protease tests (an enzyme responsible for protein metabolism in living organisms), indicating that protein was broken down into its essential amino acids;
- The forensic judgment of the appearance of various wounds and blood marks by STURP chemists;
- The matching appearance of microscopic Shroud samples with control fibers that contained known blood.[34]

Not only are all the above tests consistent with the presence of blood, but venous blood flows can even be distinguished from arterial blood flows in some of the bloodstains on the man's forehead. In general, venous blood appears denser and darker red, and it flows more slowly than arterial blood. In large wounds or wounds that puncture a vessel and produce a large blood flow, venous blood slowly thickens as it descends because it takes a few minutes for the coagulation process to begin and a clot to form. The large epsilon-shaped clot in the middle of the man's forehead is a good example of a large venous blood flow. Smaller examples of venous blood marks are found in wounds numbered 9, 12, and G3 in figure 7. In contrast to blood from a vein, arterial blood spurts from a wound, driven by the pumping action of the heart. Wound number 1 in figure 7 is a good example of an arterial blood flow.[35]

Dr. Rodante, who has made one of the most extensive studies of the forehead wounds to date, has identified the origins of many of the head wounds based on the size or coagulation pattern of blood flows on the skin. (The arterial or venous origins of blood flows matted in the hair, and not free-flowing on skin, are impossible to determine.) As examples, the epsilon-shaped forehead clot lies exactly over the frontal vein, while the arterial wound numbered A1 in figure 21 precisely corresponds with the frontal branch of the superficial temple artery.[36] According to Rodante, "The perfect correspondency of the forehead clots imprinted on the [Shroud], overlaying as they do the vein and the artery in mirror image, gives us the certainty that the linen covered the corpse of a man, who, while living, suffered the lesion of these blood vessels."[37]

Running from the side of the head toward the middle of the forehead (see fig. 22), the frontal branch of the superficial temple vein appears alongside the superficial temple artery. Since the wound numbered A3 on the right side of the man's forehead lies slightly higher and closer to the center of the forehead than wound A1 does, A3 must correspond to the superficial temple vein.[38] These examples of distinctly venous and arterial wounds indicate that the injuries evident on the man's image could have occurred only on an actual human body.

Regardless of technique, no artist, especially one working in the Middle Ages, has ever represented the distinction between venous and arterial blood so accurately. In comparison to the Shroud's realism, fig. 23, a medical illustration of wounds drawn in the 1400s, shows how poorly blood flows were understood at that time. In fact, the difference between arterial and venous blood was not even discovered until 1593, more than 230 years after some allege that the Shroud image was painted.

The episilon-shaped clot on the man's forehead contains another realistic detail. As the blood flow descended, it broadened and changed course twice. Physicians believe this was because forehead muscles spontaneously contract when they are injured. The forehead, temple, and scalp contain a web of nerves that is highly sensitive to pain.[39] Thus, contracting forehead muscles would be a natural reaction to the intense pain caused by having more than thirty head wounds.

The strikingly realistic scourge marks, too, could only be the result of actual wounds inflicted on a real human being. When photographed, then enlarged and studied under the microscope, each mark reveals a slightly depressed center and raised edges.[40] Even more startling was the discovery that a "halo" of lighter color surrounds the scourge marks when they are viewed under ultraviolet light.[41] When STURP scientists John Heller and Alan Adler chemically tested the material within these halos, they found it to be blood serum.[42] Furthermore, when these lightly colored halos were photographed under ultraviolet light, they fluoresced—a clear indication of the presence of blood serum.[43] According to Adler, the upraised edges, indented centers, and serum surrounding the scourge marks illustrate exact-

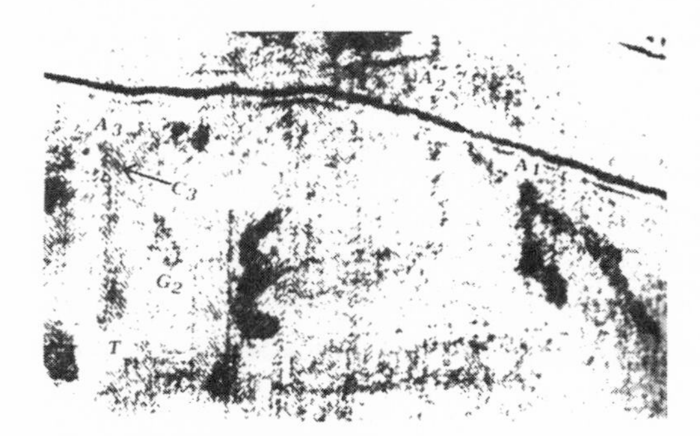

Fig. 21. Venous and arterial bleeding can both be found on the head wounds.

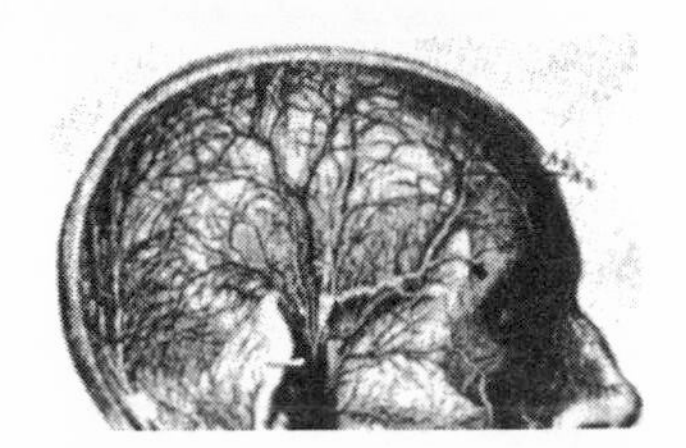

Fig. 22. Location of the superficial temple artery and vein.

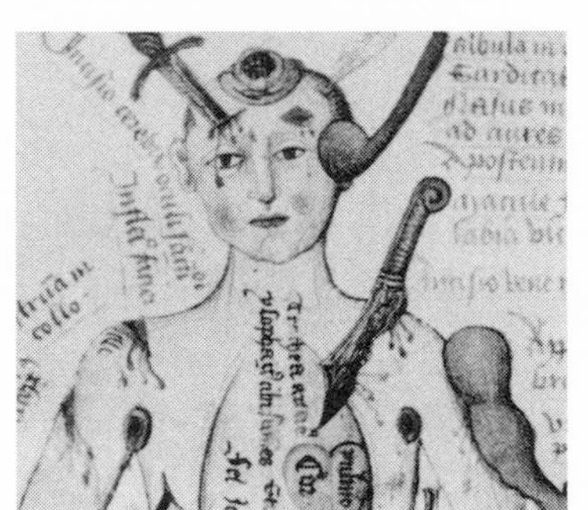

(left) Fig. 23. This medical illustration of wounds from the fifteenth century shows how poorly they were understood even a century after the Shroud was allegedly painted and first known in Europe.

ly the process called syneresis, which happens when a blood clot forms and then retracts.[44] Blood remains whole for only a few minutes after it emerges from the body and then quickly begins to coagulate. During coagulation, red blood cells and serum separate; the red cells bond to form a blood clot, which then retracts in the wound. As the clot shrinks to a slightly smaller size, the serum is squeezed out and settles around the edge of the wound.

None of the unique scourge mark features are visible on the Shroud with the naked eye. The indented centers and raised edges are perceptible only when these areas of the body image are photographed and enlarged, then examined under a microscope. Ultraviolet light and fluorescence testing are needed to show the serum-fluorescing borders and to observe that scratches and cuts invisible to the unaided eye accompany the scourge marks. These findings are highly significant because they prove that the Shroud could not have been created by an artist in the Middle Ages. A medieval artist would not have had access to photographic equipment, a microscope, or an ultraviolet light source because none of the tools would be invented for several more centuries. There are more than one hundred scourge marks on the man in the Shroud. The inaccurate representation of just *one* of them would reveal an unnatural physiological reaction and expose the work as an artistic creation! Why would an artist bother to encode such minute details anyway, since no one else could know they were there and appreciate his cleverness? We can only conclude that the scourge marks are not the product of an artist but are, instead, evidence of the natural processes of an actual body.

In addition to the halos around the scores of scourge marks covering the body, lightly colored and fluorescing borders surround many of the major bloodstains, specifically those wounds on or near the front and back of the head, the lower lip, the lower left wrist and arm, the side of the ribcage, the small of the back, and both sides of the feet.[45] These halos around the larger wounds have also been identified by scientists as blood serum. The presence of serum around the wounds means the Shroud's bloodstains are composed not only of real blood, but of whole blood. Other studies of bloodstained fibrils confirmed the presence of bile pigments, serum-type proteins (such as albumin), and nonheme proteins adjacent to blood stains—all of which are indicators of actual bleeding wounds.[46] Even more recent tests by Fourier Transform Infrared (FTIR) microspectrophotometry and ultraviolet-visible spectrophotometry confirm these findings by identifying the presence of bilirubin in blood samples and yellow serum–coated fibers from the Shroud.[47] Dr. Adler states that the ultraviolet photographs of the man in the Shroud reveal that "every single blood wound shows a distinct serum clot retraction ring."[48]

From the tiny scourge marks to the large lesions, the bloodstains and wounds evident on the Shroud demonstrate the characteristic syneresis process that occurs when whole blood coagulates as a person bleeds. Many years before STURP went to Turin to study the Shroud with modern scientific equipment, physicians had suspected that the wounds and blood flows were comprised of actual coagulated blood.[49] For example, in the 1930s, Dr. Barbet emphasized: "*The thing which immediately strikes a surgeon . . . is the definite appearance of blood congealed on the skin,* borne by all the blood-marks" (italics are Barbet's).[50] Barbet and many others hoped that one day rigorous testing could be performed on the Shroud to confirm his suspicions; specifically, he called for chemical testing and studies with spectroscopy, photography in light from different regions of the spectrum, and radiography.[51]

In 1978, STURP scientists conducted these tests, plus many others Barbet could not have imagined, which did much more than confirm Barbet's observations. Spectrophotometric studies, for instance, revealed that the bloodstained areas possess the spectral characteristics of human hemoglobin.[52] Drs. Adler and Heller obtained positive results when they tested bloodied fibrils for human albumin and human whole blood serum.[53] Working in Italy, Dr. Baima Bollone and colleagues used fluorescent antibodies to demonstrate the presence of human blood on threads removed from the Shroud. They even concluded that the blood is type AB. Using antiserums to test trace materials left in the fibrils' test tubes, Bollone also learned that the Shroud contains human immunoglobulins.[54] It was announced in 1997, after examining two blood samples taken from the Shroud at the back of the head, that Dr. Victor Tryon of the University of Texas found human DNA with both X and Y chromosomes present in the samples. This also confirmed that the samples were those of a human male. In addition, he found that the DNA was very degraded, which is consistent with ancient DNA.[55]

One other feature of blood chemistry pertinent to the Shroud is noteworthy. When blood is exposed to open air, it is normally converted to methemoglobin, a compound that results when hemoglobin is oxidized. This process—witnessed by anyone who has ever removed a bandage—turns the color of blood from red to brown. Blood will turn black in a matter of days. Blood that is at least six hundred years old should be very dark! How, then, do we explain why the blood on the Shroud is reddish in color? Simply, it is more red than brown because the blood comes from a man who has been severely scourged, beaten, and crucified.

Most, if not all, physicians would agree that crucifixion victims have great difficulty breathing, which causes their blood to become oxygen depleted. Hemoglobin that has already been oxidized (methemoglobin) has an orange-brown color. Most, if not all, physicians would also agree that the extreme pain throughout the body from a crucifixion would cause the victim to suffer "traumatic shock." Traumatic shock leads to hemolysis, a condition in which red blood cells are broken down and their hemoglobin is released into the surrounding fluid or plasma. Hemolyzed blood appears more transparent; has a richer, redder color; and contains high levels of bilirubin picked up as blood passes through the liver. Bilirubin, the principal orangish pigment in bile, is a breakdown product of hemoglobin produced in large quantities in cases of violent deaths.[56] When this hemolyzed blood then clots, its hemoglobin is exuded into the serum that seeps around the clot; red blood cells that have not been broken down stay behind in the clot itself.[57] What we see on the Shroud is a blood exudate that has a very high content of bilirubin (orangish-yellow) and methemoglobin (orangish-brown). When blood of this composition ages over time it darkens to red.

The blood chemistry of the man in the Shroud, therefore, corroborates what the body's pathology tells us: he was beaten, severely scourged, crucified, and eventually entered a state of traumatic shock. Dr. Heller summarized the findings of STURP and other scientists this way: "It was evident from the physical, mathematical, medical, and chemical evidence that there must have been a crucified man in the Shroud."[58]

UNIQUE METHOD OF TRANSFER OF BLOOD FROM BODY TO THE CLOTH

The man in the Shroud suffered well over 130 different injuries, which can be categorized into five main groups: scourge marks, side wound, head wounds, feet wounds, and wrist wounds. The injuries were inflicted at different times, with the scourging first, followed by the head wounds, the crucifixion nail wounds in the wrists and feet, then the postmortem side wound. The direction of the blood flows from the side and feet wounds was further influenced when the body was carried in a horizontal position after death.

Since the crucifixion alone would have lasted a few hours, physicians conservatively estimate that the bleeding from the various wounds went on for a period of four to twelve hours. This means the blood transferred to the Shroud varied in age from relatively fresh postmortem coagulated blood to blood that was up to twelve hours old. Yet, amazingly, all of the bloodstains transferred to the cloth in a way that has never been recorded anywhere else in history. The blood marks correspond perfectly with the natural physiological reactions of a bleeding human body that has been in changing positions. Ranging from the large side wound to the tiny scourge marks, the realistic wounds represented on the Shroud are anatomically flawless.

After examining the Shroud, Dr. Barbet was among the first to notice the unique transfer of bloodstains to the cloth. As a battlefield surgeon, Barbet had seen countless dressings removed from soldiers' wounds; these dressings covered injuries in various stages, ranging from fresh to several days old. In all of his experience, never had he removed a dressing that showed the exact form of the underlying wound with such clean outlines. Anyone who has ever removed a tissue or bandage from a cut can attest to this same phenomenon: the blood on the bandage is not a mirror image of the corresponding lesion. Commenting on the unusual correspondence between the blood on the Shroud and the man's injuries, Robert Wilcox summarized efforts to reproduce the kind of blood marks seen on the Shroud:

> Paul Vignon and Pierre Barbet found, after many attempts, that it was impossible to transfer blood to a linen cloth with anything like the precision shown on the shroud. If the blood were too wet when it came into contact with the cloth, it would spangle or run in all directions along the threads. If it were not wet enough, it would leave only a smudge. The perfect-bordered, picturelike clots on the shroud, it seemed, could not be reproduced by staining.[59]

The full significance of this unique transfer of bloodstains will be discussed in chapters 10 and 11. For now, it is sufficient to state that the bloodstains could not have been encoded on the Shroud simply by direct contact between a bloody body and a linen cloth surrounding it. While this may seem to point away from the Shroud's being an actual burial cloth, it may, in fact, point toward something truly miraculous.

DEATH BY CRUCIFIXION

As stated earlier in this chapter, pathologists and other physicians are convinced the man in the Shroud was crucified. The wounds in the wrists and feet, along with the parallel blood flows on the arms produced as the man pulled himself up to breathe, confirm this. The identification of serum around the vertical blood flows at the wrist wound and arm also indicate the blood bled and coagulated while the man was in the vertical position. Other pieces of evidence contained on the Shroud also corroborate a crucifixion: the man's abnormally expanded ribcage and enlarged pectoral muscles appear drawn in toward the collarbone and arms;[60] the upraised left leg was bent so that the left foot was apparently placed on top of the right foot; and the man was beaten and whipped.

Several physicians have discussed the exact cause of death resulting from crucifixion. Most believe it involved asphyxia—that is, difficulty in exhaling—or some other respiratory problem caused by a lack of oxygen, accompanied by or related to cardiac failure or to complications from shock and pain. Muscle spasms, progressive rigidity, and the inability to exhale would also contribute to the cause of death. When a man is hung by his arms, the pectoral muscles contract around the lungs; to exhale, the victim must strenuously and continuously work to raise himself up and expand those muscles. The pain from this pushing would be intense and constant, especially for the wrists, arms, and feet. Tremendous pain can exhaust a person and is often accompanied by excessive sweating, which causes the body to lose vital fluids and minerals that regulate heartbeat. Eventually the victim goes into a state of shock. Hence, a series of different physiological events contributes to death by crucifixion. Everyone who has written about this form of execution describes it as a gruesome and horrible type of torture. If the victim were lucky, death would come within a few hours, but some lasted as long as two or three days before dying.[61]

All medical authorities who have studied the Shroud agree the man died while on the cross. This is first indicated by the side wound, which we know was inflicted after death. The blood from this wound shows patterns that indicate the man was still in a vertical position when he received it. The side wound blood is darker and more copious than that from the other wounds, and it is interrupted by patches of a clear, watery fluid.[62] Although they differ somewhat in their explanations of this blood and water, most experts believe the blood came from the heart and the watery fluid from the pleural cavity in the chest.[63] Dr. Bucklin, who summarizes the slightly divergent opinions of pathologists, thinks that the most realistic interpretation is that the watery fluid came from the pleural spaces and, perhaps to some extent, from the pericardial sac surrounding the heart, while the blood came from a piercing wound to the right side of the heart.[64] The side wound is located between the fifth and sixth ribs. An instrument such as a *lancea* thrust upward into this ribcage region would have pierced the right auricle of the heart, which is only about three inches from the chest surface and fills with blood on death (fig. 24).

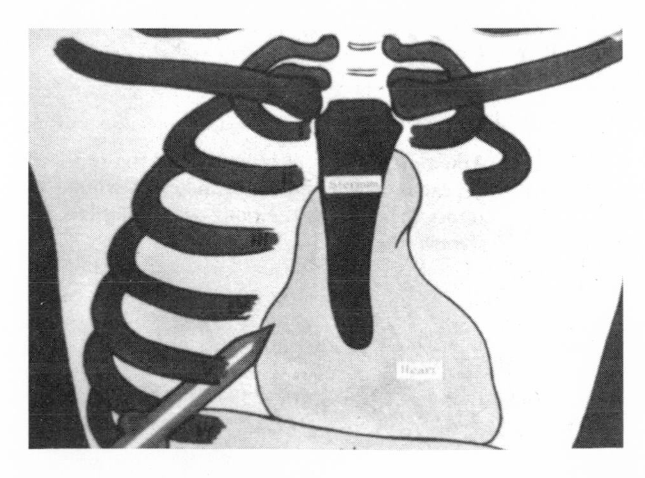

Fig. 24. Location of the spear wound on the man in the Shroud.

Further evidence of the man's death on the cross is found in the numerous identifications of rigor mortis apparent on the Shroud image.[65] Rigor mortis develops because of complex chemical processes that cause all body muscles to stiffen. The actual stiffening typically begins four to six hours after death and continues for another twelve hours. Once complete, rigor mortis gradually declines over the next twelve to twenty-four hours and the muscles relax again. The onset of rigor mortis can be accelerated by muscular exertion before death, an elevated body temperature, or warm weather.[66] In cases where physical activity has been strenuous and intense, as would be the case in a crucifixion, rigor mortis can set in immediately after death, especially in a hot climate.[67] If the corpse were then placed in a cool environment, such as a tomb, rigor mortis would tend to remain longer.

When looking at the back of the man's legs and feet, we see that his left leg is raised slightly and that both feet, especially the right one, are flat and pointed down. For the lower extremities to have remained in such an awkward position indicates that rigor mortis set in while the man remained crucified.[68] Moving up the back of the man, we notice that the thighs, buttocks, and torso are not flat, but instead are stiff and rigid. If rigor mortis had declined and the muscles had relaxed, these parts of the body would appear flatter and wider.[69] On the frontal image we see the chin drawn in close to the chest and the face turned slightly to the right. For the head to remain in this position inside the burial cloth without rotating further to the side requires the presence of rigor mortis.[70] The man's expanded ribcage is a sign of asphyxia, and the enlarged pectoral muscles drawn in toward the collarbone and arms provide evidence that the man had been pulling himself up to breathe.[71] That these parts of the body remained in such positions further indicates that the onset of rigor mortis occurred while the man hung suspended.[72] Rigor would also maintain the thumbs in the positions held during crucifixion.[73]

When we compile all the information about the wounds on the man in the Shroud with the knowledge that they were inflicted over a period of several hours, we can reconstruct what happened to him with some accuracy. Most likely, he was first beaten about the head, which caused swelling, bruises, and lacerations on his head and face. The scores of scourge marks all over his body attest to a whipping. Something made of sharp, thornlike objects placed over his head caused numerous piercing wounds on the front, top, and back of the head. Some of these wounds could have occurred from being struck on the head after the thornlike objects were placed over it. Other head wounds may have resulted from falling and being struck in the head by the crossbeam often carried by victims to the execution site, or from scraping his head against the cross when he pushed himself up and down to breathe. The shoulder abrasions could also have been imposed as he carried the crossbeam or later scraped his back during the up-and-down breathing motion. If some of these injuries were suffered while the man carried the crossbar, they may have occurred at the same time the victim apparently fell, as evidenced by the dirt in the nose and knee areas, as well as the scratches and cuts detected on his nose, cheek, knee, and leg. Such dirt and scratches suggest the man had been unable to break his fall with his hands. With a crucifixion crossbeam weighing an estimated one hundred pounds, it is reasonable to assume that victims fell frequently, especially since they were no doubt in a weakened condition already.[74]

The man's foot and wrist wounds were next inflicted by large nails driven through his flesh between the metatarsal and wrist or forearm bones to anchor him on the cross. The crucifixion alone would have taken several hours. From the two parallel streams of blood flows, the expanded ribcage, the enlarged pectoral muscles drawn in toward the collarbone and arms, and the taut legs and buttocks, we can observe this victim was pushing and pulling himself up to breathe. In addition, he likely incurred two more especially painful experiences. The median nerve, which controls the thumb, is not only a motor nerve, but is also a great sensory nerve.[75] It would be excruciating for this nerve to be lesioned by a nail, and the pain would only be aggravated further when the man pulled himself up to breathe.[76] Also, the forehead, temple, and scalp contain a rich supply of nerves whose sensitivity is among the most painful in the body.[77] The infliction of more than thirty different wounds about the head would have contributed even more pain and agony than a typical crucifixion. After he was dead, a spearlike weapon thrust into his right side pierced his heart, causing blood and watery fluid to escape.

All of the data gleaned from extensive study of the pathology evident on the Turin Shroud tells us this piece of linen was wrapped around the corpse of a man who was crucified and died while still nailed to a cross. We also know that the man's corpse lay inside the burial linen for no more than two or three days. Had he been there longer, decomposition stains would be present on the cloth, but the Shroud contains no signs of bodily decomposition. It is also noteworthy that the bloodstains remain unbroken and unsmeared and appear to be exact mirror images of the man's wounds. This tells us that the blood marks could not have been transferred to the Shroud through direct contact between the body and the cloth alone; instead, some other process must have been at work.

CHAPTER THREE

THE IMAGES ON THE SHROUD

Largely due to the efforts of STURP members prior to and since the days in Turin, the majority of the Shroud's unique features have been unlocked. The enigma of the body image could be penetrated only when technological advances supplied the required equipment and a collection of talented scientists was organized to wage the campaign needed for such extensive investigation. STURP's findings will be the focus of this and the following two chapters. At the center of this discussion lie the unprecedented characteristics of the body images and blood marks. These features combine to make the Shroud of Turin a truly extraordinary object that is far more complex than history ever imagined.

THE BODY IMAGE IS A NEGATIVE

Before explaining the image characteristics uncovered with modern technology, we have to back up one hundred years. While the scientific equipment available today could not even have been imagined in the last century, one important imaging instrument did exist—the camera—though it was primitive by our standards. In 1898, the first remarkable discovery pertaining to the Shroud accidentally emerged when Seconda Pia, an award-winning amateur photographer and attorney, was allowed to photograph the Shroud. When Pia developed his film, he became so startled by what he saw that his trembling nearly caused him to drop the heavy photographic plate. He was the first to witness the "negativity" of the body image. This stunning revelation provided the earliest clue that a great deal of information invisible to the human eye was somehow encoded within the cloth. It would be another eighty years before scientists began to realize just how much hidden evidence the Shroud actually holds.

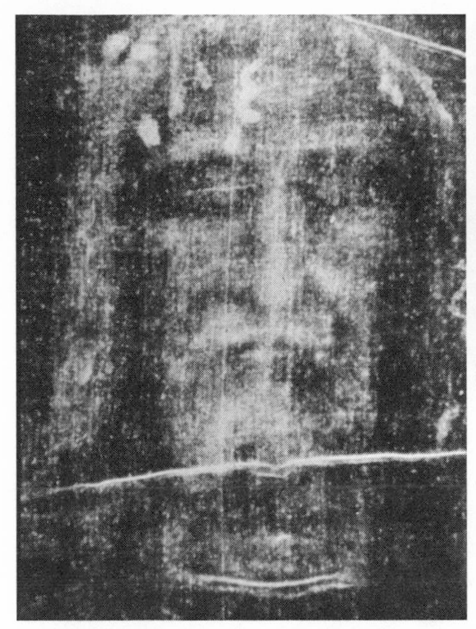

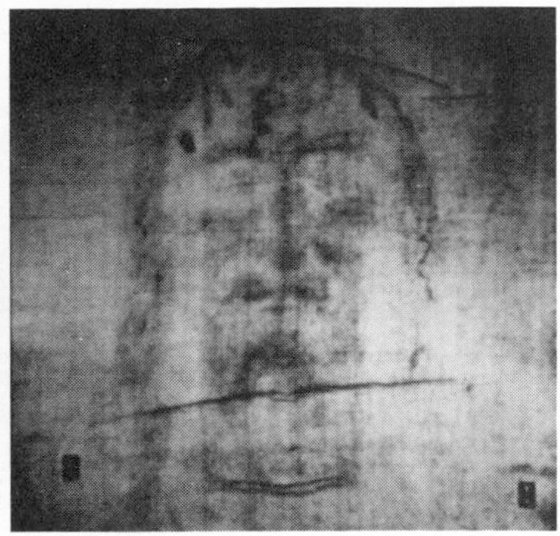

(above) Fig. 26. Negative image of the face on the cloth.

(left) Fig. 25. Positive image of the face revealed by the photographic negative.

As Pia watched the photographic chemicals do their work, a highly detailed picture of a man emerged, an image comprised of precise shadowing and light/dark contrast that suggested depth. When the Shroud is seen with the naked eye, the body image looks blurred and lacks clarity. Its straw-yellow color blends in with the background linen so the eye sees very little contrast, especially if the observer is standing near the cloth. Only on a photographic negative do the sharp details leap out to compose the realistic picture of a crucified man, a picture so distinct that the man's features, posture, and wounds are unquestionably visible.

It has long been an accepted scientific principle that a negative of a negative always produces a positive. With Pia's discovery, Shroud scholars realized the body image seen with the unaided eye is itself a negative that becomes a positive image only when photographed. On the photo negative, light and dark areas, as well as left and right sides, are reversed when compared to the Shroud image that is naturally seen. Since all photographs work this way, we can easily visualize this by thinking of any photographic negative. Blond hair appears black, and objects on the left side of a photographed scene appear on the right side of the negative. Although most photo negatives appear ghostlike and unnatural, the Shroud photo negative discloses an image of realistic clarity, as with a photo positive. As a specific example, many of the scourge marks that cover the man on the Shroud are invisible to the naked eye, but we can see scores of them on the photographic negative because it contains such minute and specific details.

This century-old discovery raises important questions. How did this phenomenon happen? Since this concept of photonegativity would be recognized only when the science of photography was invented in the mid–nineteenth century, how could a medieval (or earlier) artist be responsible for the image on the Shroud? How could a medieval artist grasp such a concept, and for what reason? Or could an artist possibly create this negative image by mistake? He would have to produce, on the naked-eye negative, the proper density and shading, as well as the fuzziness that hides the actual clarity of the image, and do it all in reverse. In addition, he'd have to imagine and then encode the numerous but minute details that would remain invisible for centuries to come. To make matters even more difficult, since the alleged artist couldn't see the real positive image that appears only through the photographic process, he'd have no way of knowing whether he got it right.

THE BODY IMAGE IS SUPERFICIAL

One of the discoveries that emerged and astonished scientists in the early 1970s was the superficiality of the image. The Shroud cloth consists of countless woven threads, each one consisting of countless fibrils. The Shroud body image lies only on the very topmost fibrils of the threads. The thread fibrils composing the body image are straw yellow in color, yet the color does not penetrate into or between the threads. Instead, the color resides only on the top two or three fibrils that make up the threads. To use an analogy, if you view your arm as a single Shroud thread, the yellow-colored body image would rest on the hairs of your arm. Further, where two fibrils cross, one over the other, there will be a white spot on the lower fibril. This tells us the mechanism responsible for encoding the color did not penetrate down to underlying fibril surfaces.

Another unusual characteristic related to the image's superficiality is the uniformity of color: body image fibrils are basically identical in terms of the degree or intensity of straw-yellow color they possess (see fig. 27).[1] Any differences in intensity between different parts of the body image are due solely to the number of yellow-colored fibrils concentrated in a given area. Although certain parts of the image may appear to be darker, they do not actually contain a darker yellow coloring; rather, the darker appearance results from a greater number of colored fibrils in that particular location.

The implications of these findings are highly significant. If a paint, dye, pigments, or any other type of painting medium was used to create the figures on the Shroud, the images could not remain on only the topmost fibrils of the linen's threads. Any liquid painting substance would naturally soak into the threads, and the fibrils would consequently become matted together. On the Shroud, however, each colored fibril is distinct from all others, and fibrils do not adhere to one another.

Another important implication of the images' superficiality also relates to the hypothesis that the Shroud is a medieval painting. Because the many fibrils that make up a single thread are too fine to be distinguished by an unaided human eye, and because each fibril has been individually encoded with color, to see them a painter would need a microscope—several centuries before the instrument was invented. And because the individual image fibrils have been superficially colored, a painter

Fig. 27. Photomicrograph of the body image on the nose.

Fig. 28. Computer scan of the face showing complete lack of directionality.

would need to use a paintbrush with a single bristle. The finest paintbrushes known are made of sable hair, but even a single sable hair is thicker than a linen fibril. In addition, a painter would need the skill not only to paint each fibril separately, but also with the same intensity.

The painting theory has another serious problem relevant to this discussion. When the Shroud is viewed at close range by the naked eye, the straw-yellow body image cannot be differentiated from the lighter yellow background cloth. To see the man's figure, a viewer must stand six to ten feet away from the Shroud. This phenomenon is due to the fact that the human eye automatically enhances the edges of objects to help the brain make sense of the multitude of things it observes. Since the Shroud image lacks sharp boundaries and is so faint, the human eye and brain cannot compose a picture of the image until they move some distance from the subject and can enhance the image's edges. Thus, our poor artist would have had to stand six feet away even to see what he was doing!

THE BODY IMAGE IS DIRECTIONLESS

One piece of high-tech imaging equipment used to study the Shroud is called a microdensitometer. This instrument digitally measures the density of details contained in a photographic negative. STURP members Don Lynn and Jean Lorre, working at the Jet Propulsion Laboratory in 1976, used a microdensitometer to scan photos of the Shroud. The scanning data was then processed through a computer and displayed on a television screen. The image resulting from this technique highlighted additional subtle details invisible to the naked eye.

These density studies revealed another image characteristic that impressed scientists: The Shroud body image appears to be directionless.[2] When the computer-displayed image was viewed at high resolutions, the only directional features found on the Shroud were in the weave of the cloth; in the body image areas, the color was randomly oriented.[3] This absence of directionality (up and down, side to side, etc.) is noteworthy because one would expect the microdensitometer to expose the presence of brush strokes running in obvious directions if the image were painted. The microdensitometer testing, therefore, offered a method for demonstrating that the Shroud is indeed a painting, yet it did not prove this. Instead, the image's directionless nature provides one more piece of evidence consistent with the theory that the Shroud image did not result from an artist's application of some foreign substance but rather was encoded directly from a body lying underneath the cloth.

THE BODY IMAGE IS THREE-DIMENSIONAL, FORMED THROUGH SPACE

Unquestionably, the Shroud of Turin's most extraordinary feature thus far disclosed is the three-dimensionality of its body image. This characteristic was revealed in 1976 when scientists examined a photograph of the Shroud with a VP-8 Image Analyzer, another piece of space-age imaging equipment.

VP-8 imaging equipment is designed not to interpret conventional photographic images, but to interpret radar-type data. For example, if a spacecraft were traveling by the planet Saturn and its rings and moons, it could send out electromagnetic impulses, such as light or radio waves, to these various features. Depending on the distances these impulses traveled to the objects and back to the spacecraft, true and accurate three-dimensional images of these features could be generated by a computer. That is because the closer a feature is to the spacecraft, the more light it will return. Only when there is an actual and direct correlation between lightness and distance will the computer generate a true three-dimensional image. The VP-8 Image Analyzer constructs images on this same principle. Without this correlation, a distorted image will result.[4]

The Shroud is a flat piece of linen, a two-dimensional object. Photographs and paintings are also two-dimensional. A regular camera does not work like radar. Like a one-eyed man, it has no depth perception. Likewise, when artists want to create an impression of distance or three-dimensionality, they use color and shadowing to trick the eye into "seeing depth." But artists can only *suggest* the dimension of depth. Consequently, two-dimensional photographs processed through the VP-8 Image Analyzer will yield distorted images. Figure 5 shows a typical example of a normal photograph of a face; the nose is flattened, and the mouth and eyes are distorted and sunken. Since no significant distance differences are represented in two-dimensional photographs, the VP-8 cannot produce an accurate reconstruction of the dimension of depth. The only way an artist could "fool" the VP-8 would be to paint, in effect, a tyopographic map of a face, each shade of paint corresponding to a certain distance. While the computer might be taken in by such a painting, you or I could easily see it for what it would be—a *diagram* of a face, not a picture.

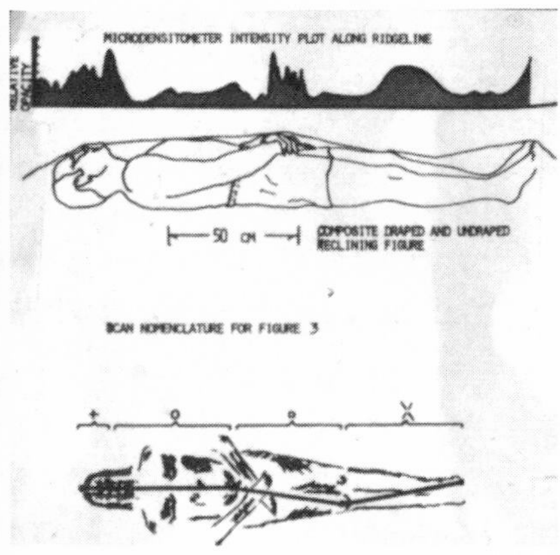

(above) Fig. 30. Microdensitometer plot.

(left) Fig. 29. Full body three-dimensional image.

As early as 1902, French biologist Paul Vignon observed that the most intense areas of the Shroud image (i.e., those areas that contain the greatest number of colored fibrils) correspond with those body parts that would have been in closest contact with a cloth lying over a body: the nose, forehead, chin, hands, etc. Areas of the body that would have been farther away from the cloth left a lighter impression.[5] This observation was not even possible until it was revealed by the detailed photographic images with such precise resolution and shading. Working on the theory that three-dimensional information might be encoded in the Shroud image, physicist John Jackson contacted Bill Mottern, image specialist at Sandia Laboratory, who had access to a VP-8 Image Analyzer. After using the machine to process a Shroud photo, both men were stunned when a correctly proportioned three-dimensional image of a man appeared on the computer screen. The details revealed by the VP-8 analysis confirmed the theory that the light and dark areas of the Shroud image do, in fact, represent actual distance information.

Because all physicists are driven by the need to support theories with mathematics, Jackson and his colleagues then set out to calculate the probable distances between a draped cloth and a body lying beneath it. They found volunteers of similar height and weight as the man in the Shroud and draped over them a piece of linen with a herringbone weave and thickness similar to the Shroud's. Using a cloth with the body image's major features traced on it, they tried to duplicate the Shroud's original drape as closely as possible. Based on photographs of the draped volunteers, they constructed a drawing of a covered, reclining man and could then easily measure the distance between the draping cloth and the underlying body.[6]

With this information in hand, the team obtained microdensitometer readings from an original slide made by Giuseppe Enrie, official photographer of the 1931 Shroud exhibition. Measurements of the Shroud photo's intensity values were then directly compared to measurements made earlier of the distances between the draped cloth and the underlying body. When calculated, the results showed "a definite correlation between image intensity and cloth-to-body distance."[7] The mathematics, therefore, support the theory that the image is less intense where the distance between the cloth and the body would have been greater; conversely, the nearer the body to the cloth, the more intense the image. Studies by other scientists working at the Jet Propulsion Laboratory,[8] as well as researchers at the University of Turin,[9] have independently confirmed this three-dimensional distance information. Interestingly, the image encoded even in places where the cloth could not possibly have been touching the body. Whatever the image-forming mechanism was, it also acted through empty space.[10]

When reconstructed by computer, the Shroud image's three-dimensional characteristic is more noticeable on the face than elsewhere on the body. That's because the face contains greater variations of depth—more rises and dips. If we think of a face in topographical terms, it looks more like a mountain range than a flat plain. Although close together, the forehead, nose, eyes, cheeks, lips, chin, and so forth all have differing "elevations." The face has a greater variety of relief than any other comparable area on the Shroud images.[11] Now recall the earlier discussion of the image's superficiality. All fibrils are encoded with the same intensity of color. If one image area appears darker than another, that area contains a larger number of colored fibrils. Thus, the number of colored fibrils somehow represents the three-dimensional distance information.

Whatever mechanism created the image on the Shroud, scientists have determined that it operated uniformly over the entire body and acted independently of the surface material on the body, encoding the presence of different types of organic matter, such as skin and hair, and perhaps encoding inorganic material as well. Close-ups of the three-dimensional face with suppressed relief suggest the presence of small, solid, round objects lying over the man's eyes. More detailed study of the eye area and its possible significance follows in chapters 6, 9, 10, and 11. What's important to understand now is that an image-forming mechanism that encodes three-dimensional distance information and is also capable of operating uniformly (no matter what body materials are involved) places certain limitations on what that mechanism might be.

THE BODY IMAGE WAS FORMED ALONG VERTICAL PATHS

From the high resolution found on the Shroud body image, the three-dimensional information that is contained within it, from the fact that neither the sides of the body nor the top of the man's head were encoded, and from certain cloth drape effects present on the body image, scientists were able to learn of and demonstrate another exceptional feature of the body image. The image-forming mechanism on the Shroud acted in vertical, straight-line paths.[12] If the transference did not operate in this way, an undistorted, highly focused image would be impossible to obtain with the VP-8 Image Analyzer.[13] Remember, the VP-8 works like radar, relying on signals from a single, unidirectional source.

The only parts of the supine body that transferred to the frontal image of the Shroud were those parts facing upward with no obstructions to the cloth draped over them. Using computer-modeling techniques and actual-size, three-dimensional models of the man in the Shroud, scientists attempted to determine the direction or path along which this transfer of information took place. After measuring various directions that the image-encoding mechanism could have taken to the cloth, scientists found that only the vertical, straight-line direction from the body up to the cloth would correspond consistently between the observable points on the flat cloth and the underlying body. Any other paths resulted in blurred or distorted images. Similarly, computer modeling and scientific calculations revealed that only a vertical encoding direction from the body up to the cloth resulted in an anatomically reasonable, three-dimensional body image such as that on the Shroud.[14] This consistent directionality was found throughout the length of the frontal image, whether the cloth had lain flat or sloped upward or downward while it was draped over the underlying body. STURP scientist John Heller noted, "It is as if every pore and every hair of the body contained a microminiature laser."[15]

This explains why the sides of the man's body, as well as the top of his head, are absent. Since these parts of the body are in vertical positions themselves, a vertical mapping process would not accurately transfer them. In an earlier study consistent with this, scientists had identified subtle cloth drape distortions along the sides of the body image that could be explained only by a vertical mapping process.[16] In the same study, STURP scientists also identified distortions in the length of the arms and fingers. While these and many other image characteristics will be accounted for in chapter 10, at this point we will note the scientists also found that only an image-encoding process that operated in a vertical direction could explain these features.[17]

DISTINCT DIFFERENCES BETWEEN THE BLOOD AND THE BODY IMAGES

The figure of the man on the Shroud is composed of two distinct substances: the straw-colored fibrils that form the body image, and the actual blood that corresponds to the man's wounds. Both the blood and the body image, are so remarkable that neither history nor science has seen anything like them before. Yet each is quite different from the other and may have been transferred to the cloth through different, although related, processes. When STURP members chemically removed small samples of blood and serum to microscopically examine the underlying fibrils, they discovered that the blood-removed fibrils look more like those of the background

Fig. 31. Photomicrograph of a Shroud bloodstain.

Fig. 32. Photomicrograph of a Shroud burn mark.

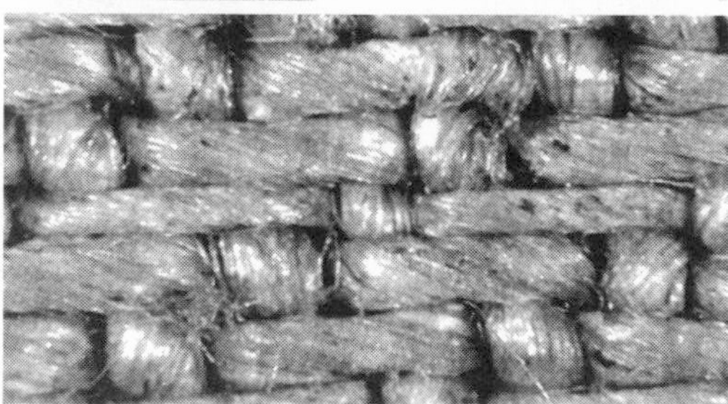

Fig. 33. Photomicrograph of dirt found at the man's heel.

cloth than the body image. In addition, in the published photomicrographs of the bloodstains, white fibrils appear where the dried blood has broken away from the cloth, suggesting that the bloodstains were transferred to the cloth before the body image was encoded.[18] Scientists also observed an absence of body image on the margins of the man's wounds where serum has been identified. This, too, suggests that the wounds, present on the cloth first, coated or sealed the underlying fibrils from the process that formed the body image.[19] Still, it is unlikely the bloodstains were fixed on the linen by simple direct contact alone, as the man was wrapped in the Shroud, because the blood marks are too clear. Instead of looking like the indistinct blob of blood you'd see on a bandage or dressing when you remove it from a wound, the marks on the Shroud duplicate the exact shape and appearance of actual blood as it forms, flows, and congeals on human skin.

One of the distinct differences between the blood and body images exists in the sharpness of their edges. Distinctly bordered edges surround the bloodstains, while the outline of the body image is vague and difficult to distinguish when viewed with the unaided eye. Because the body image fibrils are all colored with a uniform intensity, an observer can visualize the body image edges only when standing six to ten feet away. In contrast, the bloodstains vary in intensity, with the color most bold in the regions of the side wound, the head, the hands, and the feet. In the scores of scourge marks, the blood coloring is paler, so that some are visible only on the photographic negative or under ultraviolet lighting.

In addition, the bloodstains vary in depth, with matting and capillarity evident in the bloodied thread fibrils. As any liquid would, the blood has penetrated the fibrils and caused them to stick together. Blood is present between the threads of the weave and, in at least some places, it has gone through to the back of the Shroud. Unlike the saturated blood areas, the body image is superficial, and the image threads are completely separate and unmatted. The straw-yellow coloring is encoded only on the topmost linen fibrils and does not penetrate into or between the weave.

The blood and body images on the Shroud are also composed of different substances. STURP members have concluded that the blood images on the Shroud are composed of actual blood and blood products. They have also determined that the straw-yellow body image fibrils have been colored by oxidized and degraded cellulose.[20] Cellulose is an abundant natural material that is found in all plant tissues and fibers. In a sense, cellulose is the raw material from which linen is manufactured; because of this, cellulose is present throughout all fibrils of the Shroud.

Yet if all linen fibrils are composed of cellulose, and if no substance has been applied to the Shroud to color the body image fibrils, what makes the body image visible? According to STURP scientists, the cellulose in the body image areas has degraded, or decomposed. This accounts for the somewhat corroded appearance and loss of strength evident in the fibrils that compose the image (see figs. 34 and 35). Several conditions, including light, heat, and even exposure to air, will eventually cause linen to dry out. As cellulose dehydrates, it combines with oxygen. Such dehydrative and oxidizing processes normally happen when linen ages, and they produce the same chemical changes and altered structure of degraded cellulose as that found in the Shroud's body image areas. However, while all linen will naturally turn yellow with age, something has accelerated this process on the topmost fibrils of the body image threads, so they are in a more advanced stage of aging, or yellowing, than the nonimage or background threads. This slight difference in color makes the body image visible; at the same time, the color contrast is so minimal that the man's figure fades into the background cloth when the Shroud is viewed at close range.[21]

Of all the differences between the blood and the body images, one of the strangest is visible on photographs of the Shroud. Whereas the body image is a negative in its natural state and becomes a positive only on a photographic negative, the bloodstains appear as a positive image when viewed with the naked eye.

Two other seemingly minor features need to be mentioned that will have significance in later chapters. While the vast majority of all blood and scourge marks on the Shroud consists of actual blood exudate encoded by a process involving intimate contact with the body, a few of the blood images on the arms and, possibly, scourge marks on the hips and pectorals are composed of degraded cellulose, like the body image.[22] These particular marks would be more visible on the photographic negative of the body image than with the naked eye.

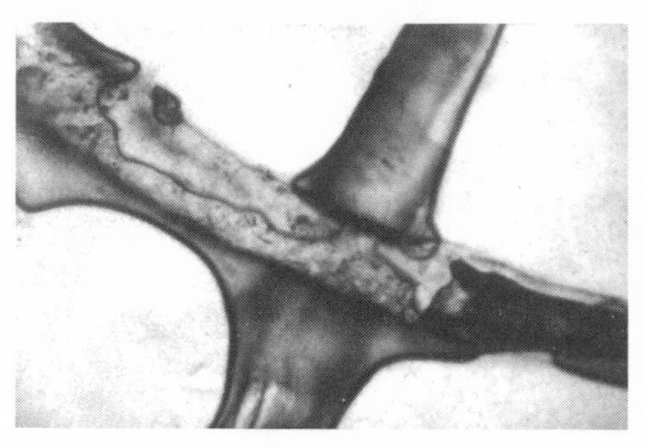

Fig. 34. Body image magnified 400 times. Note its distinctive "corroded" appearance.

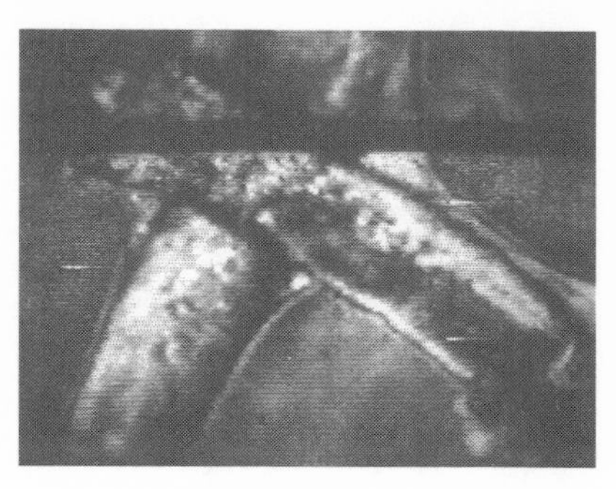

Fig. 35 Non-body image magnified 400 times. Note its intact and noncorroded appearance when compared to the body image fibril.

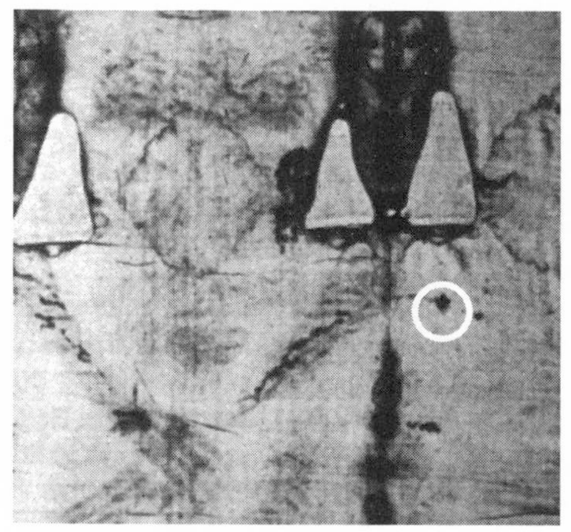

(above) Fig. 36. A continuing flow of clotted blood that ended in a rounded clot (circled) off the left elbow can be seen on the frontal image.

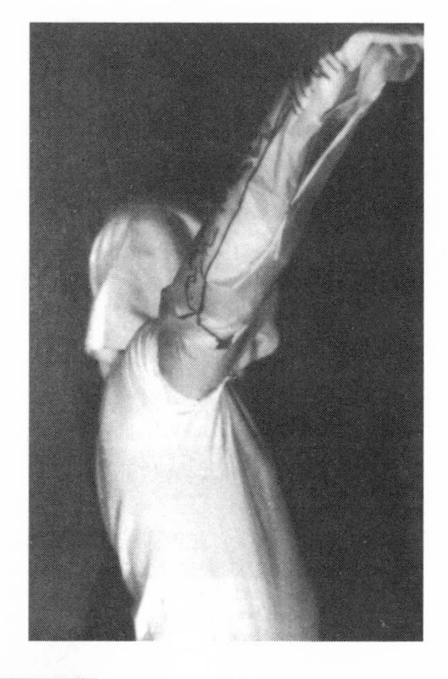

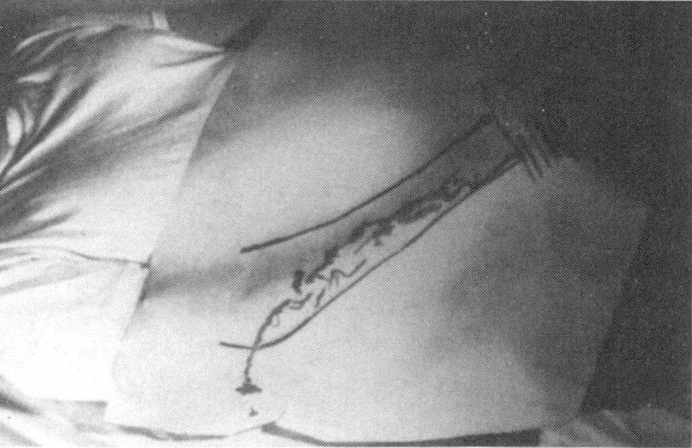

(above) Fig. 37. Tracing wrapped around arm indicating how the blood clot formed at the location off the left elbow while in the vertical position of crucifixion.

(left) Fig. 38. Illustration of how the burial cloth picks up the blood clot off the back of the elbow when the victim is lying down.

In addition, a blood mark can be seen on the cloth off the left elbow in the frontal image (fig. 36); also, an unbroken blood line runs from the elbow to this blood mark. In 1983, Dr. Gilbert Lavoie discovered that this flow originally ran from the top frontal part of the man's elbow to the back of his upper arm, as can be seen in figure 37. After the cloth was laid horizontally over the supine victim, it draped along the side or back of the man's upper arm and picked up this entire blood flow and blood mark (see fig. 38). No body image of the upper arm was encoded at this location.[23]

TECHNOLOGY AND THE SHROUD IMAGE

During much of the Shroud's existence, its most impressive features were hidden from view. With the 1978 STURP investigation, however, existing technological advances enabled scientists to visualize the Shroud more completely. As they saw details never before suspected, the public gained greater insight into the Shroud's truly extraordinary nature. The human mind finally pushed science and technology far enough so people could begin to unlock and peer into the depths of the Shroud. And the further people looked, the more remarkable the body image became.

For centuries, the Shroud of Turin was perceived as a strange yet simple piece of linen imprinted with the indistinct figure of a man. Along came the camera and a sharply focused figure jumped into view. With this technological leap, which delineated the image's boundaries and shadowing, a startlingly detailed and natural-appearing figure of a man, who had suffered the agonies of crucifixion, was unveiled. Photographic work in 1978 led to further revelations. Photos taken under ultraviolet fluorescent lighting indicated the existence of serum around the wounds and showed the faint scourge marks, along with other scratches and cuts that could never be observed with the naked eye.[24] The photographic enlarger, coupled with the microscope, disclosed the absolute realism of the indented centers and upraised edges of the scourge marks that result from the natural process of clot retraction. Photos enhanced by computer digital image analysis and display system techniques identified the actual separation of serum from red blood cells, as well as the migration of cells toward the outside edges of the bloodstains.[25] Similar technology also uncovered the narrow lesion present at the top of the side wound's blood flow.[26] Small, round objects over the eyes became apparent with advanced photographic processes such as isodensity. These objects were seen in even greater detail with three-dimensional analysis.

Other imaging and computer studies have enabled people to see the Shroud images' characteristics. Microscopes exposed the matting and capillarity of thread fibrils in the bloodstained areas, as well as the superficiality of the body image, encoded only on the tops of the fibrils and with a uniform intensity of color. Microdensitometer studies revealed the non-directionality of the body image. Perhaps most astonishing of all are the findings of the VP-8 image analysis: Three-dimensional information has been encoded on the Shroud's two-dimensional surface. This discovery has advanced understanding of the image-forming mechanism and helped to determine that it operated through space and along vertical paths as it transferred information pertaining to distances between cloth and body, information that directly corresponds with image intensity.

Yet the most remarkable aspect of the Shroud image is its ability to actually make quantum leaps. The figure on the Shroud has developed from an image on cloth, to a photographic image, to a three-dimensional image. This is unprecedented in all of history. No other relic or work of art has taken such quantum leaps. If works of art are well cared for, their quality can be preserved, but works of art cannot *improve* with age. In the case of the Shroud, as technology improves, its image actually improves. This remarkable quality, of course, is not caused by developments in technology, but results from the unique way in which the image was encoded centuries ago.

CHAPTER FOUR

PAINTING THEORIES EXPLORED

We have learned most of what we know about the Shroud only since 1978, when STURP first used science and technology to examine the cloth itself. Skeptics' attempts to explain the image, however, date back to the beginning of this century, soon after the astonishing first photographs of the Shroud were taken in 1898. Further insight came from the exhibitions in 1931 and 1933, as well as through Giuseppe Enrie's photographs. Those and others taken by visitors to the exhibits finally put to rest the totally unfounded accusations of fraud and tampering against Secondo Pia.[1] These accusations arose in an environment of startling reaction to Pia's accidental discovery by a public that could not explain this phenomenon, and who had been informed since the Middle Ages that the Shroud was a painting. Enrie's photographs would stand unsurpassed until STURP arrived in Turin with its eight tons of high-tech imaging equipment and its modern scientific methods.

In light of the previous chapter and its explanation of many of the image characteristics discovered by STURP, we now examine the historical argument contending that this image of a crucified man is a medieval painting. Yet the theory that a medieval artist created the Shroud is immediately confronted with a very thorny question: how could an artist working in the Middle Ages have encoded, on a piece of fabric, image characteristics that are so unusual and complex that only late-twentieth-century technology could begin to decipher them?

Recently, the most vocal advocate of the painting theory, and perhaps the best known among modern Shroud skeptics, is microscopist Walter McCrone. After examining fibril samples from the Shroud, McCrone announced his opinion that an artist had painted the body image by using an iron oxide (Fe_2O_3) pigment (called red iron earth pigment or jeweler's rouge) suspended in a gelatin binding medium.

He further asserted that mercuric sulfide pigment, usually called red vermilion, was added to the blood mark areas.[2] This or any other painting theory, however, is not supported by the data gathered through STURP's rigorous testing. As we will see in this chapter, McCrone's suspect methods have hurt the cause of science—and not just in the case of the Shroud. In fact, the theory that any painting medium or technique could be responsible for the extraordinary images contained on the Shroud seems impossible.

STURP scientists decided that a detailed study of the body image, background fabric, burn marks, and water stains seemed the most appropriate place to begin testing the historical painting theory. They first investigated the results of a "natural experiment" that had taken place centuries ago. In 1532, the sanctuary in Chamberry (Chambéry), France, that housed the Shroud was destroyed by fire. The cloth had been folded and stored in a silver container, or reliquary, which was heroically rescued from the burning building and then doused with water. Although the Shroud survived, it was damaged—by both fire and water. The parts of the folded cloth that were most in contact with the hot silver lining were burned completely through, while other parts were scorched. Part of the silver lining inside the container may also have melted, and a piece of molten silver fell onto the linen and burned holes through parts of it. Fortunately, the burning did little damage to the man's image. It's also fortunate that the accidental trial by fire and water provided important scientific information regarding the Shroud's physical and chemical properties.

Through direct observation and high-tech imaging studies, STURP scientists concluded that those areas of the body image closest to the fire marks are actually no different, in color or composition, from those image areas farthest away from the scorching. If either inorganic pigments or natural organic stains or dyes were responsible for the Shroud images, some evidence of their chemical alteration should be observable in and around the fire marks left in 1532. The cloth was apparently affected by a varied range of temperatures since some parts seem not at all touched by the heat, some areas were scorched, and some sections contain holes where silver burned all the way through the linen.

Most, if not all, organic media and the color resulting from them would have been altered by the heat, and most inorganic pigments available in the fourteenth century would also have changed, depending on the pigment's distance from the hottest parts of the cloth.[3] Vermilion, for example, would likely have darkened when exposed to such high temperatures. Iron oxide, too, would have become darker near the burn marks, as would a gelatin binding material. Yet no change occurred in the color of image fibrils adjacent to areas that were burned or scorched. Furthermore, most inorganic pigments and natural organic dyes are water soluble. If iron oxide, vermilion, or a gelatin binder were used to create the image, the color should have run, or migrated, in those areas where water stains are present.[4] It did not. Hence, this natural experiment tells us that the Shroud image possesses two unusual properties: It remains stable even under high temperatures, and its material composition is not soluble in water. McCrone's proposed paint pigments fail to meet these criteria.

SPECTRAL STUDIES OF THE SHROUD

While the natural experiment of 1532 provided important information about the Shroud's composition, STURP members have studied more than five thousand photographs taken at different wavelengths, and have performed many hundreds of scientific tests to search more thoroughly for pigments and try to determine conclusively whether the Shroud is a painting. These experiments, which were designed on the solid foundations of chemistry and physics, led STURP investigators to conclude that neither the Shroud body nor blood images could possibly have been painted by an artist.

To understand the many spectral tests performed by STURP, a brief explanation of the electromagnetic spectrum is necessary. Made up of both the light we can see and the light invisible to our human eyes, electromagnetic (EM) radiation is everywhere throughout our world and the entire universe. Each type of EM radiation possesses a specific wavelength and frequency that distinguishes it from all other types. Since our eyes are sensitive to a very narrow range of EM frequency, those light waves we can see are called visible light. The wavelengths of the great majority of EM radiation lie outside our visible range. EM waves with shorter wavelengths and higher frequencies are called ultraviolet, X rays, and gamma rays, while those with longer wavelengths and lower frequencies are infrared, microwaves, and radio waves. All material in the universe (solids, liquids, and gases) is composed of particles, which make up atoms that are constantly moving about. Although we can't readily see its effects, this continual oscillation of particles and atoms causes all objects to emit, absorb, and reflect different forms of EM radiation.

Every chemical element (from hydrogen and helium, which are the lightest elements, to uranium and plutonium, which are among the heaviest) has a distinctive spectral "fingerprint" that can be measured. Because the spectrum of each element is unique to it alone, spectral readings can accurately identify, and distinguish between, all the chemical elements. Spectra are quite easy to obtain from gaseous objects such as stars because they are powered by intensely hot, internal nuclear furnaces and emit radiation in such huge quantities that we can measure it even at immense distances. From these spectra, astronomers can determine what stars are composed of. When compared to a star, the Shroud is a very cool object (room temperature in Turin, Italy), and spectra of its elements cannot be obtained unless the cloth's atoms are sufficiently excited to emit radiation that can be measured in visible light. One way to excite the atoms, and thereby obtain spectra of the cloth's components, is to use fluorescence techniques.

In fluorescence testing, the object is bombarded with a focused beam of either ultraviolet light or X rays. Some of the incoming radiation is reflected, and some is absorbed by the cloth's atoms. The absorbed radiation excites the electrons of the atoms into a higher-than-normal energy state, but these electrons must quickly return to their normal (ground) state. As they jump back to the ground state, the electrons emit the radiation they absorbed in the "excitement" process. This emitted radiation will have a lower frequency and therefore be measurable within the range of visible light. Materials that absorb light of one wavelength (such as X ray or ultraviolet) and then emit EM radiation at another wavelength (such as visible light) are said to fluoresce. In short, the fluorescence process produces radiation that can be measured by sensitive spectral instruments. And once the spectrum of an object is obtained, its chemical components can be identified.

Ultraviolet fluorescence studies can detect organic and inorganic compounds by their emission spectra. Using spectrophotometry equipment to measure the fluorescence and reflectance of materials, scientists learned that the body image itself does not fluoresce at all, in contrast to the fluorescent, nonimage background cloth. This is significant because gelatin, as well as other medieval pigment binders, is composed largely of animal protein; protein contains amino acids, which fluoresce noticeably under UV light. STURP's ultraviolet fluorescence photography also confirmed that the body image is nonfluorescing. Examination of bloodied areas under ultraviolet light revealed that the bloodstains do not fluoresce, with the exception of the light outline surrounding the large side wound; interestingly, blood plasma stains on linen tested in the laboratory fluoresced in the same way.[5] A related study of the Shroud's reflectance showed that the reflectance curve characteristic of Fe_2O_3 (iron oxide) is quite different from the curve obtained from the Shroud body image.[6] Thus, STURP's UV reflectance and fluorescence photography and spectroscopy studies found no evidence that McCrone's proposed iron oxide and vermilion suspended in a gelatin binder are present in the body image or blood marks on the Shroud.

X-ray fluorescence is considered the most important test for detecting inorganic pigments, yet these studies also showed no evidence of a painted image. X-ray fluorescence studies can identify inorganic (metallic) elements with atomic numbers greater than 16, many of which are typically found in paint (including antimony, arsenic, cadmium, chromium, cobalt, lead, manganese, mercury, nickel, palladium, tin, silver, and zinc). However, no differences in metallic element concentrations were found between the image areas and the nonimage background cloth.[7] This means these elements are not present on the Shroud in sufficient quantities to be seen by the naked eye; consequently, they could not account for the body image.

While the X-ray fluorescence studies did identify the spectral fingerprint of iron, the concentrations were small and measured essentially the same in the body image areas and the background cloth. In other words, iron is evenly distributed all over the linen. If McCrone's proposed iron oxide pigment had been used to paint the body image, a spectral difference would have been clearly manifested between areas marked with the body image and the nonimage background linen.[8] Although iron's spectral fingerprint was weak in the body image areas, it was quite evident in the regions of the wounds and blood marks. This makes sense because blood, of course, contains iron.

Other X-ray tests were also used to evaluate the historical painting thesis and to gain insight into the Shroud's composition. X-ray absorption studies illustrate the respective densities of elements because more dense materials will absorb more of the X rays directed at them. Thus, materials with heavier atomic weights (i.e., denser materials) will be more visible on radiographs than those with lighter atomic weights. Since mercuric sulfide (vermilion) has an especially high density, it would be sharply visible on X-ray absorption radiographs. With this in mind, STURP studied the Shroud's side wound area where blood marks, body image, burn and scorch marks, water stains, and repair patches are all found in close proximity. If vermilion were present in the blood mark areas, it would be easily detected by the absorption tests. It was not.[9]

To the naked eye, the blood marks on the Shroud are much more visible than the water stains, with the side wound blood flow appearing especially dense. Yet the only features visible on the X-ray absorption radiographs were the water stains. When these stains were chemically analyzed, they were found to be rich in calcium and iron, which is typical of common iron- and calcium-rich hard water.[10] If the blood marks were painted with iron oxide and vermilion pigments, they would also be expected to leave a dense and distinct imprint. In fact, iron oxide and vermilion are more dense than the iron and calcium found in the water stains and should, therefore, show up even more clearly on the absorption radiographs. But in the absorption studies, the blood marks were not visible at all. (Blood does contain some iron, but it was below the detectable threshold for this test.) Hence, iron oxide and vermilion pigments could not have been used to create the bloodstains.[11]

Thermal photography techniques (which detect EM radiation in the long-wavelength, far-infrared region of the spectrum) were also used to study the Shroud. Since thermal photography exposes the presence of elements used in most oil, acrylic, tempera, or watercolor paintings, it is often used to investigate suspected art forgeries and identify the components of paint. Yet, again, no traces of painting media were found.[12]

To be as thorough as possible, STURP members performed other microscopic spectral studies as well. A mass spectrograph can determine the exact masses of individual atoms by photographing the spectra produced when a beam of ions strikes a material. Laser-microprobe Raman spectroscopy can ascertain molecular structure and chemical composition by measuring the spectrum of light scattered when a laser beam strikes a material. Neither of these studies indicated the presence of inorganic substances or any other type of foreign material on the body image fibrils that could account for the visible body image color.[13]

To summarize, inorganic elements in paint (such as iron oxide and vermilion) as well as organic compounds (including gelatin) should have revealed themselves through sharp and identifiable features in STURP's many spectral studies. None were found—not in all the ultraviolet, X-ray, and visible light studies that measured fluorescence, absorption, reflectance, and emission features of the Shroud body and blood images. As STURP members put it, the tests detected "none of the spectral characteristics expected from normal dyes, stains, and pigments. . . ."[14] This important evidence confirms the fact that neither organic nor inorganic compounds can be responsible for the body and blood images on the Shroud of Turin.[15]

CHEMICAL TESTING FOR PAINT

In addition to the sophisticated spectral investigations, STURP performed more than one thousand chemical tests and microscopic evaluations to search for possible paint pigments—both inorganic and organic. McCrone states that gelatin was not simply the binding medium for the iron oxide and vermilion pigments, but that it also accounts for the yellow coloring of the Shroud body image fibrils.[16] Gelatin is an organic (protein) substance, obtained by boiling in water the ligaments, bones, hooves, and skins of animals. If gelatin were used as a medium to bind any kind of pigment, or if it were responsible for the body image color, it should be easy to identify in the many different protein-detection studies conducted.

To search for protein, STURP investigators first used the Biuret-Lowry test, but each of the body image fibrils studied yielded readings that were negative for protein. The fibrils containing blood did test positive, but this is consistent because blood contains protein.[17] Such results support the conclusion that actual blood is present in all of the Shroud wound areas, not paint pigments in a gelatin binder. To test further for McCrone's proposed binding medium, a 10 percent gelatin solution was mixed and examined under ultraviolet light. The fact that the gelatin fluoresced contrasts sharply with STURP's finding that the Shroud body image does not fluoresce at all under ultraviolet light. These experiments were repeated for solutions containing as little as one half of one percent of gelatin, although such a solution is little better than pure water as a binding agent for a heavy pigment like iron oxide. Yet even in a gelatin solution this dilute, protein was still detectable by the Biuret-Lowry test, and the solution also fluoresced under ultraviolet light.[18] Recently, FTIR microspectrophotometry was applied to Shroud body image fibrils and to serum fibrils. While the spectral presence of protein in its characteristic regions was readily observed on the serum fibrils, it was not detected on the Shroud body image fibrils.[19]

Chemists also tried the approach of using a powerful solution of proteolytic enzymes, compounds that attack and destroy proteins. A 0.01 percent gelatin solution was applied to fibrils from other pieces of similar linen that were used as controls in many of these experiments. Even though an artist's use of such a dilute solution as a pigment binder would be ludicrous, STURP chemists tried it anyway to test for gelatin as thoroughly as possible. At this dilute level, the proteolytic enzymes still found and destroyed the gelatin protein on the control cloth fibrils. When applied to Shroud body image fibrils, the enzymes had no effect at all.[20]

Because the question of gelatin is important, the search did not end there. As the most sensitive (by a factor of ten or more) of all protein-detection methods, a fluorescamine test can identify protein at the level of one nanogram—or one billionth of a gram. But when Shroud body image fibrils were subjected to fluorescamine tests, no gelatin or protein of any kind was found.[21] Taken together, the Biuret-Lowry, proteolytic enzyme, fluorescamine, and ultraviolet fluorescence tests all conclusively prove that neither gelatin nor protein is present on the Shroud body image fibrils. Thus, the color of the body image cannot possibly be a protein compound, nor was McCrone's proposed gelatin used as a binder or suspending agent for any paint pigment.

STURP members not only concluded that protein (including gelatin) was not responsible for the image; they also determined that no other organic materials, dyes, or stains were added to produce the straw-yellow coloring of the body image fibrils. In his book, *Report on the Shroud of Turin,* STURP scientist John Heller explains that all organic colors fall into one of three categories: (1) those that change color in an acid or a base (alkali), (2) those that change color through either oxidation or reduction (the opposite of oxidation), or (3) those that can be extracted with an organic solvent.[22] Heller and STURP member Alan Adler investigated all three categories in their many attempts to alter or remove the color of the body image fibrils.

For example, when concentrated solutions of hydrochloric and sulfuric acids were applied to body image fibrils, the color did not change. Alkaline solutions also failed to alter the fibril color. Hydrogen peroxide (an oxidant) and ascorbate (a reductant) should have bleached the fibrils white, yet neither had any effect on the image color. Many different types of organic solvents, covering the entire range of solubility classes, were applied in attempts to remove the yellow coloring.[23] Again, the color proved to be nonextractable. If any organic stains or dyes were present on the Shroud, they would have to fit into one of the three categories described above. And the many efforts to remove or alter them should have succeeded.[24] The fact that STURP investigators could not modify or extract the color, nor dissolve the fibrils, supports the conclusion that no organic dyes or stains are responsible for the straw-yellow body image. In addition, both oxidative and reductive reversal of color back to "new" linen white demonstrated that the presence of iron oxide could not be the cause of the color of the body image fibrils.[25]

No *inorganic* pigments, such as iron oxide or vermilion, produced the straw-yellow coloring of the Shroud image either.[26] Because the images on the Shroud are visible, any pigment used would have to be present in amounts sufficient to be seen by the naked eye. These elements therefore have to be detectable by some means. As discussed earlier, X-ray fluorescence studies are one of the best ways to identify metallic (inorganic) elements with atomic weights greater than 16. But to be certain the X-ray fluorescence tests didn't miss anything, STURP was especially thorough in its chemical testing for inorganic paint pigments.

Chemical investigators Heller and Adler performed an extensive battery of tests to search for those inorganic or metallic elements most likely to be present on the Shroud. Almost all of these studies yielded negative results. The only metallic elements that could be identified were calcium, iron, and strontium, which are found everywhere on the Shroud, not just in the image areas.[27] The presence of these elements is best explained by the retting process of linen manufacture.[28] When linen is retted, the original flax plants are laid in a body of water, such as a pond or lake, for a lengthy period of time. As the plants soak, the useless part of the flax rots away, leaving the fibrils of linen that are then spun into thread. During retting, the fibrils absorb calcium, strontium, and iron from the water, and these inorganic elements become chemically bound with the linen. This explains why these elements are everywhere on the Shroud, just as they are everywhere on other pieces of linen used as control samples in STURP's experiments. (Control samples included three hundred-year-old Spanish linen, Coptic funerary linen from about A.D. 350, and Pharaonic linen dating from 1500 B.C.) It also accounts for the iron, calcium, and strontium being bound to the linen's cellulose and contained within the fibrils, not applied externally.[29]

In addition to this iron bound within the linen fibrils and present everywhere on the Shroud, STURP scientists identified two other forms of iron. Heme-bound iron, which is a component of blood, was found in the bloodstain areas. Iron oxide particles were also detected, but the Fe_2O_3 present on the Shroud has nothing to do with the body image, as McCrone asserts. Instead, iron oxide was found only in the edges of the water stains and in the scorch marks adjacent to the bloodstains.[30] This is no doubt because of an effect similar to "retting," which occurred during events surrounding the fire of 1532. As mentioned earlier, the water used to douse the Shroud's extremely hot or melting container would have reached high temperatures. When this heated water spread through the cloth, it carried some of the iron bound within the linen fibrils and deposited that iron at the water stain edges. There, the iron separated from the water and became iron hydroxide; as the water evaporated, a dehydrative process converted the iron into iron oxide.[31] STURP members confirmed this process by measuring the levels of iron present on the Shroud. The edges of the water stains were found to contain readings of iron oxide higher than the pure iron present inside the water-stained areas. The same kinds of measurements would result if the water used to put out the fire had a high iron content to begin with; as this water spread through the cloth, it would eventually migrate to the stains' edges.[32] The iron oxide found in areas where some of the scorch marks intersect with bloodstains has another simple explanation: As blood is heated and charred, it produces iron oxide, a common process that has been observed for centuries and can be easily reproduced.[33]

None of these types of iron found on the Shroud has any relationship to the yellowed body image. All are pure forms of iron, unadulterated with any other elements that have historically been associated with red iron earth pigment, the sort used by artists for centuries and alleged by McCrone to have produced the Shroud body image.[34] From art historians, Heller and Adler learned that iron oxide pigment is almost always impure and invariably contaminated by other elements, usually manganese, nickel, cobalt, or aluminum. Working with this information, they used a scanning electron microscope to determine the exact composition of the iron particles present on the Shroud. In this test, a beam of electrons hits a particle and produces X-ray emissions that can identify the chemical elements contained in that particle. All of the tested iron particles turned out to be pure and uncontaminated with elements such as manganese, nickel, cobalt, or aluminum.[35]

Heller, Adler, and other STURP chemists also thoroughly examined Shroud samples through high-powered microscopes. Microscopic examinations, including magnifications up to one thousand times, revealed that the image fibrils appear completely uncoated; even the joints of the fibrils can be clearly distinguished. As discussed in the preceding chapter, the Shroud body image is composed of degraded cellulose, and cellulose is made up of millions of repeated linked units, each of which is a relatively light and simple molecule. Under a microscope, cellulose resembles a series of bamboo stalks laid end to end and connected with band-like joints. If the fibrils had been coated with anything, these joints would not be distinctly visible.[36]

Through the high-powered microscopic examinations of sample slides taken from different areas of the Shroud, STURP investigators discovered and identified many minute particles. These include droplets of wax; pollens; dust spores, insect parts; modern fly ash; animal hairs; starch; and threads of cotton, silk, wool, elastic pink nylon, and other modern synthetic fibers.[37] Even a felt-tip pen marking was evident (hardly the instrument of a medieval artisan). But none of these contaminating particles are present in every area of the Shroud and its image, so none can be responsible for the man's figure. When McCrone found a trace of red vermilion on one of the sample slides he examined, he concluded that it accounts for the Shroud's blood marks—despite the fact that vermilion was absent on all other samples taken from bloodied areas. The trace of vermilion McCrone saw surely belongs in the category of contaminating particles.

The Shroud cloth holds a wealth of contamination because it is many centuries old and has been kept in different locations. It has therefore picked up an assortment of foreign particles along the way. It is well documented that throughout its history artists and the public have laid paintings, as well as a multitude of other objects, on the Shroud in order to sanctify them.[38] For this reason, minute amounts of painting pigments should be present on the Shroud. But to say that vermilion is a primary component of the bloodstains makes about as much sense as saying the body image is composed of insect parts.

PAINTING THEORY DISPROVED

When you add all these test results together, they point to one overwhelming conclusion: The body image fibrils were *not* colored with either organic stains or dyes or metallic substances. To put it simply, an artist would have to add some sort of foreign material to paint the Shroud or create the image by any other artistic method, such as block printing or transfer-rubbing over a bas relief. Yet, as Eric Jumper and colleagues affirmed: "We conclude that *no material has been added* to these yellowed fibrils to produce the color" (italics added).[39] Similarly, Heller and Adler summed up their many chemical studies: "In view of the range of our chemical testing for metal [inorganic] pigments and organic stains and dyes, we found no evidence for the application of any such known materials on this cloth."[40]

As this chapter has shown, the extensive scope of the STURP testing conclusively refutes any and all painting theses. It was only because the painting theory was widely accepted for more than six hundred years that the scientists directed so much time and energy to evaluating this possibility. Article after article published by STURP members in peer-reviewed scientific journals has been forced to conclude that McCrone's thesis is faulty. A look at his methodology shows that his thesis is faulty because his methods were flawed.

To begin with, McCrone used an amido black test to check for protein on the yellow body image fibrils. Since amido black, a reagent, stained the tested particles, he concluded that the image fibrils contained protein. However, as a basic dye, amido black will also stain oxidized cellulose in the same way, so it cannot unequivocally demonstrate the presence of protein.[41] With this one uncorroborated test in hand, McCrone searched no further for protein. Not only did he use the wrong test for protein, he also failed to perform the many, more accurate protein-detection studies. The Biuret-Lowry, ultraviolet fluorescence, proteolytic enzyme, and fluorescamine tests are much more reliable—and all those results agree there is no protein on either the body image or nonimage fibrils.

Perhaps McCrone's most serious error was his failure to recognize the existence of actual blood on the Shroud, an error that resulted because his testing for blood was equally incomprehensive. The scientists who conclusively determined the presence of blood performed over a dozen different studies, each of which confirmed this finding. In contrast, McCrone merely looked through a microscope and decided no blood particles were on his sample. Most likely, the error occurred because (as noted in his self-published articles) the blood sample particles that McCrone examined microscopically were still stuck to the Mylar backing tape that had been used to remove them from the Shroud.[42] He concluded that the red particles could not be blood because they were birefringent; in other words, they gave off a double refraction, as do most crystals. Red iron earth pigment would demonstrate birefringence since iron oxide is crystalline; in contrast, blood is not crystalline and will not show birefringence. The problem with McCrone's observation is that red particles stuck to Mylar tape will always appear birefringent because light passes through both the sticky tape and the red particle.[43]

McCrone always seems to interpret ambiguous data in his own favor, without following up with the tests that might prove him wrong. He immediately determined that his amido black test indicated the presence of gelatin protein, without realizing that amido black would result in a false positive for fibrils consisting of oxidized cellulose, or that amido black is considered an accurate indicator of protein in blood. Again, even though he identified several elements on the Shroud—sodium, magnesium, aluminum, silicon, phosphorus, sulfur, chlorine, potassium, calcium, iron, and copper—all of which are contained in whole blood,[44] he failed to follow up these clues with more extensive tests that could have proved its presence.

Pertinent to this discussion is the fact that particles from the Shroud bloodstains have flaked off from the wound areas and adhered to other parts of the cloth at some distance away. This process, described as "translocation," occurred because the Shroud has been folded and unfolded many times over the centuries.[45] Dried blood deposited on other sample linen cloths used for test purposes has translocated after only a few years. Most of the red particles identified by McCrone as iron oxide pigment may well have been translocated blood flakes. Also, as we have already discovered, the existence of different forms of iron compounds on the Shroud can be explained by the linen retting process, as well as physical and chemical changes resulting from the fire and water of 1532.[46]

To summarize, McCrone performed the wrong tests for protein; measured for birefringence while the fibrils containing red particles still adhered to the Mylar tape; failed to identify the existence of blood, even though his own work produced clues to its presence; was unaware that blood particles have translocated to all areas of the cloth; misunderstood the meaning of the different kinds of iron present on the Shroud; and was unable to recognize the degraded cellulose. McCrone did not perform the many basic experiments that can be and were used to test the painting theory: X-ray, ultraviolet, and visible light reflectance, absorption, and fluorescence spectral measurements; ultraviolet fluorescence testing of gelatin; Biuret-Lowry, fluorescamine, and proteolytic enzyme tests for protein; a wide variety of chemical tests designed to detect specific metallic elements; various methods for identifying chromophoric organic compounds, stains, and dyes by using solvents, acids, bases, oxidants, and reductants; and mass spectroscopy, which yields information regarding an object's composition. These are just some of the basic scientific experiments available to test the painting thesis. Each disproves McCrone's theory, and all were ignored by him.

It cannot be denied that McCrone's work has damaged the Shroud's reputation among many intelligent people. It would be tragic if this damage were the result of an incompetent scientist. But the reality is even worse. Walter McCrone seems to be a loose cannon with an agenda all his own. Before the STURP team could publicize the complete findings from its examination, McCrone resigned from STURP and grabbed center stage with his own uncorroborated "evidence," declaring that the image on the Shroud was nothing more than a "painting."

McCrone, a specialist in the field of microscopy, had never examined the Shroud in person, but examined only fibrils that were removed from it. When he made his announcement, he had examined those fibrils only through a microscope. He never performed any of the thousands of other scientific tests that many other STURP scientists had performed on the Shroud and its fibrils.

Besides making presumptions based on the narrow scope of his work, McCrone also preempted STURP because he alone possessed the fibrils for more than one year. This prevented other STURP team members from conducting any analyses of the fibrils and particles from the Shroud. Members of STURP called him several times requesting the return of the fibrils, but he repeatedly put off their requests. It was only after several of the scientists paid the doctor a surprise visit that the fibrils were returned to them (returned, we should note, in chopped, scarred, nicked, and generally poor condition). Unlike all of STURP's scientific articles, McCrone published most of his findings in his own non-peer-reviewed literature.

Unfortunately, when the popular media began reporting on the Shroud, following the 1978 examination, the full range of scientific findings was neither adequately understood nor fully expressed. The findings from the STURP team usually received less attention than the more limited—and more sensational—work of McCrone. As a result of Dr. McCrone and his media coverage, much of the scientific and other information learned from the 1978 examination of the Shroud has gone unnoticed, and the advancement of the public's knowledge of this subject has been hindered.

And this is not the first time Dr. McCrone has obstructed the public's grasp of history due to his questionable methodology. Dr. McCrone is known to the general public for having declared two famous historical objects—the Vinland Map and the Shroud of Turin—as forgeries. While the focus of this book is the more significant of the two objects, a brief study of Dr. McCrone's work on the Vinland Map bears strong witness to the errors in his methods.

It was his 1974 pronouncement that the historical Vinland Map was a forgery that first gained Dr. McCrone widespread public attention. This seemingly ancient map depicts an Atlantic Ocean island west of Greenland, which approximately occupies present-day Newfoundland. Donated to Yale University and studied extensively by scholars, their published results ascertained that the map was drawn about fifty years prior to Columbus's arrival in the New World in 1492. As such, it would constitute the earliest known cartographic evidence of North America in existence and would help establish the historical claim that Norsemen had reached North America before Columbus. The discovery of this archaeological treasure could prove to be a key historical finding. In fact, subsequent archaeological excavations in Newfoundland and elsewhere have now convinced most scholars that, though their attempts to settle the continent may have failed, the Norse actually preceded Columbus by as much as five hundred years. However, many of McCrone's assertions, later proven to be erroneous by scientists at the University of California at Davis and other experts, exiled the famous map to the realm of forgeries and historical obscurity until the mid-1990s.

In 1995, Yale University Press published a new edition of *The Vinland Map and the Tarter Relation,* which contained a thorough refutation of McCrone's work. (A summary of these refutations is in appendix A.) This book, along with several other scientific publications, clearly shows numerous errors in the identification and analysis of the chemical constitution of the map's ink by McCrone and his associates. However, McCrone's most fundamental error may lie in his overall method. While the University of California scientists based their findings on sophisticated, nondestructive testing of the entire map, McCrone and his associates limited themselves to a minute portion of the map. By his own assertion, "The total weight of all the samples was much less than a microgram, and if they had all been put together, they would scarcely have been visible to the naked eye."[47]

Furthermore, an analysis of the authenticity of the Vinland Map needed to be based not only on scientific testing but also on a series of cartographic, linguistic, and historic evidence. The scientists at Cal–Davis recognized that their own scientific work was only part of the total evidence and so did not judge the map's authenticity on that basis alone. They assert that "such a judgment must be based on all available evidence, cartographic and historical as well as compositional."[48]

All of these critical errors by McCrone relating to the Vinland Map correlate to his findings on the Shroud of Turin. In both instances, he based his scientific study only on minute particles of a much larger object, but not on the examination of the larger object. In both instances, by so severely limiting his techniques he ignored a broad range of relevant evidence and misled the public for decades with his erroneous conclusions. Even so, the lion's share of the blame should be heaped not on McCrone, but on the news and media organizations that allowed him to lead them by the nose. Reporters—even more than scientists—are supposed to form opinions by gathering all the evidence into one big picture.

The public has a right to have important historical objects, as well as historical events, analyzed comprehensively. This information or evidence can come from historical, archaeological, scientific, medical, or any other relevant sources. In fact, proper analysis requires that all relevant evidence should be considered when making such a determination.

McCrone made a mistake in calling the Shroud a fake from the basis of his own limited testing. The media duplicated this mistake not only when it gave McCrone so much press, but years later when it unskeptically accepted the flawed carbon-dating test results. If, like Dr. McCrone, we focus on only one aspect or part of this evidence and ignore the entire body of evidence, our analysis will be not only incomplete, but likely incorrect.

We cannot afford such errors in a study of the Shroud of Turin. The subject of this book may actually concern and relate to the most important events in all of history.

CHAPTER FIVE

ATTEMPTS TO REPRODUCE THE SHROUD IMAGE

Before we turn to the archaeological and historical evidence that will round out our picture of the Shroud's origin, let us consider one last time the theories put forward by Shroud "skeptics." None is consistent with the image on the Shroud, and some are so outlandish that it is hard to see how they could be proposed with a straight face. What these theories seem to corroborate is that the method of image formation on the Shroud is not reproducible by human or natural means. All of these proposed methods fail to account for the many unique characteristics of either the body or blood image. To be accepted as credible, any proposed theory regarding the transference mechanism responsible for encoding the images on the Shroud must be able to account for all the characteristics of the body images and blood marks.

Some of the proposed image-transference methods can be called naturalistic, while others are artistic. In general, theories pertaining to naturalistic mechanisms are based on the theory that an actual body was wrapped in a linen burial cloth, encoding the image with some natural process. Advocates of artistic methods, however, assume the Shroud was created in some manner by an artist. Based on the information acquired from testing and examining the cloth itself, the frontal and dorsal body images, the various wounds and blood marks, and the numerous fibril samples taken from all over the cloth, we are able to draw a number of conclusions about any proposed methods to duplicate the Shroud image.

STURP scientists have performed the experiments that test most of the methods claiming to explain the images on the Shroud; they have even designed new theories along with new tests, in an effort to verify the naturalistic and artistic theories. Yet none of the proposed mechanisms discussed in this chapter can replicate all the unique characteristics of both the body images and blood marks on the Shroud. In fact, none even comes close. Although the theories described below have not garnered as much support as the painting thesis, they are nevertheless important to understand—and then discard. The evidence provided by the Shroud itself discredits them, just as it proves the image did not result from the application of paint.

NATURALISTIC THEORIES

Vapograph (Diffusion) Theory

Paul Vignon, one of the first scientists to study the Shroud, made one of the first attempts to explain its images when he proposed the vapograph theory around the turn of the century, long before any scientific investigation of the cloth had been undertaken.[1] His theory is based on the scenario that the man buried in the Shroud was covered in perspiration prior to his death. The urea contained in his sweat then fermented into carbon dioxide and ammonia, the latter of which would have diffused into the cloth as a vapor. If the burial linen were soaked in aloetine (an embalming formula made of aloes and olive oil), the ammonia would react with the aloetine-soaked cloth, the linen would become stained, and the body image would thus be produced, Vignon theorized.

This notion contains numerous problems. In the first place, scientific examination has revealed that the Shroud body image resides only on the topmost fibrils of the linen threads and is completely superficial. Ammonia vapors released from a body would penetrate into and between the threads of the linen. It would be impossible for a vapor to diffuse onto and stain only the topmost fibrils of a woven cloth.

Another problem with the vapograph theory is that, according to Vignon, the darkest stains would be found at points where the cloth was in closest contact with the body. Consequently, the image of the man's back should be darker than the frontal image because the back of the body would have been in firmer contact with the linen, due to the body's weight. Yet this is not the case with the Shroud. Moreover, as STURP member Eric Jumper points out, any stains produced in a scenario involving vapors would tend to reach a "saturation point"; they would not get any darker with longer exposure to the diffusing vapor or firmer pressure.[2] Since neither fibril penetration nor saturation is present on the Shroud image, the diffusion of vapors as the image-forming mechanism must be discounted.

Further weaknesses in the vapograph theory were revealed by the three-dimensional studies conducted by Jackson, Jumper, and Ercoline.[3] They found that diffusion of vapor molecules is significantly disturbed by small convection currents, which would alter the vapor's path to the cloth. The images produced by their experiment were blurred and lacked the high resolution of the Shroud image. Additionally, the experiment yielded highly distorted three-dimensional images when photos were processed by the VP-8 Image Analyzer. Figures 39 and 40 show

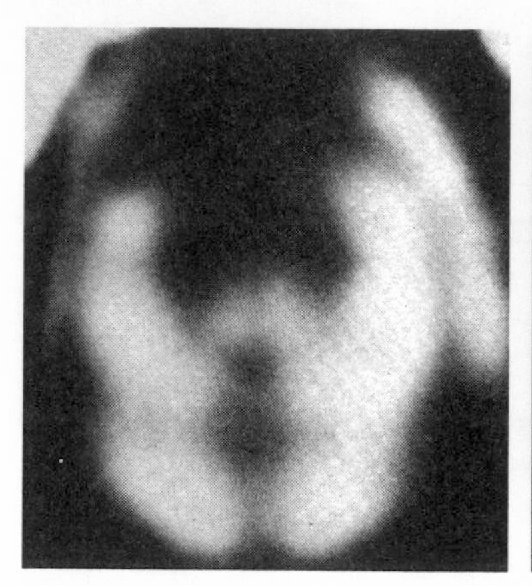

Fig. 39. Diffusion method.

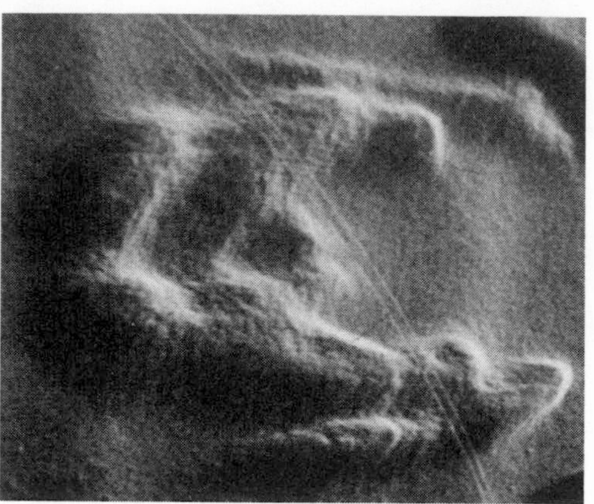

Fig. 40. VP-8 relief of diffusion image.

the photographic and three-dimensional images produced from experimental attempts to duplicate an image encoded with possible evaporation or diffusion mechanisms.

The vapograph theory fails to account for other image characteristics, as well. An organic stain mechanism would not act the same, if it acted at all, on such diverse materials as hair and skin. Even though the hair may have come in contact with parts of the body, hair itself does not perspire. The presence of images formed over different materials illustrates that an image-forming method involving the diffusion of molecules (such as ammonia) cannot account for what we see on the Shroud. Moreover, all of STURP's testing that looked beyond what the naked eye can observe—with spectrography, photometry, chemistry, and microscopy—detected none of the molecules that might be involved in the vapograph theory. While it is theoretically possible that such molecules could have decomposed or oxidized over time, some residual traces of the responsible elements should still be detectable.

It also seems unlikely that a sufficient amount of ammonia could be produced from a body's perspiration to create a satisfactorily intense image through the process of evaporation/diffusion.[4] A living body, let alone a dead one, could probably not secrete sufficient liquid. Since perspiration evaporates over a brief period, it seems improbable that a body would still be covered in copious amounts of sweat at the time it was prepared for burial.

While Vignon's vapograph theory was considered a possible explanation by many, Vignon failed to produce credible results when he conducted experiments to test his hypothesis. He reported that he reproduced an image of a hand, but he also said it was visible on the back of his test cloth, which clearly violates the superficial characteristic of the Shroud body image, and he never published photographs of his experiment. We have to conclude that his test results were not very impressive or supportive of his theory. According to French scientist Dr. Jean-Baptiste Rinaudo,

"Experiments with ammoniac impregnated plasters gave such poor images that he [Vignon] finally concluded that the image looked as if it had been the result of some irradiation."[5] Using scientific methods and equipment now available, STURP members failed to duplicate the characteristics of the Shroud image based on Vignon's theory and were forced to conclude that molecular diffusion cannot be the mechanism responsible for formation of the body image.

Direct-Contact Theories

Several methods somewhat similar to Vignon's vapograph theory have also been advocated as the mechanism that produced the Shroud image. A leading proponent of a direct-contact theory, STURP scientist Sam Pellicori, suggests the image might have developed not from diffusion but from the cloth coming into direct contact with a body covered in perspiration; body oils; and/or liquid solutions of myrrh, aloes, or olive oil. No traces of these types of organic liquid substances (or any others) have been detected on the linen.

A fundamental problem with all direct-contact theories can be demonstrated in a simple experiment by rubbing charcoal over a person's face and then draping a cloth over it. The resulting image will possess no three-dimensional information, appear grossly distorted, and bear scant resemblance to a human face.

STURP scientists Jackson, Jumper, and Ercoline attempted to reproduce direct-contact models capable of matching the image characteristics found on the Shroud.[6] They used a plaster mold of a face that was treated with different combinations of liquids and oils at different temperatures and draped with linen resembling the Shroud. This experiment identified several weaknesses in the direct-contact theory: When the cloth was draped over the face mold that had been coated with ink, two different types of shading effects were produced. An image was imprinted only where the cloth and face made contact; no impression was left if the cloth and face were separated by space. Yet, as discussed in chapter 3, the Shroud body image was imprinted even where there must have been some distance between the cloth and body (for example, the sides of the nose). When the investigators pressed the cloth onto those surface points where contact was not made during the natural draping over the face mold, the resulting image was grossly distorted (see fig. 41). The VP-8 analysis of the direct-contact photographs yielded even more pronounced image distortions (fig. 42), which means that direct-contact methods cannot produce the correct three-dimensional information.

Direct-contact theories fail to explain the Shroud images in other ways. If the images resulted from contact between linen and a body soaked with perspiration, body oils, and liquid solutions of myrrh, aloes, or olive oil, it is doubtful they would remain confined to only the topmost fibrils of the linen threads, especially on the dorsal image, where the full weight of the body would be pressing on the cloth. In addition, a liquid coming into contact with the cloth would be drawn into the core of the fibrils and spread to adjacent areas in a capillary action. In Pellicori's

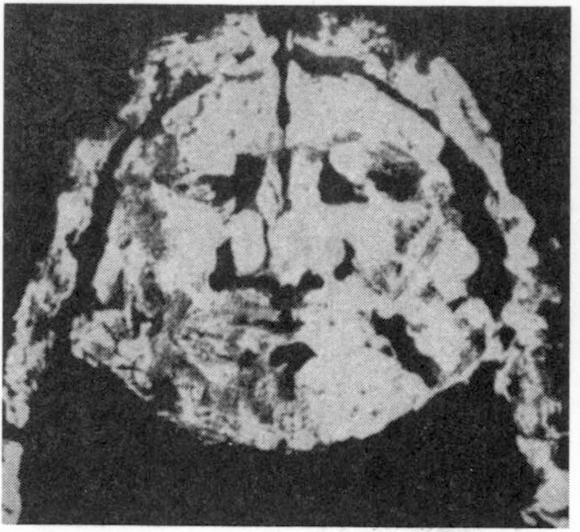

Fig. 41. Direct contact image.

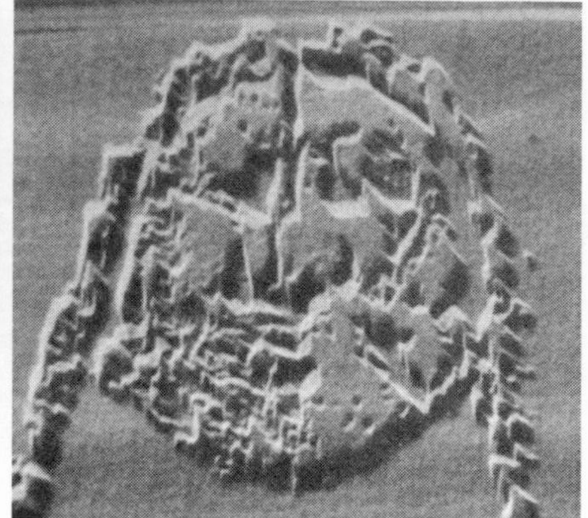

Fig. 42. VP-8 relief of direct contact image.

attempt to reproduce the image, he coated his fingers with oil, lemon juice, and perspiration and gently pressed them against the cloth. However, even when gentle hand pressure was applied (in contrast to the full weight of a body), capillary action was observed.[7] Thus the direct-contact theory cannot account for the characteristics of superficiality and lack of capillary action on the Shroud body image fibrils.

Another difficulty inherent in direct-contact theories is that the actual image-transfer mechanism must have operated independently of contact pressure because the frontal and dorsal images on the Shroud appear to have nearly equal intensities.[8] In the direct-contact scenario, the dorsal image (a result of the full weight of a body lying on the cloth) should be much more intense than the frontal image (a result only of the weight of the cloth on the body). This did not happen with the Shroud.

Producing the image on the Shroud via a direct-contact mechanism would also require a uniform application of liquids over the entire frontal and dorsal areas of the body. It is doubtful that a uniform application could be achieved everywhere. Equally dubious is the likelihood that an organic stain mechanism would operate equally (if at all) over skin and hair. Although Pellicori has addressed some of these criticisms,[9] he has never produced any facial images, let alone full-length body images, that come close to possessing all the image characteristics present on the Shroud.

STURP scientist John German has proposed an interesting variant of the direct-contact method,[10] but he too has failed to publish pictures that confirm a successful image possessing the characteristics of the Shroud. German hypothesizes that the Shroud initially touched only those high points of the underlying body that a naturally draping cloth would contact. Over time, because of fibril deformation (or sagging) and moisture in the air, German believes the cloth gradually conformed to more of the body's surface area. According to this theory, the parts of the cloth in contact with the body for the longest time (e.g., the highest points of the profile) would leave the darkest impression, while those cloth areas in contact the least amount of time (lowest profile points) would leave a lighter impression.

To be credible, German's theory has to overcome most of the problems with the direct-contact methods discussed above. Along with proponents of those theories, German assumes that moisture would be a factor in the sagging of the linen cloth around the body. However, the problems of capillarity, diffusion of liquids into the cloth, and saturation of image fibrils would be greatly compounded if the cloth, in addition to the body, were moist.[11]

German's theory cannot explain the fact that the Shroud image is continuously shaded everywhere. With the exception of the top of the man's head and sides of his body, the image does not "drop out" anywhere; instead, all body surfaces were encoded, even the sides of the nose. For the entire body image to be encoded via German's "sagging cloth," the linen would have to conform completely to every body area represented on the Shroud. However, linen is a highly elastic, cellulosic fiber that retains its shape far better than most fabrics; it is, therefore, implausible that the cloth could sag to the degree necessary to correspond with the complete body figure lying below.[12] Even if it did conform completely, the resulting image would have the same gross distortions as those found by Jackson and colleagues when they directly pressed their test cloth against every part of their experimental face mold.

Perhaps the most vocal advocate of a natural mechanism (i.e., direct-contact theory) is Dr. Eugenia L. Nitowski, a Middle Eastern archaeologist who has tested direct-contact processes most fully.[13] She tried to duplicate, as much as possible, the actual conditions involved in the burial of a crucifixion victim in ancient Jerusalem. Accordingly, she and her colleagues traveled to Jerusalem to perform a series of experiments designed to explore possible naturalistic image formation methods. They used the tomb complex at the École Biblique, which is part of the same rock shelf as that in the Holy Sepulcher and the Garden Tomb—the most likely candidates for the actual tomb of Jesus. The date of these tests also approximated the Passover/Easter season in which Jesus was executed.

Nitowski postulated that a crucifixion victim who had been scourged would undergo hematidrosis (bloody sweating), experience a loss of body fluids (perspiration and blood) along with a lack of fluid intake and sufficient rest, and endure extreme physical exertion. The victim would, therefore, become severely dehydrated, so his body temperature would rise. She further presumed that the severe trauma and emotional stress accompanying a crucifixion would produce an acidic condition in the victim's blood and perspiration.

In Nitowski's experiments, a sweat solution of normal saline and acetic acid was applied to a mannequin that had been filled with warm water to a temperature between 110 and 115°F. A mist of the sweat solution was also sprayed on the mannequin just before it was covered with a linen shroud, which was then tied with rope around the neck, waist, and ankles so the cloth would remain next to the body. Variations of this experiment were also performed, including adding blood, myrrh, and aloes to the sweat solution and using five different types of linen. A bronze coin was also placed over the eye in an attempt to duplicate the possible coin images found on the Shroud. While the mannequin was in the Jerusalem tomb, the air's humidity was constantly monitored; when humidity levels dropped below the percentages normally found in the tomb, the air was misted to raise the moisture level.

Despite the fact that Nitowski, and other experimenters, tried different combinations of liquid substances and atmospheric conditions in an effort to reproduce the Shroud body image, none of their results even remotely approach the image characteristics found on the Shroud. Nitowski's published pictures of her work, seen in figures 43 to 45, illustrate the best results of her tests. Instead of verifying the direct-contact theories, the photographs highlight the numerous problems inherent in such naturalistic, image-formation methods.

Sebastiano Rodante attempted a similar approach to test a direct-contact, image-forming mechanism. After soaking cloths in water or oily solutions of aloes and myrrh, he draped them over a plaster head that had been sprayed with bloody sweat and placed them in a catacomb in Siracuse, Sicily, for thirty-six hours.[14] Although Rodante's experimental results (see fig. 46) were superior to Nitowski's, the image produced by this procedure still suffers serious deficiencies. Figure 46 contains positive and negative photographs of his image and the corresponding three-dimensional relief made from them. Rodante's negatives clearly lack the perfectly shadowed, highly focused, and finely resolved image found on photographic negatives of the Shroud. Furthermore, this and other direct contact methods are incapable of encoding vertical directionality onto a cloth draped in different angles over a supine body, as is found throughout the Shroud's frontal image.

Perhaps, for these and other reasons, Rinaudo and other naturalistic methods fail to produce full-length frontal or dorsal images. Such images require the production of additional features that have never been duplicated by any methods to date, such as the scourge marks and their lines, many of which could not be seen until the development of modern technology. It is also difficult to imagine how realistic shoulder abrasions could be encoded under this method, which Rodante's image necessarily lacks.

Another obvious weakness of Rodante's experimental results is the color of his cloth fibrils. He describes the image color produced by his method as a brown-sepia (dark brown), which is inconsistent with the straw-yellow color revealed through microscopic examination of Shroud image fibrils. Other problems with his experiment are the same as those inherent in the diffusion and direct-contact methods already discussed. If the image were formed from contact with, or diffusion from, organic substances such as perspiration, blood, or oily or watery solutions, the image fibrils would become saturated. Also, liquid substances would penetrate into the fibrils and threads, which means the image could not be confined to only the topmost fibrils. Capillary action would also occur with direct-contact and diffusion processes. These problems of capillarity, penetration, and saturation are especially pronounced in experiments involving a wet cloth, like Rodante's.

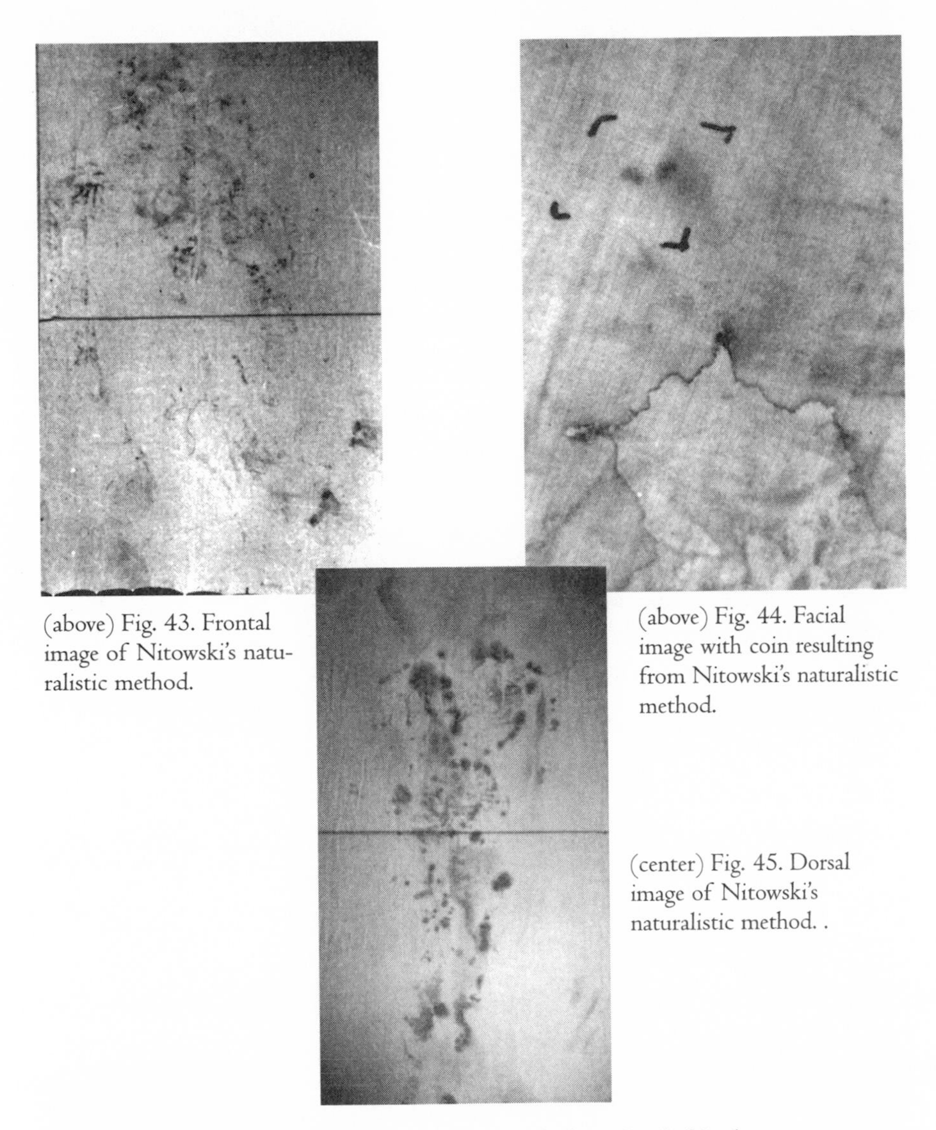

(above) Fig. 43. Frontal image of Nitowski's naturalistic method.

(above) Fig. 44. Facial image with coin resulting from Nitowski's naturalistic method.

(center) Fig. 45. Dorsal image of Nitowski's naturalistic method. .

Fig. 46. Images produced by Rodante's method, with the 3-D relief in the center.

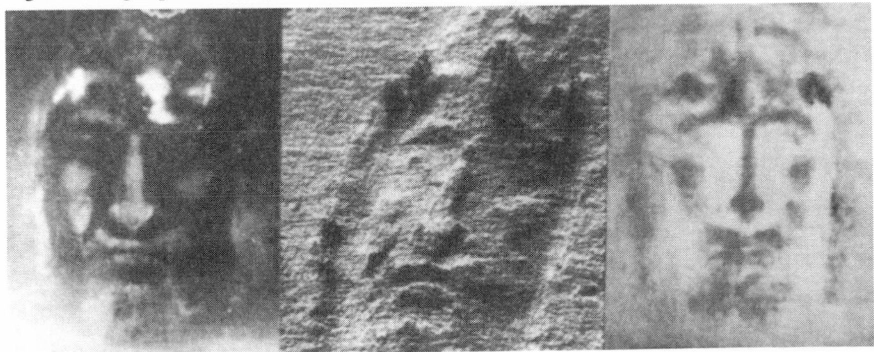

In Rodante's test, the cloth was soaked in liquid solutions of myrrh or aloes. "Soaking" means the entire cloth was covered with these substances. Yet no traces of aloes or myrrh have been detected on the body images or the cloth through the ultraviolet reflectance and fluorescence spectra obtained from the entire Shroud. Nor has any evidence of such solutions been found in the materials vacuumed from large areas of the cloth.

The work of Dr. Gil Lavoie and colleagues further discredits all the direct-contact and diffusion models of image transference, including those of Vignon, Pellicori, German, Rodante, and Nitowski.[15] From the study of a blood mark transfer near the right elbow of the man on the Shroud, Lavoie showed that the cloth and body in this area must have been in direct contact with each other (see figs. 36–38). This mark was caused by blood running down the man's arm—while he was in a vertical position—and pooling behind his elbow near the triceps muscle. When the victim was then laid horizontally, the draped or wrapped cloth came into contact with this part of the arm. Yet no body image discoloration was encoded at this location on the Shroud. Neither Rodante's direct-contact/diffusion method nor any other can account for the lack of a body image in the elbow area. Thus, Lavoie's observation directly contradicts the notion that a contact or diffusion mechanism could have generated the Shroud body image.

Volkringer Method

In 1942, Dr. Jean Volkringer, a pharmacist at St. Joseph's Hospital in Paris, discovered that when certain plants are pressed between the pages of a book, a highly detailed negative image appears on adjacent pages. He proposed this discovery as a possible explanation of what is seen on the Turin Shroud.[16] However, such a mechanism cannot explain the Shroud image. In the first place, plants that produced such images were usually pressed in books for several decades, some for as long as a century. In the case of a dead body, assuming it was somehow "pressed" between cloth, complete decomposition of everything but the bones would occur in a year or so, and decomposition stains would appear on the cloth within two to three days. In addition, the plant imprint was brown, not the straw-yellow color that can be seen close up on the Shroud body image.

To make the Volkringer theory seem even more improbable, the image of the plant usually was imprinted several pages from where the plant lay. Formation of an image was due to the book's weight pressing on the flattened plant. It is difficult to imagine how a similar scenario could apply to a light cloth draped over an entire body. Because the resulting plant image was pressed deeply into the pages, it cannot be considered superficial, like the image on the Shroud, which was immediately next to the body. The amount of pressure that a fragile dried plant receives inside a book is roughly equivalent to what a body would receive if it were pressed between two stone slabs. Unless a similar amount of pressure can be applied to a man between two sheets of linen, a logical analogy cannot be drawn to the Volkringer method on plants. Even if you could apply such pressure on a man between two sheets of linen, you would greatly increase the problems of image distortion, three-dimensionality, uniformity, and fiber saturation. In addition, you would alter or damage the bloodstains, creating more problems.

Furthermore, the image on the Shroud was found to have been encoded vertically onto a cloth that was draped over the body. It is quite difficult to conceive of a cloth being pressed against a body while still maintaining a draped configuration. We also saw earlier, for additional reasons, that the Shroud body images were encoded independent of pressure.

Neither could such a method produce an image that contains three-dimensional information of a man's body. This method does not produce true three-dimensional information of the plant in its image: There is no correlation in the image's lightness and darkness to the actual corresponding distances that the parts of the plant were from the enclosed book pages. All parts of the plant were not only flattened but were also the same distance from the page. In addition, there were usually pages in between to interrupt this distance information correlation. Nor does this method explain the wounds and other blood marks on the Shroud. This method not only fails to encode them, but would also ruin the unbroken and unsmeared quality of the wounds, blood flows, and blood marks.

No known image of any part of a body has ever been produced by a method incorporating any of the circumstances proposed by Volkringer. For all these reasons, the Volkringer method seems incapable of producing an image on the scale of a human body; a superficial image; a three-dimensional image; an image that could be encoded vertically onto a draped cloth; an image that is pressure independent; an image that could be produced within a few days; or an image that could encode the wounds and blood marks.

Singlet Oxygen Theory (Lomas and Knight)

Dr. Alan Mills has recently advocated a theory similar to the vapograph theory that attempts to incorporate elements of the Volkringer method and the Russell effect.[17] He ponders that, if a traumatized victim has a cloth draped over him in a thoroughly stable, enclosed area, that his image would be left naturally on the cloth by a combination of these methods or effects. The Russell effect, named for W. J. Russell, is the name given to a nuisance in the early history of photography whereby certain materials would leave an unwanted image of themselves on sensitive photographic plates. Although Russell was never able to explain the effect, he believed it was caused by the release of traces of hydrogen peroxide. The problem was easily solved when early photographers learned to restrict the response of their emulsions on the sensitive photographic plates. Obviously, a piece of linen cloth is not going to have the sensitivity of a photographic plate, nor is it going to have extremely sensitive emulsions on it.

Without discussing the concerns already addressed about Volkringer's method,

Dr. Mills relies on convection currents to encode the image of the victim onto the linen shroud. As was demonstrated by scientists who tested this method, these currents do not go straight up, but spread in all directions. This alone eliminates the Shroud's image-encoding mechanism as resulting from any type of diffusion. Dr. Mills seeks to overcome this fundamental problem by theorizing that a gas could be released that would provide such a necessary upward trajectory; however, a gas would most likely work the same way as a vapor. This gas would have to work in an absolutely vertical 180-degree direction from each and every point on the entire frontal length of the body. If its direction were altered at all, it would not obtain the effect required for the image on the Shroud. Dr. Mills has not demonstrated that a gas would operate in this fashion. He theorizes that the singlet oxygen molecule would be released after the cloth has been placed over a supine victim who has died a traumatic death earlier in the vertical position. He theorizes not only its release, but also its release as a gas, yet he has not demonstrated that any of these assumptions would take place under these circumstances.

Even if all of the above *could* be demonstrated, Dr. Mills theorizes that these singlet oxygen molecules in a gas would rise vertically 4 centimeters and become encoded superficially onto the cloth, including those parts of the body not touching the cloth. (He does not describe why the molecules or gas would not dissipate in the air.) However, even if that were to occur, you would receive only one solid coloring on the cloth, without any of the resolution of the detailed photographic negative, or of the image's three-dimensionality. That the gases, molecules, cloth, and body would all just naturally work together to produce these unprecedented effects may be the largest of all the many assumptions or theories that Dr. Mills makes with his method.

This method could not encode a dorsal image on the cloth; the method theorizes the release of certain molecules in a gas traveling straight up to a cloth. Such a method inherently could not account for the dorsal image. Dr. Mills does not even attempt to explain in his paper the dorsal half of the body image. Two authors have recently and erroneously utilized Mills's method without accounting for any of its many problems.[18]

Dr. Mills and his advocates also fail to account for or explain in any fashion how the blood marks, wounds, or other bloodstains could be encoded on the cloth. Neither Mills's, the vapograph, Volkringer's, the direct-contact, or any combinations of these methods can account for how the blood and scourge marks on the Shroud might have been transferred to the cloth. If a piece of linen were wrapped around or draped over a body, you would not see the detailed blood marks with serum halos that look as though they formed, flowed, and coagulated on human skin, nor the scores of tiny individual scourge marks (with their accompanying details) visible only when viewed under a microscope or ultraviolet light. To be credible, any proposed image-formation mechanism must not only account for the body images but also explain the presence of these blood and scourge marks. Also, the last two methods have failed to explain the presence of the blood mark off the right elbow, yet the absence of a body image next to it.

Bacteria and Fungi

An unusual method to account for the images on the Shroud has been mentioned in the last few years by Dr. Leoncio Garza-Valdes, who claims they were caused by bacteria and/or fungi that formed a bioplastic coating on the cloth—to quote Dr. Garza, "... a type of clear encasing that is invisible to the unaided eye... composed of millions of living microbiological organisms that have formed over time, somewhat like a coral reef."[19] However, a number of serious problems arise with such a method. First, a number of STURP scientists have examined the entire cloth, as well as fibrils taken from throughout the Shroud, and they have not found such an extensive amount of bacteria or fungi on the Shroud as Dr. Garza claims.[20] While particles of such items, along with an entire list of other contaminants or items, can be found on the cloth, this bacteria and fungi would have to be so extensive and dominant that it caused images to be seen on the cloth—yet the bacteria and fungi were discovered by Dr. Garza in a bioplastic coating under the microscope. Furthermore, the latest examination by Fourier Transform Infrared (FTIR) microspectrophotometry of thirty-four samples taken from various locations throughout the Shroud did not reveal the existence of this bioplastic coating.[21]

There are other fundamental flaws with this method. First of all, this bacteria and fungi would not result in dehydrated, oxidized linen fibrils like those in the Shroud's body image. In addition, the bacteria and fungi would theoretically attach to the Shroud from contact with the sweaty, bloodied body. However, there is no vehicle for them to get on the cloth where there was not contact with the body. The Shroud body image is highly resolved, and contains three-dimensional information even where there was not contact between the body and the cloth. Once on the cloth, the bacteria and fungi would not distribute themselves into a three-dimensional correlation or a highly sophisticated gradient of shading to result in a detailed, photographic quality. And if they did move horizontally, why wouldn't they also move vertically and penetrate beyond the first fibril or two? If the bacteria or fungi had distributed themselves so extensively in a bioplastic coating to cover more than eleven and a half feet of body image, the Shroud would be so brittle that, if you dropped it, it would shatter.[22] Instead, the Shroud is quite flexible.

Furthermore, this method would not produce an image that was encoded in a vertical, straight-line direction from all parts of a supine body onto a cloth draped over it, whether the cloth touched the body or not. Bacteria have never been known to move in such a manner. Nor would the bacteria and fungi distribute themselves in such a way that all of the thousands upon thousands of body-image fibrils on the entire frontal and dorsal images would have the same intensity. In addition, this method could not be expected to operate the same on the hair, beard, and mustache as it would on skin.

Another fundamental failure with this method is that it could not account for the shoulder abrasions. These are body-image features seen best on the detailed photographic images. Their cause is simply the skin being torn open and acting accordingly; it is not due to bacteria or fungi that are formed there in some particular or special way different from the other parts of the body. If bacteria or fungi had formed there, how could they make themselves appear different at the shoulder abrasions from the way they appear at other body-image regions on the photographic negative?

Neither could the bacteria and fungi cause the blood marks and wounds. Such a naturalistic or direct-contact method could not encode the unique, actual wounds; bloodstains; blood flows; and scourge marks. Dr. Garza states that bacteria have now taken over 95 percent of the bloodstains on the Shroud.[23] However, this result directly conflicts with thirteen independent scientific tests that established the existence of actual whole blood in these areas on the Shroud.

While writing about bacteria and cloth, textile specialist John Tyrer stated, "When bleached . . . flax has a high resistance to bacteriological attack. Under certain conditions of warmth, dampness and contamination, micro-organisms may attack cellulose, notably cotton, but flax fibers will resist damage well if kept dry."[24] When STURP scientists cut Shroud fibrils in half, they could observe the original white color of the cloth, indicating that it was most likely bleached when made. In addition, throughout the Shroud's known history since 1357, as well as throughout its proposed history to the first century, the Shroud has always been known to have been kept in dark, dry, and cool to room-temperature locations. It was hardly ever exposed to the open air, except for a few times a century. It is in excellent condition whether seven centuries or twenty centuries old. Its history is completely inconsistent with the growth or development of such an extensive amount of bacteria, as claimed by Dr. Garza.

In addition, while some bacteria could feed only on air and not the cloth, they would have to excrete. If bacteria have been on the Shroud for centuries and are so extensive as to cause its images, excretions from this extensive bacteria would have to accumulate. However, no such excretions have ever been found by any of the numerous scientists who have examined the cloth and its image and non-image fibrils.

Moreover, if something as common as bacteria or fungi caused the image on the Shroud, we should have many other examples of such images or analogous images, or even parts of images, on other cloths. In all the centuries that mankind, cloth, bacteria, and fungi have coexisted, certainly the man in the Shroud was not the only person to have died while bleeding, or sweating, or having other fluids on him, and to have been covered after death. Yet there is no such image or part of an image due to bacteria or fungi that has been remotely analogous or similar to that of the Shroud of Turin.

ARTISTIC THEORIES

Along with the painting theory discussed in the previous chapter, several other artistic theories proposed as possible image-encoding vehicles have been tested by STURP scientists and others. Like McCrone's painting notion, all the hypotheses described below are based on the supposition that the Shroud was the creation of a medieval artist—and a very clever hoax.

Nickell Powder-Rubbing Method

Joe Nickell, a former stage magician and amateur detective, has experimented with a proposed technique of image encoding. Nickell's technique involved conforming wet linen to a bas-relief while impressing all the relief's features onto the cloth. After the cloth dried, he used a cotton dauber covered with cloth to rub powdered pigment onto the impressions left on the linen.[25] The results obtained by such a technique can be seen in figs. 47 and 48, which show what happened when STURP members tested Nickell's theory. As is evident in the VP-8 photograph, Nickell's powder-rubbing method does not produce a true three-dimensional image.

A powdered bas-relief technique would involve the application of a substance to the Shroud. As discussed earlier, all the extensive scientific examinations of the cloth indicate that no powdered particles or foreign materials of any kind have been added to the body-image fibrils that could account for the coloring—and therefore visibility—of the image. Specifically, magnified photographs of the body image threads do not reveal any sort of applied particles or staining substance. The "natural experiment," arising from the fire of 1532, also disproves Nickell's theory, as it did the painting theory. Nickell promulgates that the powdered pigments used were iron oxide, myrrh, or aloes. Yet if organic substances such as these had been used to

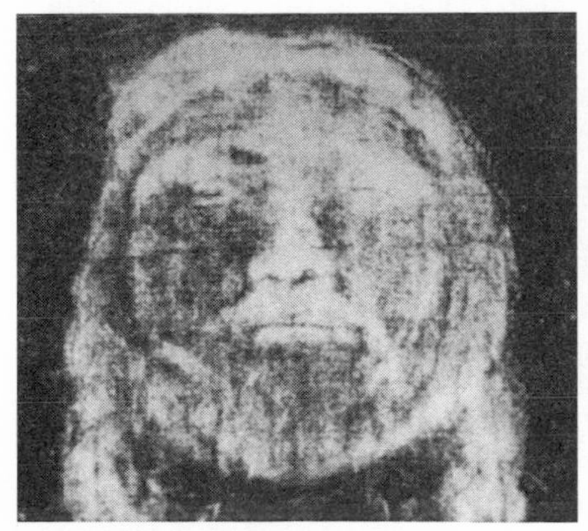

Fig. 47. Powder technique image.

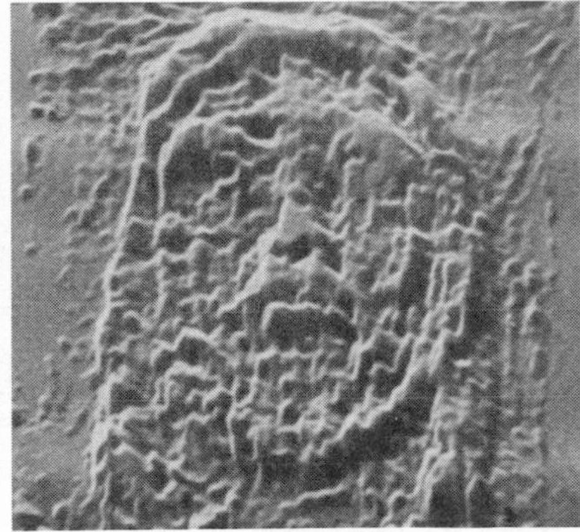

Fig. 48. VP-8 of powder technique image.

encode the Shroud image, they would have been altered or decomposed during the fire, depending on their distance from the hottest areas of the cloth. However, no such alteration in the body image can be found on the Shroud when those image areas nearer the burn marks are compared to those areas farther away.[26]

Nickell contends that his powder-rubbing method produces a superficial image. However, when STURP scientists reproduced his experiment using linen with a herringbone weave that simulates the Shroud's weave, large quantities of powder fell through the weave of the cloth and accumulated on the reverse side.[27] Shroud researcher and archaeologist Paul Maloney even tried using a piece of linen with a box weave, which is the tightest linen weave known. Maloney found that powdered particles still penetrated through to the back of the cloth.[28] Obviously, Nickell's technique fails to meet the requirement of image superficiality.

Nickell's theory seems far-fetched for other reasons. The Shroud contains, conservatively, thousands of individual body-image fibrils, but each fibril is encoded with a uniform intensity of color. Someone applying powdered pigment onto a cloth with a hand-held dauber and/or rubbing powder on woven linen could never achieve this uniform intensity on all image fibrils. In fact, experiments involving powder rubbing have shown that a uniform application of powder cannot be obtained on even one fibril. Walter McCrone, who also simulated Nickell's rubbing method, studied powdered fibrils under a microscope. Figure 49 shows a photomicrograph of a test fibril and illustrates the collection of powder at the edges surrounding a clean center. McCrone refers to this process as "snow fencing."[29]

Nickell theorizes that a likely powder pigment used to produce the Shroud body image was a form of iron oxide derived from heating ferrous sulfate (green vitriol). When a substance such as ferrous sulfate is "calcined," it is heated to a high temperature (but below the boiling point). This process, which causes the substance to lose moisture and oxidize, produces a powder or ash. Although Nickell believes such heating of ferrous sulfate explains the presence of iron oxide on the Shroud, the evidence proves him wrong. Whenever green vitriol is calcined, the contaminants of manganese, nickel, and cobalt are also produced as by-products.[30] While the process of heating may reduce these contaminants somewhat, they can always be detected with a mineral-associated iron such as iron oxide.[31]

Fig. 49. Photomicrograph of a test fibril illustrating the collection or "snow fencing" of powder.

Intensive chemical examination of the cloth by STURP scientists revealed that neither manganese, nickel, nor cobalt is found in association with the iron oxide on the Shroud. Instead, the iron oxide is pure. The chemical tests performed on the Shroud could detect any impurities above one percent. Moreover, the electron microprobe used to study the linen fibrils can detect impurities on the order of parts per thousand. None of these examinations revealed any such impurities. Calcined green vitriol produced in the fourteenth century could never be so pure as the iron oxide on the Shroud. (In the last chapter, we saw that pure forms of iron resulted from the retting process and from the water used to douse the fire of 1532.[32]

Nickell's theory is disproved by other evidence. His powder-rubbing method calls for wetting a cloth, conforming it to a bas-relief, and then letting the cloth dry before applying the powdered particles. However, the act of molding the cloth to the bas-relief would produce tension forces that would alter the cloth's configuration by stressing and elongating the linen fibrils at those points where the cloth was conformed to the bas-relief.[33] This would happen because the cellulose comprising the fibrils is composed of molecular chains; when the fibrils are stretched or straightened, the molecular side bonds between the chains become deformed and broken; these chains then adopt new configurations. The presence of moisture disrupts the molecular side bonds even more.[34] If a wetted cloth (as proposed by Nickell) were conformed to a bas-relief, the fibrils would become elongated wherever the cloth was molded to the shape of the relief. Distorted fibril configurations, or fold marks, would be evident at those locations. The process producing these elongated fibrils or fold marks would be irreversible—and, therefore, make these fold marks detectable—unless the fibrils were subjected to further stress, moisture, or temperature changes.[35] Although the Shroud was doused with water to put out the fire of 1532, the water seems to have affected only small areas of the body image, most notably the chest and knee regions. Yet the fibrils in these areas are no differently configured than those elsewhere.

Photographic examinations of the Shroud done with X ray, reflectance, and raking light (light directed at the cloth at a grazing angle) revealed various horizontal fold marks.[36] (Due to the angle of illumination used in the raking-light study, vertical fold marks would not be visible.) The locations of these horizontal fold marks are consistent with the fold configuration in which the Shroud is believed to have been stored more than a thousand years ago (to be discussed later, in chapter 7).[37] If Nickell's powder-rubbing method were responsible for the image on the Shroud, evidence of fold marks should also be evident at such places as the top of the head, eyebrow area, nose, lips, chin, hands, feet, et cetera—everywhere the linen would have been molded to an underlying bas-relief.[38] However, no fold marks in these areas can be found on the Turin Shroud.

Some obvious practical flaws are also inherent in Nickell's theory. For example, no historical evidence exists to suggest that powder-rubbing was used by artists before the nineteenth century.[39] Because Nickell used a bas-relief and not a full three-dimensional figure such as a body or statue, his technique cannot yield the types of subtle lateral distortions found on the Shroud, although this might be possible if the bas-relief itself was distorted. That a medieval artisan would accidentally use a figure that happened to be distorted exactly right is quite improbable. It's even more improbable that he would know how to distort the figure correctly or think of doing so in the first place. Yet even if an artist anticipated the need to distort the bas-relief laterally in order to encode certain deformations found on the Shroud, the artist would probably have had to sacrifice the distance information as revealed by the lightness and darkness of the body image itself.[40] Moreover, the Shroud's frontal image was encoded in a vertical straight-line direction from the supine body to the cloth draped over it. Nickell's method inherently fails to encode the frontal image in such a direction.

Nickell's attempts to reproduce blood images were as unsuccessful as his body image efforts. He asserts that the blood marks were made by tempera paint, but this assertion conflicts with the evidence showing that the blood marks were made by actual whole blood. Tempera paint cannot chemically pass for primate or human blood. In addition, the application of tempera paint could not begin to account for the serum halos surrounding the edges of the wounds covering the man in the Shroud.

Nickell has never submitted his experimental cloth to scientists for verification, but he has presented photographs showing the results of his powder-rubbing technique. On these photos his painted "blood" does not begin to approach the realism of the wounds found on the Shroud. They also do not have the shape or appearance of actual wounds that have formed and bled on human skin. The actual blood marks evident on the Shroud are mirror images on cloth of how actual wounds appear on skin. The deficiencies of Nickell's method become even more apparent when one considers the smaller wounds on the Shroud, such as the scourge marks. The slightly depressed centers and raised edges of these scourge marks have been encoded on the cloth in such a way that their characteristics are not even visible unless one examines photographs that have been enlarged and then magnified under a microscope. The fluorescing borders, composed of actual serum, around the scourge marks are also invisible until observed under ultraviolet light, as are the scourged areas that consist of only scratches or lines. None of these characteristics of the blood, serum, or scourge marks can be encoded, or duplicated, by Nickell's proposed application of tempera paint.

Craig-Bresee Method

A method similar to Nickell's has been proposed by Professor Randall Bresee of the University of Tennessee and by Dr. Emily Craig. They claim that an artist could duplicate the images on the Shroud by first drawing and encoding, by free hand, three-dimensional information of a face and body onto a paper, canvas, or other surface. The substance(s) with which the artist draws has been described as a "pigmentlike powdered aloe" and as "dry paint pigments ground into a fine powder."[41] The artist then lays a cloth over the drawing. When Bresee demonstrated the next step, he scraped or rubbed a board edgewise across the back of the cloth. When Craig demonstrated this, she took a wooden spoon and rubbed it in circles across the back of the cloth. Both techniques were done to transfer the image onto the cloth from the paper. The cloth is then heated.[42]

There are a number of reasons why this method does not duplicate the images on the Shroud. To get an image on the cloth under this method, the pigmentlike powdered substance will have to transfer to the cloth from the paper or canvas surface; yet the body-image fibrils on the Shroud do not consist of any added material, powdered or otherwise. Even if the Shroud body-image fibrils consisted of this substance, its application would not be uniform on the cloth as a whole, or on each of the individual fibrils. The scraping or rubbing of a board across the back of the cloth during the transfer process would not leave a uniform application throughout, and instances of "snow fencing" would occur on parts of the cloth as a whole and on individual fibrils. Rubbing a wooden spoon in circles on the back of the cloth would also cause a similar result.

Craig and Bresee's method resulted in a visible image on their cloth within a short distance of view. Their image did not contain the diffuseness of the Shroud image, which cannot be seen within six to ten feet.

By definition this method would fail to encode the body image in a straight-line vertical direction from the supine body to the cloth draped over it, as is found on the Shroud body images. To do so requires a body with a cloth draped over it, neither of which is utilized under this method. It's also hard to see how a forger using this method could possibly encode the intricate details found on each of the 100–120 scourge marks when many of these details would have been completely invisible to the artist. Bear in mind, if he got the invisible detail of just one of these scourge marks wrong, it would betray his work as a forgery. Moreover, this method cannot encode the other blood marks and wounds that could only have formed and flowed as a result of whole blood exuding and coagulating from real wounds on a real body.

Heated Bas-Relief/Scorch Theory

Another possible image-forming mechanism similar to that proposed by Nickell involves pressing a stretched cloth over a heated bas-relief. Such an idea was first proposed in 1961 and tested, with limited success, by placing a white handkerchief on top of a heated small medallion that bore a carving of a horse.[43] This theory is more intriguing than most because the Shroud image does appear to have many of the physical and chemical properties of a light scorch.[44] STURP scientists Jackson, Jumper, and Ercoline tried to duplicate the image on the Shroud by testing the scorch hypothesis more fully. To accomplish this, they heated a full-size bas-relief model of a face and stretched over it a linen cloth of a thickness similar to the Shroud.[45]

The image produced by this heated bas-relief method can be seen in figure 50. This photo clearly illustrates that the resulting image lacks the high resolution and

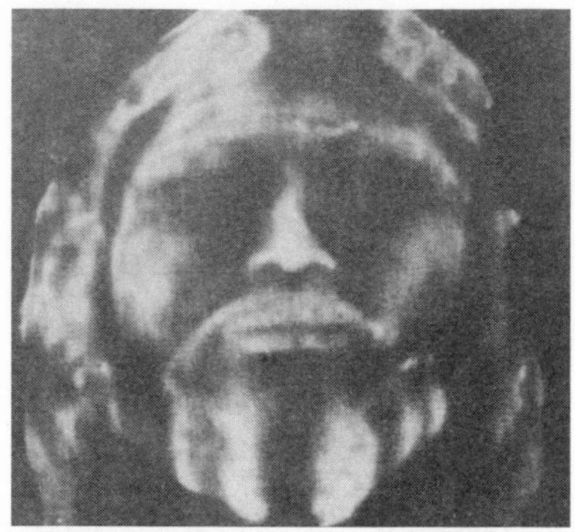

(left) Fig. 50. Bas relief on dry linen.

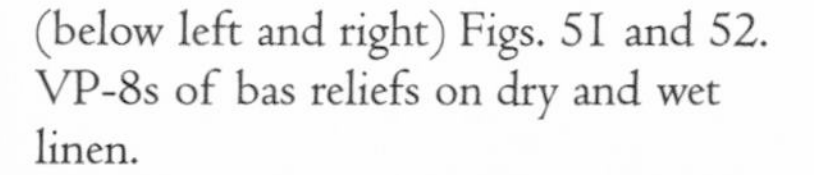

(below left and right) Figs. 51 and 52. VP-8s of bas reliefs on dry and wet linen.

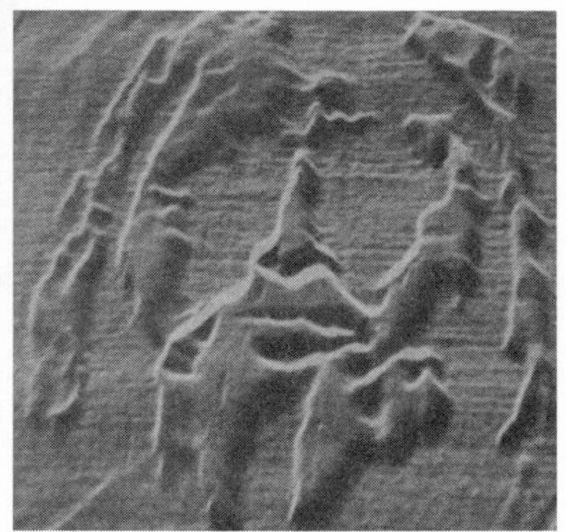

sharp focus found on the Shroud. While the bas-relief method seemingly yields a respectable three-dimensional image, problems are evident in the accompanying VP-8 relief of this image. Hollow spots below the eyes, next to the bridge of the nose, below the lips, in the beard, and on the forehead are all noticeable in figure 51. Further, a slight plateau is visible on the high spots of the VP-8 relief, similar to those produced in VP-8 analysis of results from experiments with direct-contact methods.

Even though the heated bas-relief produced better three-dimensional information than other methods, Jackson and colleagues concluded that this process could not encode many of the necessary Shroud image characteristics. For example, regardless of the temperature of the bas-relief, thermal discoloration appeared on the back side of the test cloth within several seconds after being placed on the hot bas-relief. Thus, the superficiality characteristic is violated because the image could not be encoded only on the topmost fibrils of the linen. The researchers tried to circumvent this problem by wetting the cloth, thereby extending the scorch time. When this technique was tried, new problems appeared. The image's contrast was reduced, causing more severe distortions in the three-dimensional analysis and resembling images obtained from direct-contact techniques (see fig. 52). In addition, because the cloth was essentially flat when the image was encoded, tests of this image-forming method failed to generate an image that contains the subtle lateral distortions that are consistent with the cloth-drape effects found on the Shroud.

Another flaw with the heated bas-relief image-forming method is that a body image discoloration should be produced wherever the body and cloth touched. The blood mark behind the man's right elbow near the triceps was transferred to the cloth, but no body-image discoloration representing this part of the arm was encoded on the Shroud. It is incomprehensible that a forger/artist would think to encode this blood mark from the arm without also encoding a body image from the same area.

Microscopic and ultraviolet examinations of the Shroud indicate that the blood images were transferred to the cloth before the body image.[46] If the body image were encoded through contact with a hot surface, thermal discoloration or degradation of bloodied fibrils would be evident because the blood images would have been in direct contact with the bas-relief heated to temperatures high enough to scorch linen. Indeed, this effect appeared in the experimental testing of this technique.[47] Microscopic study of the bloodstains on the Shroud, however, reveals no thermal discoloration or fusing (except in areas where the fire marks of 1532 intersected bloodstains). Furthermore, a heated bas-relief could not produce the many other aforemtioned unique features of the blood on the Shroud.

Hot Statue Method

Just as the heated bas-relief method cannot account for all the Shroud image characteristics, neither can the hot statue technique, which involves laying cloth over a full-size three-dimensional hot statue. A hot statue would produce an isotropic radiation source, which means the heat radiates the same in all directions. This type of uniform radiation could not produce the subtle cloth-drape distortions found on the Shroud because the distance information encoded onto the cloth would not be transferred along vertical, straight-line paths;[48] instead, the heat would travel in all directions and produce a blurred image. Thus the three-dimensional shading and high resolution of the Shroud image could not be encoded simultaneously if this image-forming method were used.[49]

Furthermore, the hot statue technique would scorch the image into multiple layers of the linen's threads, which means the image could not be superficial and confined to only the topmost fibrils of the cloth.[50] The many characteristics of the blood and serum marks also could not be reproduced with a draped hot statue. In particular, the blood marks would undergo thermal degradation as a result of their contact with a hot surface (as discussed above).

Another objection to the hot statue method lies in the inevitable creation of "hot spots" or well-defined regions of enhanced image density at points where the statue touched the cloth. Such spots would necessarily result from thermal conduction,[51] yet no such regions are present on the Shroud body image. As discussed in chapter 3, the entire image contains the same density of coloration.

Radiation and Electrostatic Fields (Kirlian Method or Corona Discharge)

Radiation and electrostatic hypotheses—sometimes grouped together as the Kirlian or the corona discharge method—have been proposed as possible explanations for the Shroud image. Kirlian photography involves a photographic plate or paper connected to a high-voltage, low-current electrical source. An image of the electrical "airglow" of an object—a Kirlian photograph—can then be developed. Experiments have shown, however, that the Kirlian theory has the same types of serious flaws that are inherent in other diffusion techniques. Advocates of the Kirlian method say their tests produce a brown stain on the cloth,[52] but the color of the Shroud body image fibrils, when examined microscopically, is actually straw-yellow.

Jackson and colleagues tested this hypothesis by coating a plaster face with phosphorescent paint and charging it to emit radiation. They then contoured sheets of photographic film over the face to simulate a draping cloth. When the light emitted from the charged face struck the film, an image "developed."[53] From this experiment, scientists learned that such a technique produces no facial image at all; the only visible result was a uniform discoloration that completely lacked shading variation and therefore produced no distance information.

A related hypothesis suggests that electrostatic fields, possibly associated with lightning, may have encoded the image on the Shroud. In a complex procedure, scientists simulated an electrically conductive face lying beneath a cloth that was attached to aluminum foil electrodes. They sent enough electricity through the electrodes to trigger heating under the linen, a process that transferred a "thermal map" of the face to the cloth.[54] The disappointing results of this experiment include the electrostatically produced image and the VP-8 analysis shown in figsures 53 and 54.

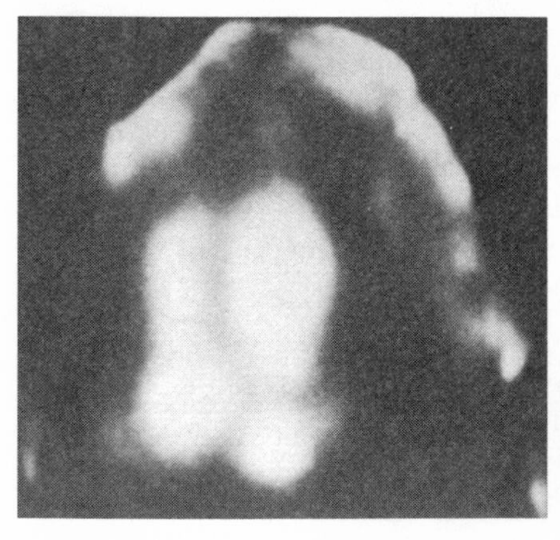

Fig. 53. Electrostatic image.

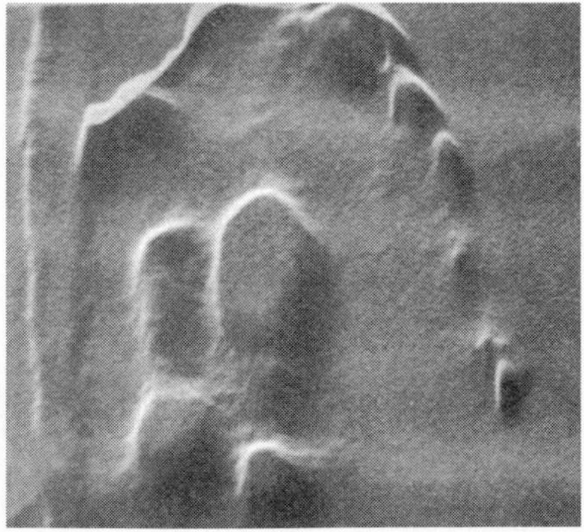

Fig. 54. VP-8 relief of electrostatic image.

There is little recognizable form to either image. Since electrostatic fields involve a diffusion mechanism, the technique also fails to transfer three-dimensional information. Further, this mechanism could not produce body image fall-off on features such as the fingers, as is found with the man in the Shroud.[55] In addition, to encode all of the body image and off-image features found on the Shroud, the amount of power and electricity that would be required would likely burn the cloth.[56] At a minimum, this would fail to produce the body image's superficiality. In addition, the scorch produced by the method would fluoresce—unlike the Shroud's body image.[57]

Scientist Giles Carter, who has experimented with cloth imaging, points out three further problems with this method: "(1) How was the dorsal image generated if the body rested directly on the cloth? (2) Would there be a sufficient potential difference between a body and a linen cloth around and about it? (The natural fibers would tend to short out the potential difference.) (3) An image has not yet been produced by this mechanism."[58]

Neither radiation nor electrostatic field transference mechanisms can encode or account for the various wounds and scourge marks on the body. Even assuming they could, they would cause thermal degradation and fusing of the blood marks and serum.

Engraved Lines Technique

Most of these artistic methods involve action-at-a-distance mechanisms (either evaporation or emission) that might encode three-dimensional information from a surface that contains an actual three-dimensional shape (a bas-relief or a sculpture). Some have speculated that a mechanism requiring distance and a three-dimensional object might not be needed; instead, an engraving technique may explain the Shroud image. For example, STURP scientists Schwalbe and Rogers have wondered whether an image with the Shroud's characteristics could be produced if a cloth were suspended above a flat metal sheet that had been heated and engraved with lines.[59] The engraved lines would change the emission of heat from point to point over the metal sheet. Radiant heat might then discolor (scorch) fibrils to produce an image that corresponded with the engraving below. With this technique, three-dimensional information might be encoded because of differences in the number (density) of engraved lines that represent cloth-body distance, and perhaps image resolution could be achieved by bringing the cloth close to the heated engraved metal plate.

As with the other naturalistic and artistic theories, this hypothesis has been tested by STURP scientists. After engraving a copper plate (a favored etching metal that has a sufficiently high melting point to scorch cloth), Jackson and colleagues studied its radiant emissions as it was heated. Although the etched lines correlated well with the three-dimensional relief of the test face used in all their experiments, the intensity of emissions that resulted from heating the engraved copper plate failed to preserve this correlation. They discovered that when the hot etching was brought near the cloth, the linen reacted to the heat and deposited "reaction products" on the metal. The variations in heat emissions caused by the different number (density) of engraved lines were disrupted by these deposited products. As figures 55 and 56 indicate, neither the high resolution nor the three-dimensionality of the Shroud image was reproduced in this radiant heat–emission experiment.[60] Because obvious problems were inherent in this image-formation theory, scientists also investigated whether a heated, engraved plate that was brought into direct contact with filter paper (instead of linen) could produce an image of the face. But the resulting images were unrecognizable.

Furthermore, if radiant or emitted heat were responsible for the Shroud image, the bloodstains and serum—which were transferred to the cloth before the body image—would exhibit thermal discoloration or degradation of bloodied fibrils, as would happen with any image-forming method in which heat was involved. In addition, if any direct contact, diffusion, or radiation mechanism—or a hybrid of these methods—had been used to encode the image on the Shroud, there should have been body-image discoloration encoded in the region of the man's right elbow near the large blood mark in this area. However, no body image is evident on this part of the Shroud. Moreover, none of the proposed image-forming methods discussed in this chapter can account for the realistic, mirror-image blood flows and wounds with serum halos, or the subtle scourge marks and other details that are invisible to the naked eye.

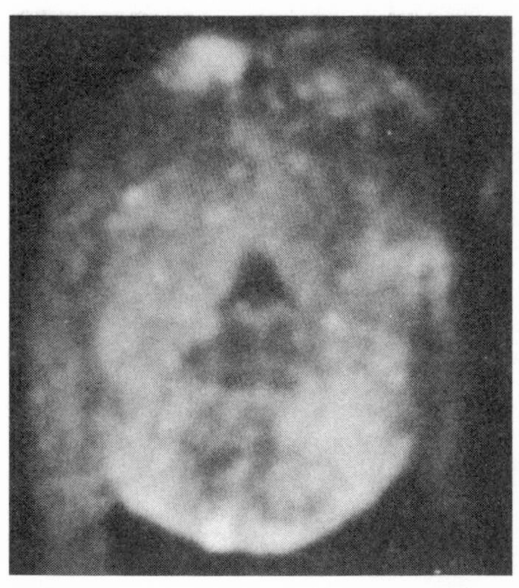

Fig. 55. Image produced by engraved lines method.

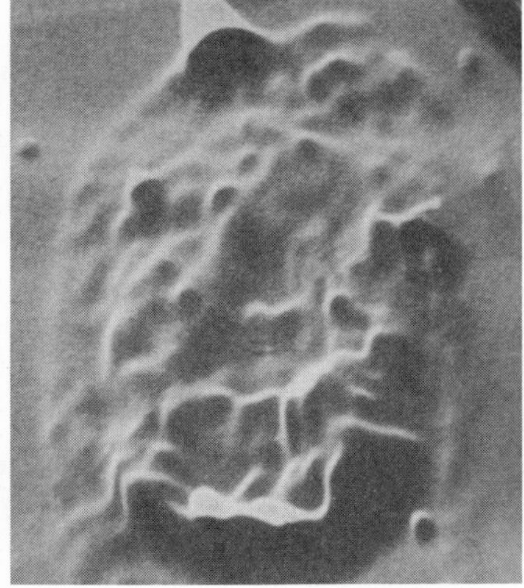

Fig. 56. VP-8 relief of engraved lines method.

Kersten and Gruber Method

Another Shroud theory that is similar to some of the previous methods has recently been attempted by Holger Kersten and Elmar R. Gruber.[61] They ground coarse-grain, crystalline substances of myrrh and aloe to a fine powder with pestle and mortar. They mixed the powders, stirred them into "a strong white wine," and let the mixture stand overnight. Most of the solid components dissolved in the liquid, forming a thick brown emulsion. This emulsion was then painted over the body of one of the authors, who had previously made several trips to the sauna in order to raise his body temperature. A cloth was laid over him while he lay supine; towels were placed on top of the linen. When the time came to remove the cloth, the emulsion stuck so much that the cloth had to be pulled off the body by another person using both hands.[62]

There are a number of reasons this method failed. First of all, the body image on the Shroud is not composed of this thick brown emulsion, and tests on the cloth and the fibers clearly reveal that the body-image fibers are oxidized, dehydrated, degraded cellulose. In addition, such a method would not leave a uniform image on the thousands of fibers in the frontal and dorsal images, as well as on each part of each individual fibril, since the cloth would have uneven contact on the body coated with the thick emulsion. The fibers would stick together and the image would penetrate beyond the topmost superficial fibrils. In some of the authors' experiments with the same pulverized substances, Kersten and Gruber admitted "the coloring penetrated the sheet and could be seen clearly on the reverse side as well."[63]

Their image does not show any shoulder abrasions on the body, and such an emulsion applied to the body would certainly cover them up. If one tried to paint around them, they would not encode onto the body image. This technique also fails to encode the scourge marks and all the other bloodstains. Because the bloodstains were encoded onto the Shroud before the body image, even if the authors were to attempt to duplicate them, they would have to encode the bloodstains first on the cloth, paint around them on the body, lay the cloth back over the body in the exact location without altering the bloodstains, and then attempt to encode the body image.

This method does not produce an image on cloth so subtle that it cannot be perceived within six to ten feet of it. Since this image is produced like a direct-contact method, it contains noticeable distortion in width throughout the image. The authors did not produce any kind of a photographic negative image with any kind of detail or resolution for any part of the body image under their method. Nor do they produce or even attempt a facial image of any kind. Since the face contains the largest variation of contours within the smallest distance on the body, if Kersten and Gruber had attempted to produce a facial image under their method, it would have contained the most distortion.

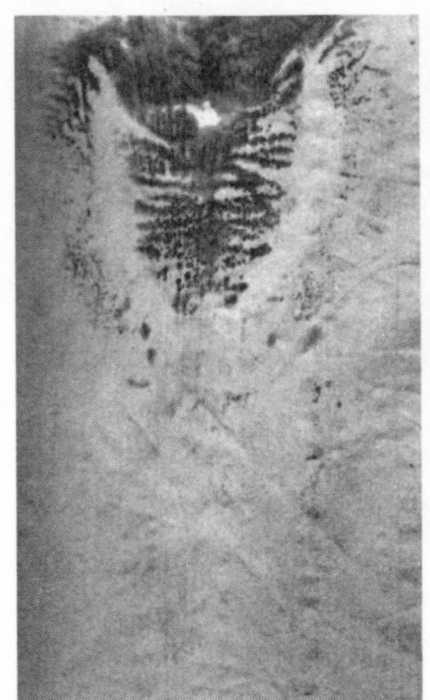

Figs. 57 and 58. Images from Kersten and Gruber's method.

Furthermore, this method cannot encode three-dimensional information onto the body image. Since this image is encoded by direct contact, it also fails to encode the image onto a draped cloth in a straight-line, vertical direction. The authors state that a modified version of this method would overcome this problem by having vapors rising straight up from the body onto the cloth. However, as was demonstrated by scientists testing the vapograph and diffusion methods, vapors do not travel in this fashion but go in all directions.

Medieval Photography

Two other very similar proposals have recently been offered—one is by Lynn Picknett and Clive Prince; the other by Professor Nicholas Allen[64]—which allege that the images on the Shroud could have been encoded by a form of photography in the medieval ages. These proposals claim that the images on the Shroud could have been created by standing a body in the sun across from a cloth placed in a dark room with a lens located in a wall of the dark room equidistant between the two objects. However, both proposals produce inferior results, and even to get these inferior results, the authors have to employ equipment, technology, and materials that did not exist in medieval times.

Under the medieval photography methods, solutions such as egg white and ammonia bichromate (Picknett and Prince's method) and silver nitrate or silver sulfate (Allen's method) are first brushed onto the cloth and allowed to dry. Afterward, these solutions are removed. The cloth is then stretched upright more than fourteen feet away from the body. An image projected onto the cloth under such conditions would lack the unique three-dimensionality that is found on the Shroud, for all the points on the body would be almost the exact distance from the cloth as the other parts of the body. The approximately one-inch difference from the tip of the nose to the next closest part of the body to the cloth in this method would be about 1/1,740 closer to the cloth. There wouldn't be any noticeable differences in distance to encode. However, a one-inch difference between points on the body of a man lying on his back with a cloth draped over him would be a very great difference, since the greatest distance between the draped cloth and the body would be no more than three to four inches.

The complete failure under these methods to produce the three-dimensionality that is found on the Shroud can easily be seen in the VP-8 reliefs of their images in figures 60, 62, and 64.

The frontal body image on the Shroud shows characteristics of the cloth draping over the body. Neither of the photographic methods could possibly reproduce this unique characteristic of the Shroud frontal body image, for in their methods, the cloth is flat and apart from the body. They would not contain image information that necessarily derives from a draped cloth over a supine body.

In addition, neither of these methods has been able to produce even half of the Shroud *body* images—no visible or satisfactory dorsal image has been produced under these methods. Perhaps that is because the dorsal image consists largely of the scourge marks. We saw earlier how intimate contact between the cloth and the body had to have occurred in order to have intricately encoded these scourge marks. Under these methods the cloth and the body do not come close to making any contact. For this same reason, these methods would not be able to encode any of the numerous and realistic blood marks that formed, flowed, and congealed on human skin.

Allen acknowledges that STURP came up with convincing evidence to support the notion that the blood and scourge marks are formed from real blood and that these important findings cannot be ignored.[65] However, his image lacks not only scourge marks but also wounds and blood marks of any kind. Picknett and Prince's method also lacks scourge marks, and while they attempted to place blood marks on a part of the image, they failed to obtain any of the realistic characteristics found on the Shroud.

Both of these proposals call for applying the wounds and blood marks after the body image was encoded. However, the blood marks on the Shroud shielded the cloth fibers under them from the mechanism that encoded the body image. Yet, with both these proposals, encoded body-image fibers and threads would be lying underneath the blood marks and flows, since their blood marks are applied after the body images are encoded.

On Allen's full-length body image, the man's beard, nose, and the hair on his head are the most prominent facial features. This makes it obvious that such a method could not encode a real body. Hair simply will not reflect sunlight in the same manner as skin, if at all, and to be actually encoded, it would have to be added afterward. Picknett and Prince freely discuss this problem with their facial image and admit that the hair on the head, beard, and mustache had to be added.[66] Allen does

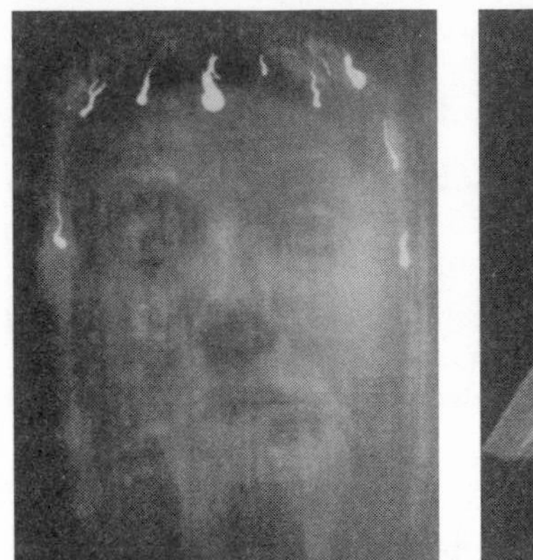
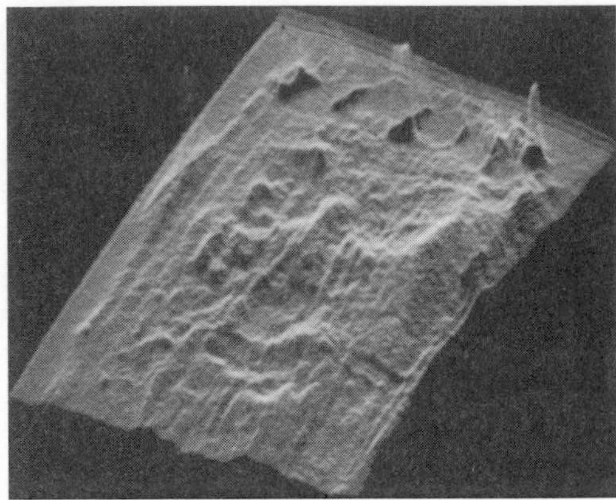

(above) Figs. 59 and 60. Image produced by Picknett and Prince and VP-8 relief of it.

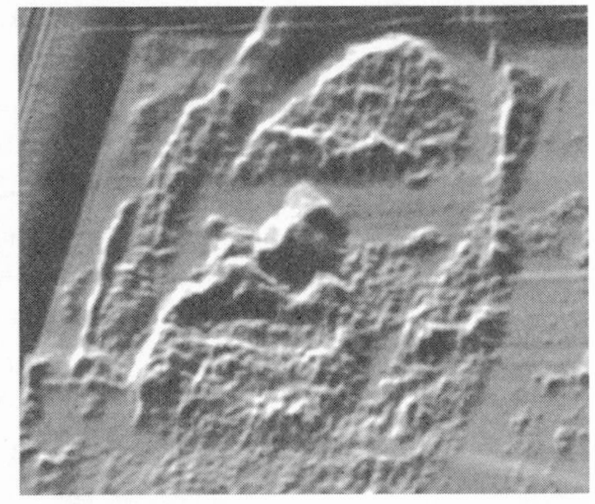

(above) Figs. 61 and 62.Facial image produced by Nicholas Allen and VP-8 relief of it.

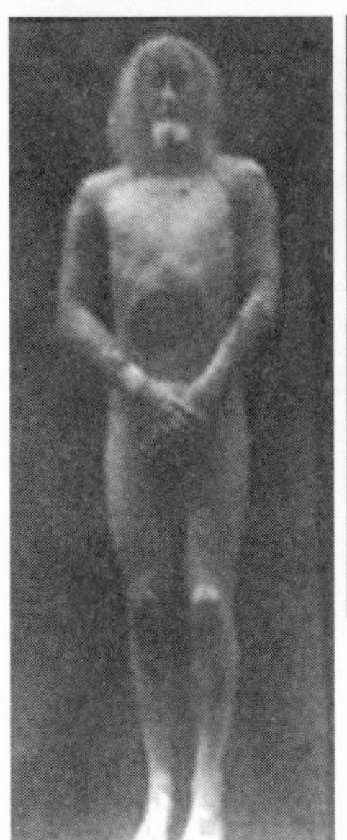
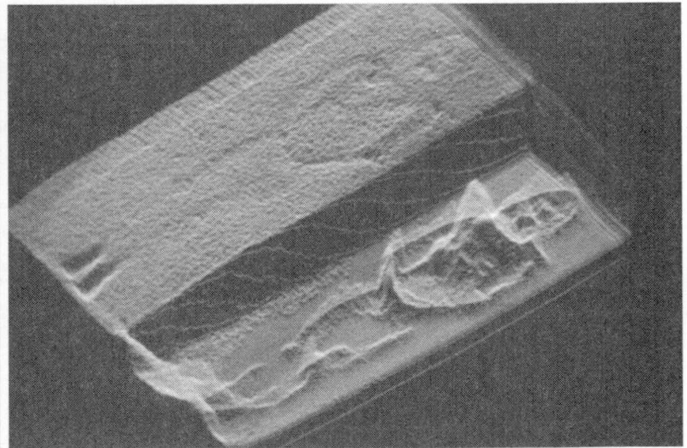

(left and above) Figs. 63 and 64. Full-length frontal image produced by Nicholas Allen and VP-8 relief of it.

not discuss the problem of the lack of reflectivity of the hair, or whether these items were added, or how they were encoded in his images. Regardless, the nose, beard, mustache, and hair at the top of the head is much brighter than the rest of the facial features in his image and, therefore, fail to duplicate the Shroud image in this respect. Unlike these two methods, the mechanism or process that created the image on the Shroud acted similarly over both skin and hair. STURP photographer Barrie Schwortz observes obvious shadowing in Allen's facial and full-length images. He notes that this, too, differs from the Shroud, which lacks shadowing.[67]

These two methods have difficulty conveying a body image at all. This can be seen in Allen's frontal body image, where a large gap appears from the man's ribcage to the middle of both upper legs along with a somewhat brighter square area on a flat chest; both feet appear much brighter than the rest of the body. You can make out the beard and nose most easily on the face, but there are blank spots over the eyes with the rest of the face being difficult to make out. What you can make out of the face does not appear to match the face of the model that he uses, nor does the hair. (Picknett and Prince do not even attempt to produce a body image—frontal or dorsal.) The authors of these methods do not provide any excuse for these problems. Aside from other inherent problems with these methods, part of the explanation may be found in another problem, use of the necessary light-sensitive pretreatment solutions of silver nitrate and silver sulfate. Allen states that "only two light sensitive 'substances' could have been employed by these hypothetical 'photographers' (medieval or Renaissance for the matter), viz.: silver nitrate and silver sulfate."[68] However, he also notes elsewhere that he has "found that both woven linen and cotton material which contains size, repels both silver nitrate and silver sulfate, and results in the formation of areas that either do not hold an image or hold an image of less intensity."[69] For other reasons, Allen writes that he had trouble keeping the pretreatment solutions on and holding the image on large cloths.

Allen produced a facial image separate from the full-length body image, which minimizes the above problems with a large image and large cloth. However, even here, the beard, mustache, nose, and hair on the top of the head appear much whiter than the rest of the face, thereby failing to duplicate the Shroud image and causing great distortion when its three-dimensionality is measured on the VP-8 Image Analyzer (figs. 61 and 62). In addition, both eyes and the areas around them are completely missing and consist of a blank area.

These brighter areas appear to have been added. Perhaps Allen had to augment his image as Picknett and Prince freely admit they did. The plaster cast used by Allen had a cloth placed over the area where the hair would be on the head. (On Allen's full-length image, the "hair" does not seem to match the model's.) In one of his articles, Allen seems to be saying indirectly that he had to repeat the process for the hair.[70] In addition, the nose is not only much brighter than the rest of the skin, but it does not match the nose of his model. The eyebrow ridge of his facial image also does not match the model's eyebrow ridge.

In order to finally fix the image and remove the light-sensitive solutions after exposure to sunlight, Allen soaks or immerses his cloth in ammonia (5 percent) or urine. One of the purposes for this soaking is to remove all traces of silver from the pretreatment solution on the cloth. Yet even this may not accomplish the purpose. Allen does not state that all of the silver is removed. He merely states that "*most* of the silver is removed" or that his immersion "would have *ostensibly* removed all silver" (italics added).[71] If any silver remained, it would most likely be on the image features and would constitute another defect in the effort to duplicate the Shroud images.

Picknett and Prince state they obtained "the most Shroud-like results" when they added urine to the chemical mixture applied to their cloth and held it in front of a fire to literally burn the mixture off and scorch the body image.[72] However, an image produced by these methods would not match the properties of the Shroud body image, which does not fluoresce under ultraviolet lighting, although its scorch marks fluoresce orange-red.

According to Picknett, Prince, and Allen, their methods result in a "scorched" body image on the cloth.[73] Yet such processes would also likely scorch, singe, or otherwise alter the blood marks already on the cloth. To duplicate the Shroud's blood marks which do not contain any such alterations, the artist using medieval photography methods would have to encode the body image around the numerous blood marks all over the body without interfering with the blood marks. This seems impossible to accomplish.

The most fundamental deficiency of these methods in attempting to reproduce or duplicate the image on the Shroud is that both proposals fail to use a human being. From the scores of medical experts who have examined the Shroud and its image, and from the wealth of medical and anatomical data, clearly the cloth wrapped the body of a crucified man. This evidence is further confirmed by the chemical, physical, and mathematical evidence. While referring to the detailed photographs, Allen acknowledges "the image in the Shroud . . . displays a degree of anatomical/medical/pathological knowledge that simply was not available to even a prominent medieval natural philosopher let alone a medieval artist or forger of relics."[74] If one is going to duplicate the images of the man in the Shroud, he is going to have to encode the images of a real human, who received real wounds, from which real blood flowed. However, a human body will not convey an image under these proposals. In order to test this, the authors would not have to crucify and kill a person. They could simply stand a human in the sun to see if he reflected the amount of light needed for their proposals.

Physicians and scientists have demonstrated that the Shroud image is of a dead human whose body is in rigor mortis, without the appearance of any decomposition at the time of image formation. Both medieval photography methods require the body to be standing upright in the sun, receiving equal parts exposure of sunlight in the mornings and afternoons. While the authors of these two methods are vague about such details, Picknett and Prince state that their forger would have used the "hot Italian sun," while Allen informs us that it takes "at least four days" to imprint the frontal image on the cloth.[75] To encode both sides of the images on the Shroud under these conditions would have taken at least eight days. Under warm conditions, it is extremely difficult to think that decomposition of the body would not occur within a period of eight days, or that the body would stay in rigor mortis as long. If the sun does not shine each consecutive morning and afternoon, the number of days increases. To overcome these difficulties, our medieval forger would have to move to a cold climate to preserve the rigor mortis and lack of decomposition.

If the forger were in the far north or the top of a mountain range with his corpse, it still would not solve his problem. The skin consists mostly of moisture, and ice crystals would develop on the corpse, which would refract and disperse the light and make it impossible to encode the image. Windblown snow, as well as any kind of precipitation, would also be a problem. The pretreatment solutions applied to the cloth by Allen consist of 99.5 percent water and .5 percent silver nitrate or silver sulfate. These would freeze on the cloth and refract and disperse the light also. A fire to keep the solutions or the cloth warm would ruin the effect of the light-sensitive solutions.

In addition, the man in the Shroud's left leg is clearly upraised. If standing upright, the man's right leg would be closest to the ground with his toes pointing down. If you were encoding this man's image while he was standing upright under these two methods, you would not be able to put any weight on his legs or feet and still have them projected the way they are on the Shroud. If you slung him under the armpits with all the weight under his shoulders, the rigor mortis would likely give out there before at least eight days, and he would start to slouch. This would be all the more compounded by the fact that the man's hands are not tied and his left shoulder is already clearly dislocated.)

As previously stated, Picknett and Prince never tried their tests using a real human. Not only that; they did not even use sunlight. They simply placed a plaster model about a foot or so from a small wooden box that contained their stretched and treated cloth, with two ultraviolet lamps placed next to the box and facing the model at approximately 45 degree angles. The authors could have had a human sit in a chair to test his or her reflectivity compared to the model's. If they had used a human, it would have been the *only* element of realism in the test of their own proposal. Picknett and Prince wildly claim that Leonardo da Vinci left his own image on the Shroud. Instead of claiming that a person who wasn't even born until one hundred years after the Shroud was first exhibited in Europe left his image on the cloth, they should see if a human image can even be encoded using their method.

Allen, for his part, seems to think that a corpse alone would not reflect enough light to encode a good image. The bright white body cast used by Allen can be seen in fig. 65. Presumably to overcome this problem, Allen alludes to painting a corpse under this method in order to increase its reflectivity.[76] However, this would interfere with encoding a number of the wounds on the Shroud that were found to be part of the body image, that is, they consist of degraded cellulose.[77] Furthermore,

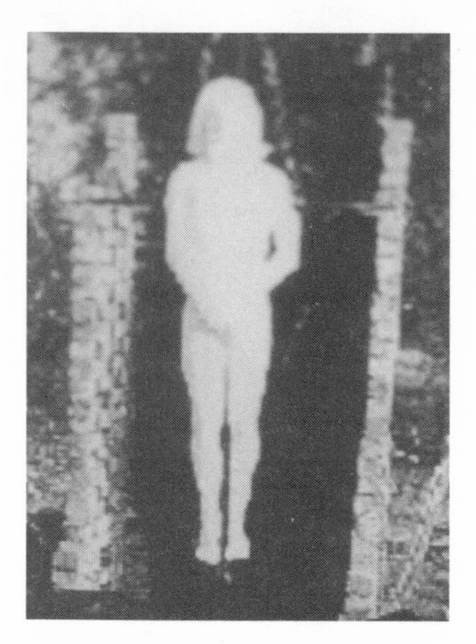

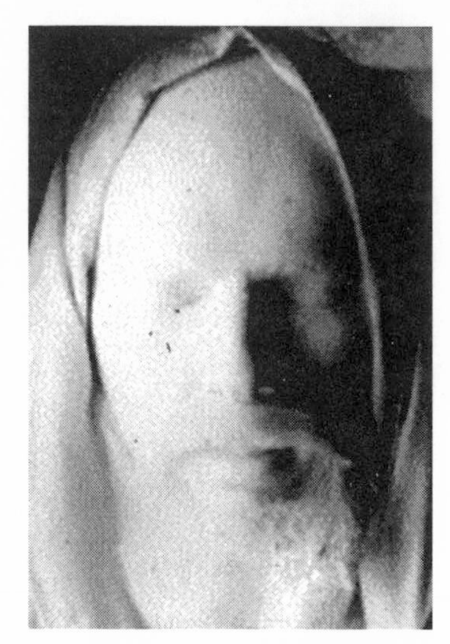

Figs. 65 and 66. Allen's full-length and facial models.

scratches, cuts, and abrasions appear on the man's leg, knee, and shoulders. If any of these items had been painted over, they could not have been encoded.

There are several other reasons why the methods, technologies, or experiments employed under these proposals could not have been used in medieval times. Both methods call for placing the cloth in the back of a dark room (camera obscura). While very rare accounts of the camera obscura can be found from medieval times, they were made only by piercing a small hole or pinhole in the front wall of the dark room to view an object, upside down, on the back wall. In fact, for centuries the camera obscura was known as the pinhole camera. Figure 67 shows a drawing of the camera obscura by Reiner Gamma Frisius in 1544, illustrating how to use it for viewing a solar eclipse. It was not until 1550 that the mathematician Girolamo Cardano suggested replacing the hole with a lens. Even then, its purpose was for tracing images to be reproduced right side up. Even this *limited* use did not occur until two centuries after the Shroud first appeared in Europe.The use of a lens to make a permanent image would not occur for several more centuries.

Picknett and Prince blatantly use an aperture from an early modern camera, as well as lenses from cameras and slide projectors. They use ultraviolet lamps at very close range on a model head. They also use ammonia bichromate on the cloth. They do not encode scourge marks, realistic blood marks, dorsal image, full-length frontal image, three-dimensionality, vertical directionality, or uniformity in the image. Not only do they fail to produce the numerous features required of the Shroud image, they use modern technology to create their image.

Allen refers to Picknett and Prince's effort as a "mismanaged 'experiment,'" and states that they "simply employ a technique based heavily on a standard nineteenth-century recipe."[78] Allen further criticizes their use of ammonium bichromate and

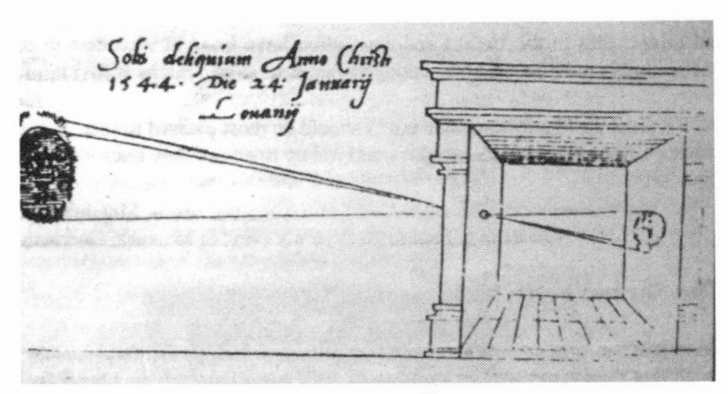

Fig. 67. Drawing from 1544 confirmating that camera obscuras still did not even have lenses two centuries after the Shroud is first displayed in Europe.

their "laissez faire manner" and attitude by quoting Picknett and Prince themselves: "We must say straightway that there is no evidence that this particular solution was used in Leonardo's day, still less by the Maestro himself."[79]

Ironically, among his very valid criticisms of Picknett and Prince's book and experiments, Allen does not criticize their use of lenses and apertures from cameras and slide projectors. Perhaps that is because he himself is using a type of lens that was not available until the nineteenth or twentieth century. He uses a biconvex, optical-quality quartz crystal lens measuring more than seven inches (180 mm) in diameter.[80] A lens made of regular glass will not transmit ultraviolet light, the portion of the EM spectrum that makes the image on Allen's treated cloth. Of course, no one in medieval times knew about ultraviolet light at all, much less what materials would or would not transmit it.

Optical-quality quartz lenses are made by first heating quartz to about 3,500°F, at which point this material becomes flexible and can be shaped. Not until the Industrial Revolution could furnaces burn that hot. In medieval times, there would not have existed a container that could even hold the quartz at that temperature, for the melting point of iron is about 285°F less than quartz.

Quartz usually occurs in fine crystals shaped like six-sided prisms topped by six-sided pyramids (figs. 68 and 69). In order to grind a lens without melting the quartz, you would have to use the long axis, or tubular part of the prism, and chisel off the pyramid ends without shattering the crystal. Cutting quartz was unknown in medieval times. Assuming you could chisel off both ends of the uneven-based pyramids without shattering the crystal, you're left with only approximately 50 percent of your original length.

You need approximately eight to ten inches of length and width before your next step. Yet, as can be seen in figures 68 and 69, there is not much width on the tubular or prism-shaped crystals. To obtain a length and width of eight to ten inches in natural quartz, the crystal would have to be very, very large—at least a foot and a half. Unfortunately, the larger crystals tend to contain fractures and inclusions (water, air, or dust particles) appear cloudy or milky. For Allen's purposes, he needs a very large, transparent, rock-crystal-quality quartz, which is very rare, just to begin his process. The largest and most abundant of these are found in the Western hemisphere, which certainly did not export them in medieval times.

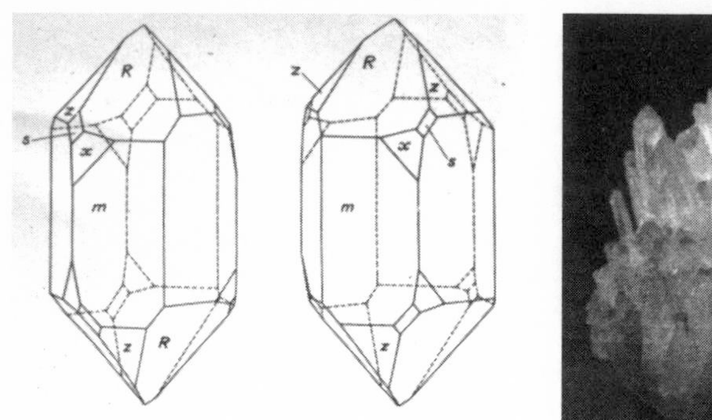

Figs. 68 and 69. Quartz in its usual construction and natural appearance.

Once our medieval forger had obtained this unlikely stone he would have to form a perfectly circular lens, with smooth, equal curves around each side of the complete circle. The convex curves of each side would have to match perfectly. If our medieval forger is off a fraction of a degree anywhere, it will throw off the highly resolved and focused image. In addition, he cannot have imperfections of any kind on the perfectly curved surfaces anywhere on the entire lens. There is absolutely no history of such skill or such a product in medieval times.

Our medieval forger would most likely have had to have done this perfect job with only his hands and a piece of cloth with some sand on it. According to Ernst Keller of the prestigious Karl Zeiss Co., perhaps the foremost lens-making company in history, while small quartz crystals are mentioned occasionally in history as crude magnifying glasses or lenses, they were probably found naturally in streams or at beaches. Keller completely confirms the historical record that optical-quality quartz lenses were not used as eyeglasses in medieval times, nor were they made with the precision or in the size that Allen requires for his method.[81]

Allen does not inform us where he acquired his lens, but it is extremely doubtful that he chiseled and hand-ground it from natural stone. Since optical-quality quartz lenses do not appear historically until the nineteenth century, Allen has the burden of demonstrating how a seven-inch, optical-quality, biconvex quartz-crystal lens without any imperfections could be made.

With his quartz lens, Allen focuses on ultraviolet rays on a cloth treated with a silver solution. The light-sensitive qualities of such solutions were not discovered for four centuries after Allen claims they were used to create the Shroud.

In addition, the pretreatment solution that works best for him is silver sulfate;[82] however, he does not say if it was ever used in medieval times. Allen goes out of his way to inform us that silver chloride was used in medieval times;[83] however, he subsequently informs us that "other substances such as silver chloride . . . are not suitable for this technique as they do not produce an image which conforms to the image formation characteristics as found on the Shroud of Turin and which were documented by the STURP committee in 1978."[84]

Neither set of authors tested their hypotheses to the furthest extent. It's almost as if they went out of their way not to do so. However, the most telling thing about their methods is that even while using eighteenth-, nineteenth-, and twentieth-century techniques and knowledge, the authors were completely unable to encode the image's scourge marks, realistic wounds and blood flows; shoulder abrasions; three-dimensionality; and vertical directionality. They were also unable to encode a human body at all, let alone one in rigor mortis and without decomposition. They were unable to encode a satisfactory dorsal image and would have serious problems encoding a uniform intensity of the whole body image and on each of the individual fibrils. The one method that attempts to encode a full-body, frontal image also has problems with silver nitrate and silver sulfate staying on the cloth and holding an image. This "full-length" figure also has noticeable problems with the feet, hair on the head, beard, mustache, and nose (with the "face-only" image having corresponding difficulties). The full-length body and facial images also contain noticeable blank gaps and shadowing in the image where the Shroud does not. According to one author, the pretreatment solution used by the other authors would not work, and his pretreatment solution would leave some remaining silver on the body image, unlike the Shroud. Lastly, according to their own descriptions of their images as scorches, the images should fluoresce under ultraviolet lighting, unlike the Shroud body images.

Other Artistic Theories and Experiments

A statement by a French aristocrat, Antoine de Lalaing, that the Shroud was "boiled in oil, tried by fire and steamed many times, without either effacing or altering the said imprint and figure"[85] has given rise to other theories of how the Shroud's images were formed. This statement related to an exhibition of the Shroud that Lalaing attended in 1503. Related theories state that the Shroud was painted and then boiled in oil or water resulting in the image we see on the Shroud. Under these theories, a chemical reaction between the paint and the fibers, or from the boiling itself, results in the images on the Shroud. First of all, Lalaing's statement contains no sources, or names of the participants or witnesses, nor does it contain any times, places, or other useful information to support it. Secondly, such theories could not produce the images on the Shroud.

If the Shroud was boiled in oil, this oil would easily have been detected in the numerous tests that STURP and other scientists have performed on the cloth and its fibrils. Boiling the cloth in oil would also have darkened or blackened the cloth.[86] If the Shroud had been boiled in water, we would not see the high calcium content that is found throughout the cloth. Furthermore, if a chemical reaction had taken palce on the fibers, it would have left signatures that the numerous chemicals and spectroscopic tests would have detected.[87] If the boiling is claimed to have somehow caused the image, this wouldn't even result in a negative image because the image-bearing fibers would have been protected the longest and reacted the least.[88]

Boiling the cloth in either oil or water would also alter or remove the Shroud's numerous blood marks, which—as we have seen—could not have been applied by an artist after the imaging occurred. Since this method requires an original application of paint, the resulting body image would not be superficial throughout. Moreover, under this method the frontal body image could not be three-dimensional or vertically directional to the cloth draped over it.

All of the naturalistic and artistic theories described in this chapter fail to duplicate the numerous unique characteristics of either the body image or blood marks on the Shroud of Turin. To illustrate the difficulty of achieving just one of these image characteristics, two certified forensic artists attempted to duplicate only the shading that contains the three-dimensional information in the Shroud body image.[89] Forensic artists were used in this experiment because they regularly compose realistic, monotone drawings that possess qualities like those of the Shroud image. The task was made as easy as possible: the artists were not required to draw on flexible, absorbent linen but to use pencil on paper; they drew only a face, not a full figure; and they ignored cloth-drape effects.

In the first test, each of the two artists produced a free-hand drawing of a shaded facial image, based on the model face used in all the experiments conducted by STURP scientists. The artists drew the facial relief as they perceived it, as a medieval artist would most likely have done. When these drawings were processed by the VP-8 Image Analyzer, both produced obvious relief deformities. In a more rigorous test, the artists were given specific relief data for fifteen anchor points on the face (nose, cheeks, lips, chin, forehead, etc.) so they could convert this distance information into shading that should more accurately represent the relief of the facial model. In addition, the artists were allowed to erase their shading if they felt it was too dark for a particular facial area (something a medieval artist could not do if he were painting on cloth).

These "rigorous" drawings containing specific relief data showed no significant improvement over the free-hand ones (see figs. 70–73). In addition, conspicuous deformities are still apparent in the VP-8 image analysis, despite the artists being given precise distance information. For centuries, artists have tried to duplicate the images on the Shroud; however, none ever came close. Even though these specialized forensic artists were given many advantages that a medieval artist would not have had, they could not reproduce this one image feature of three-dimensionality. Based on this relatively simple drawing experiment that attempted to duplicate a single body-image characteristic found on the Shroud, some scientists have concluded that the Shroud image could not be the result of any artistic technique involving eye/brain/hand coordination.[90]

None of the proposed image-forming theories detailed in this chapter comes close to explaining or reproducing all the unique features of the Shroud's images. In fact, many of these image characteristics cannot be duplicated by any known method. Some proposed theories can partially reproduce some of the image characteristics but only at the expense of others. In its review of the mystery of the Shroud's body

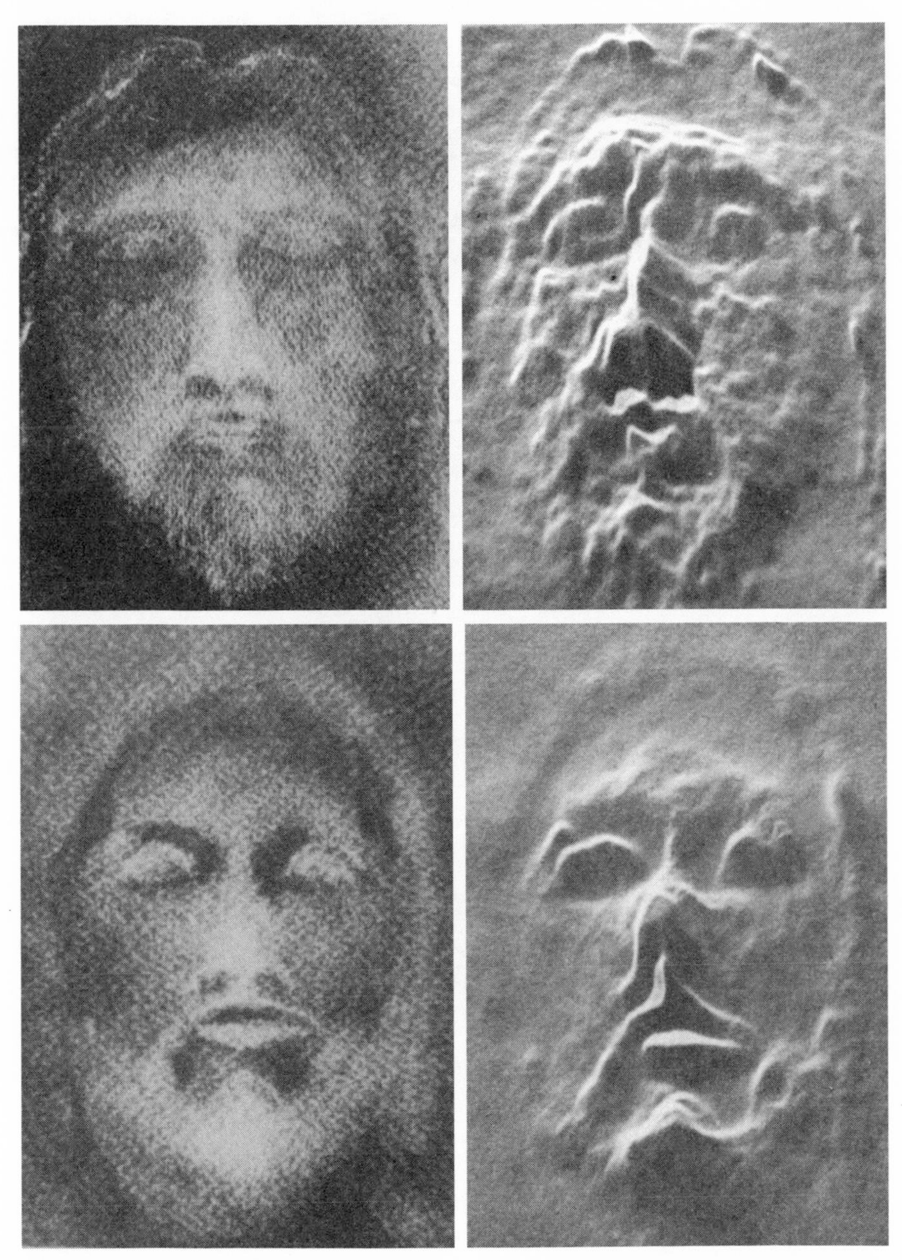

Figs. 70, 71, 72, and 73. Free hand images and their VP-8 reliefs by two certified forensic artists.

images and blood marks, STURP summarized: "The basic problem from a scientific point of view is that some explanations which might be tenable from a chemical point of view are precluded by physics. Contrariwise, certain physical explanations which may be attractive are completely precluded by the chemistry." Furthermore, "there are no chemical or physical methods which can account for the totality of the image, nor can any combination of physical, chemical, biological and medical circumstances explain the image adequately."[91]

Despite all the space-age technology available today and the extensive examination of the Shroud conducted in Turin and elsewhere, STURP member John Heller expressed the frustration his fellow scientists and he feel when pressed to explain the Shroud image: "If you were to give me a budget of 10 million dollars and told me to make a replica of [the Shroud] . . . I would not know how to do it."[92] Modern science has been able to discover the various unique characteristics that comprise the Shroud image, but it has given no concrete clues to duplicate that image. It may turn out that the image on the Shroud of Turin transcends the laws of physics as we know them. Perhaps when the image-encoding mechanism is finally understood, we will be confronted with the possibility that the transference process resides in the realm of the extraordinary and miraculous.

CHAPTER SIX

ARCHAEOLOGICAL ARTIFACTS AND OTHER EVIDENCE POINTING TO TIME, PLACE, AND IDENTITY OF THE VICTIM

TEXTILE STUDIES OF THE SHROUD

While in so many respects the Shroud is the most extraordinary cloth ever known, it is not close to being the oldest surviving cloth of its kind. If kept in dry climates and exposed to little air, textiles will survive for thousands and thousands of years. The examples are numerous. STURP scientist Dr. John Heller observed a cloth at the archaeological site of Diuropus that was five thousand years old and in good condition.[1] Many surviving mummy cloths originated two thousand to three thousand years before Christ. At the Museum of Egyptology in Turin, I observed numerous cloths from Egyptian dynasties that predated the Shroud by thousands of years, and certainly other museums in the world hold similar ancient cloths. A seven-thousand-year-old textile from the deserts of the Peruvian coastal plain, at the Paloma Village site, was reported in 1981.[2] Numerous other textile artifacts—dating back four thousand to five thousand years—have also been found in good condition in the Andean highland.[3]

Throughout the Shroud's history, the cloth has always been kept in the ideal environment for preservation: dark, arid surroundings, either folded or rolled inside a container or sealed inside a wall. With the exception of a few public viewings each century, the linen has rarely been exposed to sunlight or open air. The Shroud is much thinner and more flexible than most people realize. STURP scientists described it as being in excellent condition and having the thickness and flexibility

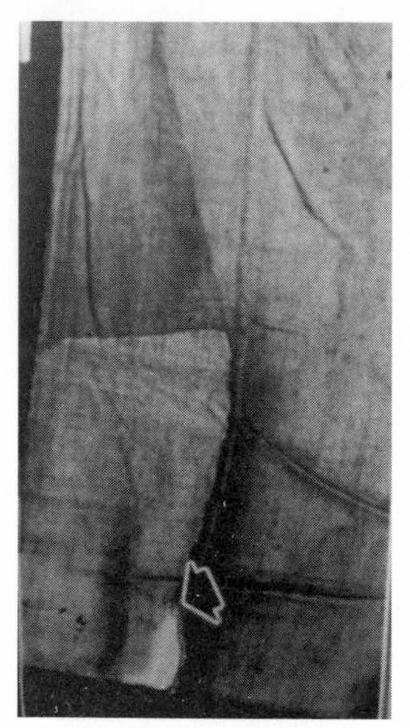

Fig. 74. The arrow indicates the location from which the Raes samples were taken.

of a T-shirt. Although there are clearly cloths older than the Shroud, the ancient ones that I observed weren't in nearly as good condition as the Shroud is today. The same observation could also be made for most other cloths from the medieval ages.

In 1973, Gilbert Raes of the Ghent Institute of Textile Technology in Belgium was allowed to examine four small pieces of material from the Shroud of Turin. Two were threads: one weft thread 12 mm long and one warp thread 13 mm long, taken from the lower left corner of the front of the cloth. The other two pieces consisted of an irregular portion of cloth, 13 x 40 mm, taken from the same area, and a piece, 10 x 40 mm, from the side strip alongside this area. Figure 74 shows the location from which these pieces were taken. He found both samples of thread and the portions from the main body and side strip to be made of linen. The cloth samples were spun with a Z twist and woven in a three-to-one (3:1) herringbone twill pattern.[4] This type of weave means that the weft thread passes under three warp threads and then over one, with each successive weft thread beginning at an ascending point one warp thread earlier (and then, in series, at a descending point) to form a diagonal "herringbone" pattern.[5] This is more complex than the plain "one over, one under" weave more commonly found among ancient linens. As such, it would have been an expensive cloth for its day.

Several authors have questioned whether the complex weaves found on the shroud were capable of being produced in the first century A.D. However, cloths with Z twist and twill weaves have been dated to times well before the time of Christ, as evidenced by a late–Bronze Age cloak found at Gerumsberg, Germany.[6] In the burial wrappings of the mummy of King Thutmes II (c. 1450 B.C.), fabrics with a 4:1 twill were found.[7] A scarf from the burial garments of King Seti I (1300 B.C.) contained a border with a 1:3 weave.[8] The disposition of threads in the frame and the operation of the loom are identical with 1:3 and 3:1 twills. A piece of fabric from the tomb of Queen Makeri (1100 B.C.) had a 1:3 twill bordered with a 1:10 twill.[9] Similarly, mummy cloth of the high priest Nessita-neb-Ashir from the same period contained weaves with 1:2 twill, 1:3 twill, and 1:6 twill.[10] One particularly striking example of fabric weaving is a linen girdle of Ramses III (1200 B.C.) This specimen, which is seventeen feet long, is woven with threads of five colors in a design composed of a 3:1 twill alternating with a 4:1 and 5:1 pattern.[11]

William Geilmann, who was a textile expert from the University of Mainz, studied pieces of linen from Palmyra, dating between the first and third centuries A.D., and one of them had the same 3:1 pattern as the Shroud.[12] Herringbone twill examples in silk, thought to be of Syrian manufacture and dating from A.D. 250 and A.D. 276, have been found in Syria and England.[13] Italian sindonologist Emanuela Marinelli claims that this weave originated in Mesopotamia or Syria and was known in the Middle East at the time of Jesus.[14] The late John Tyrer, who for more than twenty-five years was head of textile investigations at the Manchester Chamber of Commerce Testing House and Laboratories in England, wrote that "linen textiles with 'Z' twist yarns and woven three-to-one reversing twills similar to the Turin Shroud could have been produced in first-century Syria or Palestine."[15] Tyrer also was told by an experienced handloom weaver that a 3:1 fabric like the Shroud was easier for him to weave than a plain cloth of the same quality, which suggests the weaving of the Shroud cloth would not have been as difficult for a first-century handloom weaver as some have thought.[16] Italian textile expert Franco Testore says the twill was well known in Egypt as far back as 3400 B.C. and was especially prevalent for the use of mats.[17] Bands of twill weave linen from Egypt dated between A.D. 136 and 200 have also been found.[18] The pattern used to make the Shroud linen certainly is one that could have easily been produced in the first century A.D., especially in the Middle East, which stood at the crossroads of the world's major trade routes. Although Raes concluded that the pattern found on the Shroud could have been produced in the first century A.D., he could not state that it actually was, based on the pattern alone.

Using a microscope and polarized light to obtain the best possible contrast to view the thread and cloth samples, Raes made another interesting finding: He identified unmistakable traces of cotton fibrils in the portion taken from the main body of the Shroud. This indicates that wherever the Shroud linen was made, it was woven on equipment also used for weaving cotton.[19] The cotton was found to be of the *Gossypium herbaceum* variety, which is characteristic of the Middle East. Cotton was known to have been introduced into the Middle East by the seventh century B.C.[20]

Interestingly, microscopic traces of cotton were found on the Shroud, but there were not microscopic traces of wool. Jewish law (Mishnah) prohibited the mixing of linen and wool, as demonstrated by Deuteronomy 22:11 ("You shall not wear mingled stuff, wool and linen together") and Leviticus 19:19 ("Neither shall a garment mingled of linen and woolen come upon thee"). However, neither Deuteronomy, Leviticus, nor the Mishnah prohibited the mixing of cotton and linen.[21]

Another indication that the Shroud had a much older date than the Middle Ages was found in studies of the patches and backing cloth sewn onto the Shroud after the fire of 1532. In a series of transmission photographs, the Shroud was illuminated from behind and photographed from the front. The resulting photos clearly revealed that the patches were much lighter than the Shroud. While some of the difference can be attributed to the different thicknesses of the Shroud and the patching cloth, STURP scientists state, "This will not adequately explain the difference in the transmission coloration."[22] According to these scientists, "This is an indirect indication that the Shroud is a great deal older than the patches and implies that the Shroud probably has a history prior to the known A.D. 1350 date."[23]

Similarly, after studying X-radiographs of the Shroud, John Tyrer stated:

> I am very interested in the comparison that can be made between the altar cloth used to patch the Shroud, the Shroud itself, and the backing cloth. . . . The altar cloth and the backing cloth are plain woven, and are much better products than the Shroud. They seem to contain less weaving faults whilst the Shroud is a very poor product by comparison. It is full of warp and weft weaving defects, many mistakes in "drawing-in." The impression I am left with is that the cloth is a much cruder and probably earlier fabric than the backing and patches. This I think lifts the Shroud out of the Middle Ages more than anything I have seen about the textile.[24]

ROMAN EXECUTIONERS

A strong indication of a premedieval date for the Turin Shroud can be found by examining other archaeological artifacts from the first century A.D. The dumbbell-shaped marks that cover the body of the man in the Shroud from head to feet on the front and back are especially significant. As stated before, these dumbbell-shaped marks precisely match the size and shape of the marks made by the Roman flagrum, an example of which was found in Herculaneum, the sister city of Pompeii, destroyed in A.D. 79 and excavated in the eighteenth century. This instrument is occasionally depicted on Roman coins, as well. A flagrum in this form was not typical of any other culture.[25] With its dreaded *plumbatae*—balls of lead (sometimes bone) attached to the end of its thongs in groups of two or three—the flagrum dug deep, contused wounds on the flesh and caused hemorrhaging and considerable lowering of physical resistance.

Jewish law strictly limited the number of strokes that could be administered during a scourging to forty, so the Pharisees, in order to make sure this number was never exceeded, reduced the number to thirty-nine. Roman law, however, imposed no limitation whatsoever. Roman executioners were bound only by the requirement that the victim be alive for his final crucifixion. This form of scourging was so severe that the use of a flagrum on Roman citizens was forbidden.[26] Apparently the man in the Shroud was not a Roman citizen.

As discussed earlier, the elliptical lesion on the man's side also matches in detail several excavated examples of the Roman lancea, a spear of varying length, which had a long, leaflike tip that thickened and rounded off toward the shaft. The wound does not correspond to the *hasta* (a long, heavy spear with a point of various designs), the *hasta velitaris* (a short javelin with a very thin, long point), or the *pilum*, which was used chiefly by the Roman infantry and also had a long, thin point, but was much heavier and twice as long as the hasta.[27] Whereas these last three weapons were designed to break off inside the victim's body so they could not be reused against the Romans, the lancea was designed for repeated use. As such, it is quite typical of what might have been standard issue for the soldiers of the Roman military garrisons guarding Jerusalem in the first century A.D.

The blow to the right side was characteristically taught to members of the Roman militia because their adversaries usually had shields to protect their left sides. According to the second- and third-century theologian Origen, the lance thrust into the side of Jesus was inflicted according to the Roman military custom: below the armpits, the location where the Shroud image wound is found.[28]

The executioners of the crucified were recruited from the ranks of the Roman Army and comprised the Roman military guard, which was commanded by a centurion and carried out all crucifixions ordered by the government. These executioners were responsible not only for standing guard at the cross until the condemned was dead, but also for carrying out a variety of other functions. Among these was escorting the victim to the crucifixion site after affixing the crossbar to him.

There are differing schools of thought among researchers as to how the *patibulum,* or crossbar, was carried by a crucifixion victim. One school believes the beam was placed horizontally across the back of the shoulders, with the victim's arms tied to it (fig. 75). Other schools assert that either the victim's arms were tied to the cross, completely outstretched or the crossbeam was carried perpendicularly over one shoulder, as one would carry a beam. With any of these methods, the act of having to carry one's own crossbeam to an execution was probably intended to cause further humiliation as the victim suffered the taunts and jeers of onlookers along the way. Only the crossbar was carried, for the *stipes crucis,* or vertical part of the cross, was permanently affixed in the ground. The crossbeam alone weighed about one hundred pounds, and carrying it in a scourged, beaten, and weakened condition would have been extremely difficult. As evidenced by the cuts, scratches, and dirt on his face and knees, the man in the Shroud probably experienced such difficulty and suffered falls.

Fig. 75. Crucifixion victim carrying a cross over his shoulders.

Mocking and tormenting of crucifixion victims were customary and resulted from the judicially sanctioned subjection of the condemned to the military guards who performed the executions. The man in the Shroud has been beaten about the head and face, as is indicated by the swelling around the cheeks and eyes, as well as by the cuts on the face and head. Of particular interest is that he had placed upon his head a grouping of sharp, thorny objects. As seen in the earlier discussion of the head wounds, the thorny objects are believed to be in the configuration of a cap that covered the man's entire head. In the East, a mitre was traditionally used for a crown.[29] A mitre is a caplike structure that encloses the entire skull and is not like the "wreathlet" that serves as a crown in the West.

It was common for the legs of crucifixion victims to be broken so that they could no longer push themselves up to breathe and would then asphyxiate. Since Jewish law prohibited leaving bodies on a cross after sunset, the Romans instituted this measure out of deference to the Jews. The remains of a Roman crucifixion victim who was crucified in Jerusalem around A.D 50–70 had sustained broken tibias and a broken fibula.[30] However, the legs of the man in the Shroud do not appear to have been broken. This corresponds to the Gospel accounts of the execution of Jesus, which state that his legs were not broken.

While crucifixions undoubtedly occurred in history for many centuries, this particular type—involving instruments and techniques of the Roman military guard executioners—was banned by Emperor Constantine around A.D. 315. This indicates that the crucifixion of the man in the Shroud took place before then.

PONTIUS PILATE COIN INDICATES AGE AND ORIGIN

One indication of an even more specific date for the crucifixion of this particular victim may be available in the Turin Shroud image. It comes from the uncorroborated evidence of coin images found over the eyes of the man in the Shroud. The presence of coins was first suggested by the three-dimensional images of the Shroud face made with the VP-8 Image Analyzer in 1976.[31] In these experiments, scientists were surprised to discover two small objects, both nearly circular and approximately the same size, over the eyes (fig. 6). More evidence of the presence of a coin was found later when photographs were taken of an enlargement of the Shroud face made from a sepia print based on the original 1931 photographic plates of Giuseppe Enrie. These photographs suggested several features that were uniquely characteristic of a Pontius Pilate coin, or lepton, issued between A.D 29 and 32. These studies were conducted by the late Francis L. Filas, S.J., of Loyola University in Chicago, and several numismatists who assisted him.

The first of the features noted by the Loyola team were the letters UCAI appearing at the 9:30 to 11:30 clock positions on the coin over the right eye. These letters seem to be part of the inscription TIOUKAICAPOC (C was an alternate form of Σ, and U or V alternates of ϒ[32]), an abbreviation of TIBEPIOUKAICAPOC

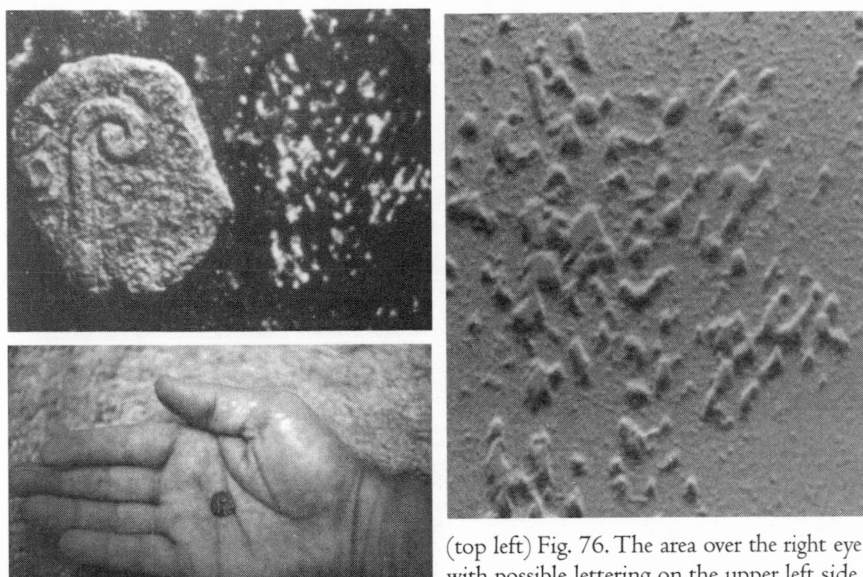

(top left) Fig. 76. The area over the right eye with possible lettering on the upper left side with staff in the center, and clipped coin margin, compared to an existing lepton with similar shape and staff, but whose letters have mostly worn off. (above right) Fig. 77. Log E Interpretation of the same area over the right eye. (left) Fig. 78. Pontius Pilate coin in hand, indicating its small size.

("Tiberiou Kaisaros," Greek for "Of Tiberius Caesar"). Both inscriptions have been identified on Pilate coins. Pontius Pilate coins that bear this first inscription have the same corresponding letters, UCAI or UKAI, appearing at the same 9:30 to 11:30 clock positions as those found on the coin over the right eye of the man in the Shroud. When a Pilate coin with this same inscription was enlarged on a screen to match the size of the enlarged right-eye area of the man in the Shroud, the size of the letters on the Pilate coin and the Shroud eye matched, with both measuring approximately 1½ mm.[33] The matching of four consecutive letters strongly suggests that this is not an optical illusion or coincidence. According to Father Filas, for these letters to have appeared by accident, or as a result of a chance pattern in the weave of the cloth, is almost impossible; the odds of all four letters appearing in consecutive order are extremely remote.[34]

An even more convincing point of authenticity to support the existence of a Pontius Pilate coin over the right eye of the Shroud image can be found in the letters UCAI. Prior to the identification of these letters on the Shroud, an interesting point concerning these Pilate coins had never been known to numismatists. The UCAI is actually misspelled and should read UKAI. The misspelling probably occurred because the pronunciations of "Caesar" in Latin and "Kaisaros" in Greek were identical, with both having the hard "K" sound (though the Greek C sounded like the Latin S). After finding this spelling over the right eye of the man in the

Shroud, Pilate coins were checked for their spelling. It was discovered that at least four Pilate coins with this same misspelling exist today.[35]

Furthermore, the letters UCAI on the Shroud face are located around the curve of an astrologer's staff, or *lituus*. This lituus is another very important point of identification, for it was used as a constant motif on coins minted by Pontius Pilate after A.D. 29. Following the rule of Pilate, this lituus was not used again by a ruler in Palestine, nor anywhere in the Roman world, as a central independent symbol.[36] Occasionally, it has been found as a small side decoration, but never more than that.

On the Shroud image, the lituus is not as clear as the inscription, but the image on the coin is completely consistent with a lituus turned to the right, or clockwise, as was the lituus on the coin with the inscription TIOUKAICAPOC. When a photo of the Pilate coin was enlarged to match the size of the enlarged coin over the right eye of the man in the Shroud, the lituus measured 11 to 12 mm from its base to the top of its curve; this is the same measurement as the lituus on the coin found on the Shroud.[37] In addition, unlike the graceful curves of the lituus stem on most Pilate coins, the coin mentioned above with the abbreviated inscription has a cruder-appearing lituus that lacks graceful curves on its stem. Again, this design matches the lituus found on the coin over the right eye of the man in the Shroud.[38]

Critics of Filas's work have asserted that the presence of all these features could be just a coincidence, an accident, or a chance pattern in the weave of the cloth. Yet these critics have not stopped to consider the extreme remoteness of such possibilities. That the weave of the cloth or the grain of the film, or something else, could accidentally produce the appearance of a Greek letter is certainly possible, but then one must consider many other features as well: that all four letters are upright and not sideways, upside down, or backward; that the letters appear side by side and match the spelling found on other coins minted by Pontius Pilate, and all the other features that make the image consistent with the genuine article. Too much is left to chance when one realizes that, of the entire 14'3" length and 3'7" width of the Shroud, this would all coincidentally be found at the precise location over the right eye of the man in the Shroud. In fact, the odds of all of the above occurring as a result of some coincidental pattern of the weave or grain of the film or some other accidental cause, and not from an inscription, are astronomical.[39]

Further comparison of the enlarged area over the right eye of the man in the Shroud with the enlarged Pontius Pilate lepton reveals even more similarities. The sizes and outlines of both are quite similar.[40] Further, the right side of the rim of each has been clipped at the 1:30 to 3:30 clock position (see fig. 76). Father Filas summarized the many points of comparison: "To sum up, there exists a combination of size, position, angular rotation, relative mutual proportion, accuracy of duplication . . . and parity [i.e., turned in the proper direction]. This combination concerns at least six motifs: a *lituus* or astrologer's staff, four letters, 'UCAI,' and a clipped coin margin."[41]

While these features are seen in varying degrees on numerous photographic negatives taken by different photographers,[42] they are most clearly visible on an enlargement of the entire two thirds life-size photograph. The Enrie photographs were taken with film that emphasized contrast, whereas subsequent photographers used improved film that tended to downplay contrasts.[43] Also, subsequent photographers secured the Shroud to its frame with magnets, which produced tiny folds or draping effects rather than the stretched tautness of the Shroud cloth that was obtained by Enrie, who is thought to have used metal tacks.[44] Unfortunately this means that STURP's many photos do little to prove or or disprove the existence of these coins. Further imaging of the Shroud should take Ernie's method into account so we may learn more about this theory.

The photographic negative from which all of the above-discussed features were found has been processed in a Log E Interpretation System, which is very similar to a VP-8 Image Analyzer. Pictures of the enlarged areas over the eyes were also processed with this system. The letters UCAI, the *lituus*, and the clipped edge at the 1:30 to 3:30 clock position are all apparent in figure 77. Furthermore, for the first time the clarity of the boundary of a coin over the left eye also became visible. These nondistorted features appear on the Log E Interpretation image in the same manner as found on the photographic negative; this only points further toward a coin with the same inscription, motifs, and designs.

While Pontius Pilate coins could have been used as currency in any part of the Roman Empire, they were most likely circulated in the area where and at the time in which they were minted: first-century Palestine. Coins used in connection with burials in ancient Jerusalem and the surrounding area have enormous relevance, though the subject has only recently begun to receive attention and is deserving of much more. There are now many known instances wherein the use of coins in burials has clearly been found. Most significantly, all examples occurred primarily in Jerusalem and the surrounding area around the time of Christ.

In 1970, it was first reported, but only in Hebrew, that excavation at the site of a fortress in the Judean Desert, at ᶜEn Boqeq, uncovered a buried man with silver coins from c. A.D. 133 placed over both his eye sockets.[45] Quite close by was a Bar Kokhba coin (A.D. 132–135) The letters and documents of Bar Kokhba were found hidden just thirty kilometers north of ᶜEn Boqeq near ᶜEn-gedi, in a region that was a traditional place of refuge, where David hid from Saul and where the last stand of Masada took place. In addition, ᶜEn Boqeq lies south of the zone around Jerusalem from which the Jews were excluded between A.D. 135 and 220.[46] In light of these factors, archaeologist William Meacham thinks there is an excellent chance the excavated man was a Jew. Furthermore, he recognizes " . . . most importantly, the ᶜEn Boqeq burial establishes that the coin-on-eye ritual was found in second-century Judea. . . ."[47]

Such a discovery was remarkable because the buried body apparently had been completely undisturbed with the skull intact after almost two thousand years. Such discoveries are fortunate and rare, allowing us to observe the exact, original use of these ancient and tiny coins. One fortunate aspect of the man in the Shroud image is that we can see his original position inside the burial cloth; if there was a coin associated with his burial, it was clearly placed over his right eye. (As noted earlier, some believe there is also evidence of a coin over the man's left eye.) The main reason it is difficult to determine precisely how coins were used in connection with Jewish burials between the first century B.C. and the first century A.D. (Second Temple Period) is that the Jews engaged in the practice of secondary burial in ossuaries. The deceased would be laid out in a shroud or coffin for approximately a year, after which time the body would have decomposed, leaving only the skeleton. The bones would then be collected and secondarily buried or placed in an ossuary, a large, rectangular chest or container usually made of limestone and often decorated and inscribed. Reburials, or secondary burials in ossuaries, were rare in Jewish tombs after the Roman destruction of Jerusalem in A.D. 70; however, during the Second Temple Period, Jews were the only group to utilize this practice.[48]

In 1979, Rachel Hachlili described her findings from excavations in the hills overlooking Jericho of tombs hewn out of rock, which dated from the middle of the first century B.C. until the first century A.D.. These burials, located outside the then city limits of Jericho as required by Jewish law, consisted of primary and secondary burials. In one of the tombs, a skull contained two bronze coins of Agrippa I (A.D. 37–44). In another tomb, a bronze coin of Herod Archelaus (4 B.C. to A.D. 6) was found in a damaged skull of a skeleton that had been interred in a coffin, and a second, earlier coin of Yehohanan Hyrcanus II from 63–40 BC was uncovered in the debris by the door. Immediately following the description of all four coins, Hachlili states, "The coins originally must have been placed on the eyes of the deceased. . . ." She added that this practice was followed often.[49]

Hachlili was probably one of few people aware that coins were found over the eyes of the deceased at ᶜEn Boqeq, because she was the first to announce this finding in a 1983 English-language publication. The first two coins located at Jericho were found together in the same skull and obviously formed a pair. Archaeologist William Meacham thinks that, since the third coin found at the site was inside a damaged skull, the fourth coin discovered in the debris of the same tomb may have been that coin's mate: The date of the second coin coincides with the date of the tomb's first use, making it unlikely to have been an intrusion after the tomb's closure.[50] These tiny coins would pass through the eye socket into the skull as a body's soft tissue decomposed.

Interestingly, after Shroud of Turin researchers discussed the appearance of a coin over the eye of the man in the Shroud and drew connections with ancient Jewish burial customs, Hachlili changed her interpretation of her findings: She stated her new belief that the coins found at Jericho had been placed in the mouth, not on the eyes.[51] However, according to analyses of an anatomist, physical anthropologist, and archaeologist, it is not possible for a coin to drop from the mouth into the skull when a body is in an ordinary, supine position because the foramen magnum—the only hole through which an object could pass—would be blocked by intact cervical vertebrae.[52] Coins placed in the mouth would ultimately fall into the throat near the cervical vertebrae, or even the upper thorax, but not the skull. Even if the head were tilted 15 to 20 degrees, the possibility of falling into the skull is still only slight. Meacham reports that dozens of exhumations of recent burials have revealed loose teeth near the cervical vertebrae, the shoulders, and even among the ribs, but that none had managed to fall inside the skull.[53] Recent experiments by Mario Maroni have demonstrated that it is possible for coins to fall into the skull through the upper eye sockets only and not through the mouth.[54]

Recently, construction workers in Jerusalem, outside the old city boundaries, stumbled on an amazing discovery: a burial cave containing six undisturbed ossuaries in their original positions. This burial cave was also from the Second Temple period. A bronze coin of Herod Agrippa I dating to A.D. 42 or 43 was found in the skull of an adult woman in one of the ossuaries. Remarkably, two elaborate and ornate ossuaries contained the final resting place of the Caiaphas family,[55] and according to Dr. Reich of the Israel Antiquities Authority, one of these "in all probability" contains the name and remains of "the high priest who presided at Jesus' trial—or at least a member of his family."[56] These two ossuaries contain three different inscriptions of "Caiaphas," the well-known family of high priests of this period. Two almost identical inscriptions on one especially beautiful ossuary read, "Joseph, son of Caiaphas."[57] Dr. Reich, who conducted the etymological study, further states that these inscriptions "may well be understood as Joseph of Caiaphas."[58] The New Testament refers to the high priest who presided at Jesus' trial by the single name Caiaphas; however, the first-century historian Josephus gives his proper name as Joseph Caiaphas and "Joseph who was called Caiaphas of the high priesthood."[59] As the high priest in Jerusalem from A.D. 18–36, Caiaphas charged Jesus with blasphemy, causing the council to attack Jesus—which started a series of bruises, wounds, and execution that can all be found on the man in the Shroud.

It is logical to conclude that a family so close to the Caiaphas family as to be buried with them would be following Jewish burial customs and not pagan customs, which called for placing a coin in the mouth. While little is positively known about the reason for laying coins over the eyes, William Meacham writes, "The ritual significance of closing the eyes of the deceased is noted in the Bible (Genesis 46:4) and in the first/second-century Mishnah. . . . The use of coins for this purpose may have had a special significance, for instance in rare types of death, or may have occurred more randomly. . . ."[60] It may very well be that Jesus' death and burial (hurriedly before the Sabbath and sundown) fit into such random circumstances, and that the use of coins by knowledgeable and respected members of the council (who buried Jesus) would not violate, but uphold, Jewish laws and rituals. The use of coins at burials by families closely associated with the family of high priests is far more likely to have followed Jewish burial customs than to have followed pagan rituals.

Many other burial sites in Jerusalem and the surrounding area at the time of Christ have been discovered with coins, but they have not been reported publicly in an extensive or detailed manner. For example, forty-two coins were found in a recess in Jason's tomb, and an Agrippa I coin dating to the year A.D. 6 was found in the debris of an ossuary tomb in Talpiot. Another Jerusalem tomb yielded a coin of

Alexander Jannaeus (103–76 B.C.) among its artifacts, as did a disturbed burial site.[61] A coin "of the Second year of the Revolt" (the first Jewish Revolt lasted from A.D. 66 to 70), was found in an undisturbed tomb, along with "a coin of Tiberius" and "two unidentified coins."[62] Perhaps most interesting of all, because it is like the one on the Shroud image, was the discovery of "a Pontius Pilate coin in a Jewish tomb on Jabel Mukaber, south of Jerusalem."[63] These brief descriptions were taken from an overview of coins found at Jewish burial sites of the Second Temple Period in and around Jerusalem by Rachel Hachlili and Ann Killebrew. At the conclusion of their overview, the authors state, "Among tombs dating to later periods, coins are only occasionally found."[64]

It is obvious from the scores of coins found at a wide variety of tombs in Jerusalem and the surrounding area that coins were used or associated with the deceased in some way at their burials during the period from the first century B.C. to the first century A.D. Whether the use was to observe a Jewish ritual of closing the eyes, paying a tribute, or some other reason, there is a common use of coins with burials in this time period and in this region. After observing the many discovered examples of coins found in Jewish tombs, Zvi Greenhut, the archaeologist who excavated the site of the Caiaphas tomb as part of the Israel Antiquities Authority, states, "I believe we must now regard coins discovered in the context of Jewish tombs from the Second Temple Period to be elements connected to the burial ceremony, despite the fact that they have not always been found in direct relation to the skulls or bodies of the deceased."[65]

All these instances of contemporaneous Roman coins (as well as some coins of Jewish rulers) found in connection with Jewish burials, in or near Jerusalem, around the time of the first century B.C.or first century A.D. are extremely significant for a number of reasons. All these discoveries have been made by archaeologists only in the last thirty years. The definite indications of such a coin over the right eye of the man in the Shroud were discovered only in the last twenty years by studying enlargements of the photographic negatives and three-dimensional reliefs. These facts, combined with the impossible problem of trying to encode such subtle features without being able to see them, rule out the possibility of forgery of the Shroud coin image. Moreover, this coin image relates to several important points of identification. The dates when this coin was minted and the predominant period of time such coins were used at burials, relates quite specifically to the first century A.D. The use of this coin in context with other evidence found on the Shroud further indicates that this victim was Jewish, that he was buried according to Jewish burial customs, and that the burial took place in or near Jerusalem.

LIMESTONE SAMPLE AND POLLEN ANALYSIS CONFIRM SHROUD ORIGINATED IN JERUSALEM

An important indication that the events depicted on the Shroud of Turin occurred not just in Palestine, but specifically in Jerusalem, is supported by an examination of the limestone in the Ecole Biblique tomb in Jerusalem. The Ecole Biblique provided researchers with access to the same rock shelf as the Holy Sepulcher and the Garden Tomb, both of which are considered the most probable choices for the actual tomb of Christ. Tombs in the Palestine/Transjordan area were carved out of limestone, which remains wet and pliable and which rubs off easily with the slightest contact.[66]

Calcium carbonate is the major component of limestone. The limestone in the Jerusalem tomb was determined to be in the form of travertine aragonite, rather than the more common travertine calcite.[67] Aragonite is less common than calcite and is formed under a much narrower range of conditions. The Jerusalem sample also contained small amounts of strontium and iron.[68] A calcium sample taken from a Shroud fiber on the foot has been compared to the calcium sample from the Jerusalem tomb. The Shroud sample was found to be in the form of aragonite, not the more common calcite, and also exhibited small amounts of strontium and iron.[69]

This match was confirmed by Dr. Ricardo Levi-Setti[70] of the Enrico Fermi Institute at the University of Chicago. Dr. Levi-Setti analyzed the calcium from both the Shroud fiber and the Jerusalem tomb with a high-resolution scanning ion microprobe. The resulting graphs show that these samples are an unusually close match, except for minute pieces of flax that could not be separated from the calcium sample taken from the Shroud fiber and that caused a slight organic variation.[71] Limestone samples taken from other tombs located at nine different test sites in Israel were also analyzed by Dr. Levi-Setti—but only the sample taken from the Jerusalem tomb matched the limestone on the Shroud.

In the 1970s, Dr. Max Frei was allowed to take dust samples from between the threads of the Shroud by means of adhesive tape. Dr. Frei, an internationally known criminologist, served as president of the United Nations' fact-finding committee to investigate the death of Dag Hammarskjöld. He founded and directed the renowned scientific department of the Zurich Criminal Police for twenty-five years. The STURP scientists, when taking cloth samples with sticky tape, used a torque applicator (which limited the pounds-per-square-inch pressure on the Shroud) and took samples only from the surface. Frei was not constrained by a torque applicator, and by moving the sticky tape laterally over the cloth, he was able to lift material from between the cloth threads.[72]

From these samples, Dr. Frei, a botanist and expert in Mediterranean flora, could identify fifty-eight different pollen grains. Pollen grains, virtually indestructible, can last thousands, even millions, of years. Although extremely tiny, they have all

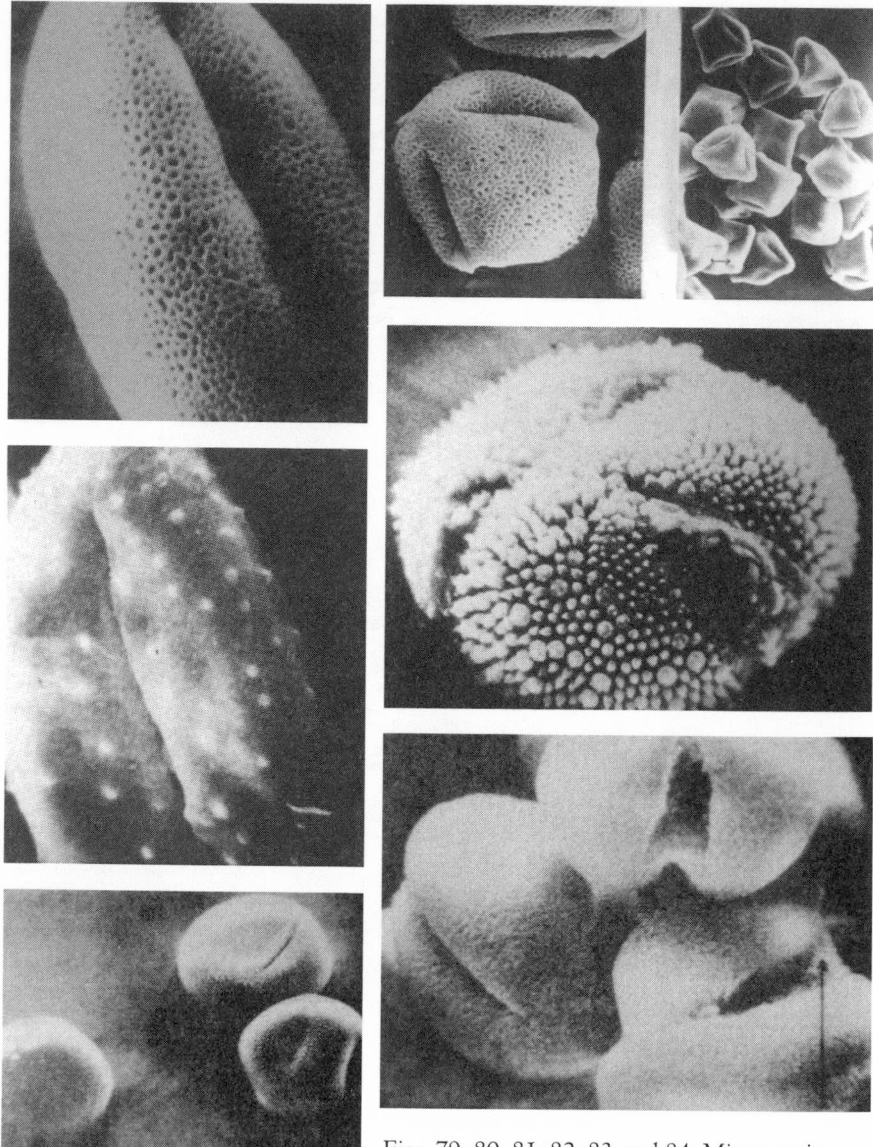

Figs. 79, 80, 81, 82, 83, and 84. Microscopic photographs of pollen indicating the individual natures of each pollen.

different shapes and features representative of the plant species from which they derive. Examples of these are shown in figures 79 to 84.

Dr. Frei spent the last nine years of his life identifying these various pollen grains and the parts of the world in which their plants are found. He also made seven different trips to the Middle East, in different floral seasons, to compare the pollen grains on the Shroud with those of ripe pollens from the Middle East. This was necessary because most of these Middle Eastern pollens were not yet registered with microphotographs in the botanical manuals nor in the herbariums. Each of his identifications was made under an optical microscope, at magnifications ranging from 60x to 1200x, and under a scanning electron microscope (SEM).

Dr. Frei discovered that pollen grains from sixteen different plant species that grow only in sand deserts or in soils with a high concentration of salt (such as those found around the Dead Sea) are present on the Shroud. These plants do not grow in France and Italy, where the Shroud has a documented history.[73] He also determined that pollen grains of seven different plants that grow in rocky hills and stony terrain, such as in Palestine and neighboring countries but not in France or Italy, are also present on the Shroud.[74] In addition, pollen grains from six other plants grown in Anatolia (a vast plateau in Turkey) and areas between Iran and the eastern Mediterranean, but not in France or Italy, also exist on the Shroud.[75] This can only mean that the Shroud has a history outside of France and Italy.

The non-European species of plants and pollen grains constitute the vast majority of samples taken from the Shroud. Perhaps the most interesting fact emerging from Dr. Frei's extensive study is that all of the non-European pollen species, except three, grow in Jerusalem. The city occupies an extraordinary geographical position situated on the ridge of the Judea mountains, between the Mediterranean area and the steppes and deserts around the Dead Sea, each of which has very different soils. Even of the three exceptions, one species grows exclusively in Constantinople,[76] and the other two are found in Edessa, Turkey.[77]

Certainly pollens have adhered to the Shroud in later centuries when it was exposed to the local atmosphere. However, they comprise a small minority of the pollens identified on the cloth. With the vast majority of pollens identified on the Shroud coming from Jerusalem and the surrounding Middle East area, the only possible conclusion one can draw is that the Shroud of Turin has a strong historic link with this part of the world.

The cloth's origin in Jerusalem is supported further by the fact that the clear majority (forty-five of fifty-eight) of pollens found on the Shroud grow in Jerusalem.[78] The number of pollens on the Shroud from plants that grow in Jerusalem is almost three times greater than the number from plants that grow in France or Italy, where the Shroud has spent its last 630 years (but rarely exposed to the open air). The majority of the Jerusalem pollens are insect pollinated and could not have ended up on the Shroud as a result of winds.[79] The only realistic explanation is that the Shroud cloth itself was manufactured in Jerusalem.

Most plant pollens typically fall onto horizontal surfaces, such as water or fields, in the process of being transported to other plants through pollination. The flax used to make linen was soaked in natural bodies of water during the retting process, which was the first opportunity for pollen to attach itself to the Shroud material. An even greater opportunity is provided during the bleaching process, when the linen made from the surviving flax was laid out in a field of grass for long periods of time and regularly sprinkled with water. Because pollens contain such features as small hooks and sticky surfaces, they would easily adhere to the textile material. The quality of the linen, as well as its whiteness, depended on the duration of the bleaching process. Since the Shroud is of very high quality for an ancient linen, it can be assumed that it received a lengthy bleaching process. If the Shroud is the burial garment of Jesus, it would have been purchased new in Jerusalem by Joseph of Arimathea, according to the New Testament.

A complementary theory of how the pollens could have gotten onto the Shroud began with the research of Dr. Whanger, who has concentrated on enhancing the off-image areas of the Shroud. By underexposing photographs of the Shroud, the off-image areas appear darker and, therefore, allow more detail to be revealed. In 1985, Whanger detected what he believed to be a clear flower image near the head of the Shroud image, whose presence was first suggested by Oswald Scheuerman's observations in 1983. After finding other similar images, Whanger thought they might have great relevance, so he acquired a six-volume set of the definitive study on the botany of Israel. Whanger spent the next four years painstakingly comparing the faint images on Shroud photographs with life-size drawings in the botany books, and using his Polarized Image Overlay Technique to check his findings. By 1989, he had tentatively identified twenty-eight species of plants that grow in Israel.[80] Although Whanger showed his findings to other Shroud researchers, he did not publish them until they could be confirmed by Dr. Avinoam Danin, professor of Botany at Hebrew University in Jerusalem and a world-renowned authority on the flora of Israel. Danin not only confirmed almost all of Whanger's identifications, but he also discovered a large number of additional flower images that were not found by Whanger.[81]

Of the twenty-eight plants, twenty-seven grow within the close vicinity of Jerusalem, where four geographical areas containing different specific climates and flora can be found. (The twenty-eighth plant grows at the south end of the Dead Sea.) All twenty-eight would have been available in Jerusalem markets in a fresh state, and most would have been growing along the roadside or in nearby fields. While three of these plants grow in France and nine grow in Italy, "half are found only in the Middle East or other similar areas and *never* in Europe" (italics added).[82] One of these plants grows only in Israel, Jordan, or the Sinai, with its northernmost boundary between Jerusalem and Jericho. Danin concluded that there is only one place in the world where all of these flowers can be collectively found—Jerusalem.[83] Furthermore, the blooming season for all these plants is March and April.[84]

There is also a strong correlation between the flower images and pollen grains found on the Shroud. For example, Danin and Whanger, along with Dr. Uri Baruch, an expert in the pollen of Israel, have also identified numerous images and pollen of three thorny plants. One of these plants left over ninety pollen grains and is found only in and around Israel. Of the twenty-eight plants identified by Whanger, Frei had previously identified the pollens of twenty-five of the same or similar plants. Dr. Frei's identifications have sometimes been criticized for identifying to species, instead of to family or genus. In fact, Danin and Baruch have not been able to make as many specific identifications in their examinations of Dr. Frei's

Fig. 85. On the left is the image of a chrysanthemum found on the Shroud; on the right is a drawing of this flower.

pollens. However, Frei was able to separate the pollen grains from the sticky tape and independently mount and rotate them under SEM at 1200x magnification. While Dr. Whanger acknowledges that Frei's pollen identification list was naturally used as a guide to search for flower images on the Shroud, the pollen grains and flower images could constitute corroborating evidence for each other.

Interestingly, one of the floral species on the Shroud that grows in Jerusalem and blooms in the Spring, *Capparis aegyptia,* provides further corroborating information of the events depicted on the Shroud. Danin, Baruch, and Whanger state: "*Capparis aegyptia* is also significant as an indicator for the time of the day when its flowering stems were picked. Flowering buds of this species begin to open about midday, opening gradually until fully opened about half an hour before sunset. Flowers seen as images on the Shroud correspond to opening buds at about 3 to 4 o' clock in the afternoon. This was confirmed by a two day experiment with, first, *Capparis aegyptia,* and later with *Capparis spinosa* Veillard."[85]

Furthermore, after examining flowers at various stages after they've been picked, Whanger concluded that their images most closely matched those that had wilted for twenty-four to thirty-six hours.[86] This gives an indication of when the flower images might have been formed. This time frame is consistent with the formation of the body images, which occurred within two to three days after the body was placed within the Shroud, due to the lack of decomposition. These flower images, like the possible coin images, do not contain all of the unique features found on the body image and are very difficult to discern. They are most likely secondary images that also formed at the time the primary images formed. Further study should be undertaken to confirm the flower images, but so far, all evidence points toward corroboration.

Danin's confirmation of, and additional findings to, Dr. Whanger's identifications are very recent and ongoing, and most scientists have not yet had a chance to thoroughly review all of them. But most scientists who have observed their presentation think that there are flower images on the Shroud. Danin has identified flowers and thorns on the photos of Pia, Enrie, and Miller, as well as on the ultraviolet fluorescent photos. He has even been able to identify two floral images on the Shroud itself with binoculars.[87] The implication of their identifications are enormous. In addition to confirming Frei's identifications, they could confirm the Jerusalem location, the period as the spring or Easter season, that different types of thorns were involved, that the flowers were picked around 3 to 4:00 in the afternoon, and that the images were encoded before two days had elapsed.

JEWISH CHARACTERISTICS OF THE VICTIM AND JEWISH BURIAL CUSTOMS

The physiognomy of the man in the Shroud appears to be Middle Eastern. After analyzing photographs, one of the world's most distinguished ethnologists, former Harvard professor Carlton S. Coon, described the man's image as definitely "of a physical type found in modern times among Sephardic Jews."[88] A study of features such as the man's hair and beard allows further deductions.

The long hair parted in the middle and falling to the shoulders is a style of particular interest. The style indicates that the man on the Shroud was not part of the Greco-Roman culture. As Professor Werner Bulst, S.J., points out, "Of the numerous portraits we have of Greek and Roman origin, there is not one of a man with hair parted in the middle and falling to the shoulders."[89] In addition, the Romans were also, with few exceptions, clean shaven, in contrast to the longish hair and beards typical of Jewish men of antiquity.[90] Perhaps the most strikingly Jewish feature on the Shroud image is what appears to be a long streak of hair in the middle of the back of the head, which falls to the shoulders, leaving an impression of an unbound pigtail. Long hair caught at the back of the head was a common fashion for Jewish men in ancient times.[91]

Another strikingly Jewish feature is the possible appearance of a chin band on the man in the Shroud. It was proper in ancient Jewish burial practices to bind the jaw to prevent the mouth from falling open.[92] Many have wondered whether the man in the Shroud had a strip of cloth running under his chin, along both sides of his face, and tied at the top or back of his head. It does look as though the beard on the front of the chin, just to the left of center, has been pushed up. The same upturned beard is visible on the three-dimensional image, where the hair on the left side of the face seems to drape over an invisible object.[93]

The burial posture of the man in the Shroud is similar to skeletons found at the c. 200 B.C.–A.D. 70 Jewish monastic Essene sect at Qumran, where the Dead Sea Scrolls were discovered. Numerous skeletons have been found there, lying on their backs, faces pointing upward, with their elbows protruding and their hands crossed over the pelvic region.[94]

The use of a single linen shroud as a burial cloth can also be found in ancient Jewish tradition. The Jerusalem Talmud, Kilaim 9:32b, states that Judah the Patriarch, the compiler of the Mishnah, who lived at the end of the second century A.D., "was buried in one linen shroud (without any other garments)."[95] From this we can infer not only that a linen shroud was used for burial, but that the body was naked when buried.

Ian Dickinson, a researcher from Canterbury, England, was struck by the fact that the measurements of the Shroud—14'3" by 3'7"—seemed odd. Research indicated that the international standard unit of measurement at the time of Jesus was the Assyrian cubit (21.4 inches). When measured in Assyrian cubits, the Shroud is 8 cubits by 2 cubits, a strong indication that this standard unit was used to measure the linen cloth.[96]

The Code of Jewish Law proscribes that if the body is contamintated with "lifeblood" or blood flowing before and after death, it should be wrapped in a sheet without washing or removing clothing and interred.[97]

One other group of writings is worth mentioning here—the Apocrypha. While these writings contain mostly legendary material and are usually not considered credible, they do reveal customs and traditions of the age in which they were written. These writings, contemporaneous with the New Testament, refer to the use of shrouds as burial cloths almost exclusively.[98]

Even as late as the twelfth century, Maimonides, a great Jewish scholar of his era who wrote about Jewish funerary practices, mentions a white linen cloth used to wrap corpses.[99] The Romans cremated their dead, and the Egyptians disemboweled and pickled them before swathing them in bandages. It is obvious that neither of these methods was used for the man in the Shroud.

The use of a fine linen cloth such as the Shroud on a crucifixion victim would indicate a degree of wealth, respect, family ties, or ranking not normally associated with common criminals. Individual burials were usually not allowed for those crucified. The general practice was to leave the crucified on the cross after death or to throw them on a heap to be devoured by scavenging animals and birds of prey. In deference to local custom, the Romans would permit the bodies of crucified Jews to be buried in a common pit. The Romans also allowed, if requested, bodies to be given to families for individual burial. According to Professor Bulst:

> The practice of burial of a crucifixion victim was confined to a very short period: In Palestine, in the first century of our era. In 6 A.D. Augustus removed the Jewish King Archelaus, son of Herod I, and installed a Roman procurator for Judea and Samaria who had the authority of the death sentence. . . . At the same time, however, the Jewish government was still in existence, which required burial before sunset according to the Jewish law. This exceptional double rule was finished with the Jewish War in 66 A.D.[100]

Professor Bulst argues that the allowance of an individual burial for a Jew crucified by the Romans in Palestine strongly suggests a date in the first century.

An indication of Roman deference to local custom can be clearly seen in the archaeological evidence of Yehohanan, the only crucifixion victim whose remains we have today. They were found in a tomb at Giv'at ha-Mivtar, located northeast of Jerusalem, which was part of a huge Jewish cemetery from the Second Temple period.

Yehohanan stood about five feet six inches (167 cm) tall, or "about the mean height for Mediterranean people of his time."[101] He was crucified sometime between the start of the first century A.D. and the destruction of the Second Temple in A.D. 70.[102] His crucifixion was undoubtedly performed by the Romans, because it was unlawful for Jews to put a man to death.[103] The Romans had militarily conquered the area during this time and had reserved the power of execution for themselves.

Both heel bones of Yehohanan were thought to be attached by a single nail, a large spike at least seven inches (17–18 cm) long, which contained fragments of olive wood (from the upright of the cross) attached to the end of the nail where it had bent in the wood. The tibia of the man's right leg had been brutally fractured, clearly the result of a single strong blow. The left tibia was not available, but the left fibula was also broken in a straight, sharp-toothed line, characteristic of a fresh bone fracture. Yehohanan's skeleton provides clear evidence of the practice of breaking the legs of a crucifixion victim in order to hasten death by preventing the condemned from raising himself up to exhale and, thus, to breathe. This, too, most likely deferred to strong Jewish custom (Deuteronomy 21:22–23), which prescribed the burial of crucifixion victims before nightfall. Unlike Yehohanan, no bones of the man in the Shroud appear to have been broken.

Some have thought the height of the man in the Shroud exceeds that for men in that part of the world in ancient times, but that is not true. His height is usually estimated to be about five feet ten inches, but when the factor of cloth drape is considered, the height of the man in the Shroud is more likely about five feet nine inches or five feet nine and one half inches.[104] (This is all the more probable as the cloth was rolled lengthwise on a spool for hundreds of years and it would have stretched.) As stated earlier, Yehohanan, at about five feet six inches (167 cm) was "about the mean height for Mediterranean people of his time."[105] In a recent excavation in Meiron, a Jewish settlement in the Galilee area with burials spanning the first century B.C. to the fourth century A.D., the average height for men, based on humerus bones, was five feet nine inches.[106] Actually, the height of the man in the Shroud corresponds to the ideal height of 4 ells (176 cm or 5'9.29"), according to an interpretation of the Talmud.[107]

Many critics of the Shroud have asserted that it could not have been the burial cloth of Jesus Christ because the man in the Shroud was not buried in accordance with Jewish burial custom. Their objection is that the body of the man in the Shroud had not been washed. Normally, when Jews buried their dead, the body was

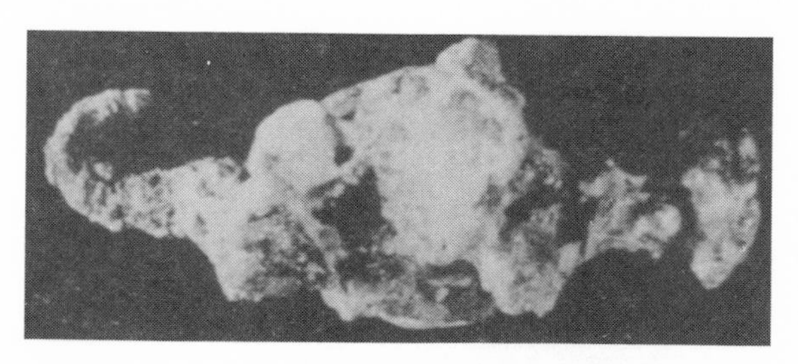

Fig. 86. A seven-inch nail with part of Yehohanan's heel bone found in an ossuary in Jerusalem.

anointed with oil to clean it, and then washed with water. Some scriptural scholars assert that Jesus' body was washed, basing their conclusion on the description of Jesus' burial in John 19:40. If Jesus' body was washed following his crucifixion, then it could not be Jesus that is depicted on the Shroud of Turin, for a large amount of blood is present on the cloth.

Several very convincing arguments refute assertions that Jesus' body was washed. In the first place, the burial that Jesus received was hurried and incomplete. It was the Day of Preparation and the Sabbath was approaching, and all work had to cease before sundown. Not only was this the Sabbath, but this particular Sabbath was also the Passover, one of the holiest of all Jewish occasions.

It was already evening when Joseph of Arimathea went to Pilate to ask for the body of Jesus (Matthew 27:57–58, Mark 15:42–43). This would have been at least a ten-minute journey (and possibly longer because of the crowds gathered for the Passover). There most likely would have been some additional delay in gaining an audience with Pilate, and even further delay as Pilate awaited confirmation of Jesus' death from the centurion (Mark 15:44–45). Since the release of the body would have involved a legal act of some importance, more than likely a document of release would have been prepared as well, especially since the Jews had sought permission a short time before to have the bodies of the crucified taken away (John 19:31). Joseph of Arimathea then had to proceed back to Golgotha, purchase a linen shroud along the way, take the body down from the cross, and transport it to the tomb. In describing Jesus' burial, Luke uses the graphic word *epephosken* (23:54), meaning that it was already "lighting-up time," when lamps were lit or when the first stars became visible.[108]

Jewish law[109] prohibited a crucifixion victim from remaining on the cross at night. The Gospel accounts clearly imply that Jesus' burial was hurried and provisional and that the Sabbath was fast approaching. Perhaps the Sabbath had already arrived by the time Jesus' body was laid in the tomb. The women had followed Jesus' body to the tomb and saw how he was laid. In fact, the reason they returned on the Sunday following the Sabbath was to take spices and ointments they had prepared to anoint him for burial (Mark 15:47–16:3, Luke 23:55–24:2).

Regarding the large and abundant quantity of myrrh and aloes mentioned in John 19:40, several writers postulate that these were in the form of dry blocks or in powdered or granulated form.[110] Such a large amount would have been used to postpone putrefaction temporarily until the washing and anointing could be completed after the Sabbath and Passover.[111] Nowhere in any of the four Gospels does it say that Jesus' body was washed. This contrasts with Acts 9:37, where Luke simply states, "In those days she fell sick and died; and when they had washed her, they laid her in an upper room." If this had been the case with Jesus, any of the four Gospel writers could have mentioned it.

Victor Tunkel, a Jewish scholar who has been a London University Faculty of Laws Member, asserts that Jewish burial rites positively and unequivocally prohibited washing a bloodstained body that died under the circumstances in which Jesus did. He regards the absence of evidence indicating that the body of the man in the Shroud was washed as evidence for the authentic "Jewishness" of the cloth.[112]

Doctors Lavoie, Klutstein, and Regan have published a well-documented discussion concerning the Jewish law dealing with the question of blood on a corpse. After detailing the positions of the Mishnah, the Talmud, and the Code of Jewish Law on these questions, and even tracing the extent to which the issue of blood is dealt with in the Old Testament, they conclude:

> The Mishnah and Talmud specifically quantitate the minimum amount of blood which is necessary to become unclean (a quarter-log of blood) [An amount equal to about one and one-half eggs] and the period at which it becomes unclean (at the time of death). This blood is described as an uncleanness of the first order and should be buried with the corpse.
>
> In the case of Jesus, who died a violent death, it can be stated that there was blood on his body that flowed during life and after death. Furthermore, as to the quantity of blood that flowed after death, an accumulation of at least a quarter-log of blood can be easily inferred by John's description of Christ's wounds, especially when considering the wound on the side. The blood on his body was, therefore, mingled blood which could not have been washed off the body because it had to be buried with the corpse in order to comply with Jewish custom.[113]

Lavoie et al. relate rabbinical comments made through the centuries, as well as more recent abridged versions of the Code of Jewish Law, as handed down through the centuries. All these sources consistently conclude that, "If a man dies a violent death and blood is shed, the blood is not washed from the body. He is simply buried in a white linen sheet with his clothes not removed for fear of losing blood that has flowed from the man at the time of death. This blood flowing at death is considered life-blood."[114]

Since full burial is not permitted on the Sabbath, the Mishnah sets out procedures that may be taken to preserve a corpse until the Sabbath has passed. The chin could be bound up, not in order to raise it but that it would not sink lower.[115] It was also permitted to place objects on the eyes to keep them closed, although one could not close the eyes.[116] There is evidence to suggest that both of these steps were taken with the man in the Shroud. Those attending the corpse could anoint and wash the body, as long as they did not move any member of it. However, this last act would have been impossible with Jesus, for he had been crowned with thorns, scourged, stabbed in the side, and crucified. He would have had blood all over his body, front and back, literally from head to feet. Whether those attending to the burial of Jesus had worked up until the Sabbath, or had continued to attend to him through the beginning of the Sabbath, they had done everything they could to properly bury Jesus according to applicable Jewish laws.

An understanding of all these various rules and their applications would have been known by Jewish religious figures or leaders such as Joseph of Arimathea or Nicodemus, who were members of the council. The detailed knowledge of Jewish laws and burial customs on the part of those who buried the man in the Shroud is just one more important point of congruence between the death and burial of the man in the Shroud and the death and burial of Jesus.

A *soudarion* is mentioned by John in 20:7 in his description of Jesus' burial cloths and in 11:44 when describing Lazarus as he exited the tomb. Soudarion, derived from Latin, refers to a handkerchief or sweat cloth. In John 11:44, it is depicted as being around the face of Lazarus; and in the account in 20:7, it is over the head of Jesus.[117] As stated by the late John A.T. Robinson, "The only position, I submit, which fits both these descriptions, assuming that they are referring to the same custom, is of something tied crossways over the head, round the face and under the chin. In other words it describes a jaw band. . . ."[118] In John 20:7, the soudarion is described as still rolled up and lying in its own place when the two apostles returned to the tomb on the Sunday following Jesus' crucifixion.

THE IDENTITY OF THE MAN IN THE SHROUD

Based on the archaeological evidence described above, the man in the Shroud appears to be a Jew, who was crucified by the Romans and buried in Jerusalem, in accordance with first-century A.D. Jewish burial customs. The next obvious question that arises is whether this crucifixion victim is the historical Jesus Christ. While absolute proof may be impossible, since we have no living eyewitnesses, we can compare the Gospel accounts of the wounds inflicted on Jesus Christ and the events surrounding his death and burial with the details visible on the Shroud. (While studying this comparison, the reader should remember that most of the evidence on the Shroud is impossible to duplicate or forge.)

Then Pilate took Jesus and scourged him (John 19:1; see also Matthew 27:26 and Mark 15:15). Jesus was scourged by the Romans; as such, there would have been no limitation on the number of strokes he received. The man on the Shroud received a severe scourging by the Romans, and scourge marks cover the front and back of his body, from head to feet.

And some began . . . to strike him . . . And the guards received him with blows (Mark 14:65). *And they struck his head with a reed* (Mark 15:19). Jesus was beaten on two occasions, once before the high priest, chief priests, and the council, and again at the praetorium by the Roman soldiers (Matthew 26:67 and 27:30; Mark 14:65 and 15:19; Luke 22:63; John 18:22 and 19:3). The man in the Shroud has been beaten about the face and head, which is clearly indicated by the swellings and lacerations evident on the image.

Plaiting a crown of thorns they put it on his head (Matthew 27:29; see also Mark 15:17–20 and John 19:2). Jesus was mocked as King of the Jews, and part of this mocking consisted of placing a crown of thorns upon his head. Such a crowning has not been found among any of the recorded tortures of the condemned prior to crucifixion, except in the case of Jesus. The man in the Shroud has numerous puncture wounds all over his scalp, and evidence strongly suggests he wore a full crown, of the type used in the East in ancient times.

So they took Jesus, and he went out, bearing his own cross (John 19:17). The man in the Shroud has abrasions on both shoulders as if a rough, heavy object had been placed across his shoulders. Based on the dirt found in the image areas of the man's nose and knees, as well as the cuts and abrasions on his face, knees, and legs, it appears that the man in the Shroud fell. This would have been expected for someone who was forced to carry a heavy crossbeam, especially if he had first received a beating and severe scourging. Jesus, too, apparently fell from this same strain (which would have been compounded by his sleepless, night-long agony of desertion and trials), for Simon of Cyrene was forced to carry the cross for him (Matthew 27:32; Mark 15:21; Luke 23:26).

"Unless I see in his hands the print of the nails, and place my finger in the mark of the nails, and place my hand in his side, I will not believe" (John 20:25). These words reveal that Jesus was nailed to the cross during his crucifixion. Some crucifixion victims were tied to the cross, while others were nailed. Jesus was nailed to the cross (John 20:20, 25, 27; see also Luke 24:39, 40). It is obvious that the man in the Shroud also has nail marks in the areas of his hands and feet.

So the soldiers came and broke the legs of the first, and of the other who had been crucified with him; but when they came to Jesus and saw that he was already dead, they did not break his legs (John 19:32, 33). In the case of Yehohanan and the thieves crucified with Jesus, *crucifragium* (the practice of breaking a victim's legs) was used to bring on death. This was not done to Jesus or to the man in the Shroud.

But one of the soldiers pierced his side with a spear . . . (John 19:34). The side wound on the man in the Shroud was determined to have come from a Roman lancea. John describes this very same instrument being used on Jesus.

. . . and at once there came out blood and water (John 19:34). This flow of blood and water is recorded only once in history, and that is with the crucifixion of Jesus Christ. Christian apologists from the time of Jesus to the fourth century, a period of frequent crucifixions until they became illegal c. A.D. 315, regarded the flow of blood and water as a miracle. The same flow of blood and watery fluid can be found coming from a postmortem side wound on the man in the Shroud.

Jesus' body was given an individual burial in a new linen shroud; however, since the Sabbath and Passover were approaching, the burial was incomplete and his body was not washed (Matthew 27:59–62 and 28:1; Mark 15:42, 46–16:3; Luke 23:52–24:2; John 19:12, 41–20:1). The man in the Shroud was also given an individual burial, in a fine linen cloth, and his body was unwashed. Both Jesus and the man in the Shroud also appear to have been buried with sudaria, or chin bands, over their heads, and those who buried them apparently possessed detailed knowledge of Jewish burial customs. Moreover, while wrapped in a Shroud, Jesus was laid in a nearby tomb hewn from the rock in either the Holy Sepulcher or the Garden Tomb, both of which consist of the same rock shelf. Invisible traces of limestone that match samples from this rock shelf, discovered only by microscopic examination, have been found on the Shroud.

The Gospels specifically state that, when the women returned to the tomb on the Sunday following Passover and the Sabbath, the body of Jesus was no longer there; it had somehow left both the tomb and the linen shroud in which it had been wrapped. The body of the man in the Shroud did not stay within the cloth longer than two or three days, for no decomposition stains appear on the Shroud. The Gospels relate that a miraculous event caused the dead body of Jesus to leave the cloth. The images of the man in the Shroud have never been duplicated despite centuries of effort from people throughout the world. The more science investigates the images, the more they seem to defy the laws of physics and science and require something miraculous to account for their unique characteristics.

CHAPTER SEVEN

THE HISTORY OF THE SHROUD OF TURIN

Prior to the twentieth century, the Shroud had been publicly displayed only nine times since its first arrival in Turin in 1578. Between 1502 and 1578, it had been kept in Chambéry, France, where the fire and dousing occurred in 1532. In the 1350s, it was exhibited for the first time in Europe as the burial cloth of Jesus, and again in 1389; both exhibitions were at a collegiate church in Lirey, France, established by Geoffrey de Charny. Geoffrey de Charny died suddenly in 1356 at the Battle of Poitiers, and the explanation of how he acquired the Shroud, or from whom, perished with him. From its first exhibits in France until today, the Shroud has a known and documented history.

The whereabouts of the Shroud between 1204 and the 1350s have been variously theorized, each of which could easily explain the cloth's transfer to Europe from the Byzantine Empire in the 1200s. There, in Constantinople, a full-length shroud with the figure of Jesus' naked, dead body was clearly identified in two unmistakable references shortly before its disappearance in 1204 at the sacking of Constantinople by the Fourth Crusade.

From 1203 to 1204, the Fourth Crusade besieged Constantinople, the great capital of the Byzantine Empire, known today as Istanbul, Turkey. The army, led by Marquis Boniface de Montferrat, was demanding payment for deposing Alexius III and seating his nephew, Alexius IV, on the throne of Constantinople. During this time, the soldiers were both inside and outside the walls of Constantinople, awaiting payment and observing the city's enormous riches, the likes of which they had never seen before. In one of the darkest hours of Christianity, the men of the Fourth Crusade attacked and sacked the city of Constantinople, stealing and looting everything of value that they could carry. The Shroud disappeared, along with most of the capital's riches. Robert de Clari, in his chronicles about this period, describes his daily visits in Constantinople while the army camped outside the walls. In describing the Shroud and its disappearance he states:

> . . . there was another of the churches which they call My Lady St. Mary of Blachernae, where was kept the shroud [sydoines] in which Our Lord had been wrapped, which stood up straight every Friday so that the figure of Our Lord could be plainly seen there, and no one, either Greek or French, ever knew what became of this shroud when the city was taken.[1]

Just three years earlier, in 1201, Nicholas Mesarites, a Greek who was keeper of the emperor's relics in the Pharos Chapel of the Boucoleon Palace in Constantinople, also wrote of the Shroud. In that year, during a palace revolution, he had to defend the chapel against a mob. In that connection he wrote:

> In this chapel Christ rises again, and the sindon with the burial linens is the clear proof. . . . still smelling fragrant of myrrh, defying decay, because it wrapped the mysterious, naked dead body after the Passion. . . .[2]

THE SHROUD, THE MANDYLION, AND THE IMAGE OF EDESSA

The Mandylion was a sacred cloth that was said to bear the divine and miraculous imprint of the face of Jesus. It was said to be "not made by hands," nor the product of painter's art or pigment. This cloth was so revered that artists came from great distances to copy it. The Mandylion arrived in Constantinople in 944 and disappeared from there in 1204, when the city was looted during the Fourth Crusade. Before 944, it was kept in Edessa, now Urfa, Turkey, since at least the sixth century. There, it was equally revered and known as the Image of Edessa. Several sources believe the cloth was first brought to Edessa from Jerusalem in the first century, either by the disciple Thaddaeus or a messenger of Edessa's King Abgar V.[3] Shortly thereafter, this cloth disappeared and did not reappear again until the sixth century. It was said to have a tremendous influence on the development of the likeness of Christ in art. Regarding the emergence of the traditional likeness of Christ in art, Wilson states:

> From the point of view of the tradition of the Eastern Orthodox Church, there is absolutely no mystery about this. The universally recognized source of the true likeness of Jesus in art was an apparently miraculously imprinted image of Jesus on cloth, the so-called Image of Edessa, or Mandylion, so highly venerated that a representation of it is to be found in virtually every Orthodox church even to this day.[4]

A study to examine the artistic depictions of Christ through the centuries was first undertaken by Paul Vignon, and then others, in the earlier part of this century,[5] and was later expanded by Ian Wilson twenty years ago.

These researchers found that the standard and conventional likeness of Christ we know today was not always the way in which Jesus was depicted. The image so familiar to us now did not begin to appear until the sixh century. Prior to this, depictions of Christ varied greatly. Nowhere in the New Testament is there a description of Christ's physical appearance. Throughout the first five centuries A.D. Jesus was usually portrayed as young, cleanshaven, and with short hair. While there were certainly some exceptions, figures 87 through 91 show how Jesus was usually depicted in the first five centuries of Christianity.

THE TRADITIONAL IMAGE OF CHRIST BECOMES ESTABLISHED

Beginning in the sixth century, this likeness abruptly changes. The likeness of Christ takes on the form of long hair parted in the middle and falling to the shoulders. He has a forked beard, with a thin mustache that droops to join the beard. His face is longer and more refined. His nose is longer and more pronounced, and his eyes are more deeply set. His whole countenance is also set in a rigidly front-facing attitude.

A consistent pattern of anomalies or oddities in the depictions of Christ also emerged in the sixth century. These anomalies include a three-sided square between the eyebrows, a V shape at the bridge of the nose, a second V within the three-sided square, a raised right eyebrow, accentuated cheeks, enlarged left nostril, an accentuated line between the nose and upper lip, a heavy line under the lower lip, a hairless area between the lower lip and beard, heavily accentuated owlish eyes, and a transverse line across the throat. These features appear regularly in pictures of Christ and are apparent in the negative image visible on the Shroud of Turin. A full listing of these anomalies and the locations of the facial markings can be seen in figure 92 from Ian Wilson.

What is most intriguing is that many of these oddities appear with no apparent artistic purpose. In fact, many of them are irrelevant to, or detract from, the naturalness of the face. While each feature does not appear in every representation, the consistent pattern of their appearances indicates that artists through the centuries were studying and interpreting the various features found from a similar source. We know artists used the Mandylion as just such a primary source. The fact that all these features appear on the Shroud makes a good case for declaring the Shroud and the Mandylion to be one and the same. This is called the "iconographic theory," which was first developed by Vignon to assert that the Shroud had a definite existence and influence on artists well before the 1300s, and that the similarity in imitations of Jesus' features could not be explained any other way.

The artists would have been working with the negative image on the cloth. They would not have had the well-focused and highly resolved positive image revealed by the photographic negative. As a result, the features on the cloth would be vague and

(clockwise from top left) Fig. 87. Christ's face from a mosaic pavement of the fourth century found at Hinton St. Mary, Dorset, England. Fig. 88. Detail from Sarcophagus of Janius Bassus, A.D. 359. Fig. 89. Portrait of Jesus, ca. 500. Fig. 90. Christ healing a blind man. St. Appollinare Nuovo in Ravenna, 510–520. Fig. 91. Christ enthroned. Apse mosaic, San Vitale, Ravenna, ca. 545.

Fig. 92. Facial markings found on the Shroud.

1. A transverse streak across the forehead.
2. The three-sided "square" on the forehead.
3. A V-shape at the bridge of the nose.
4. A second V-shape, inside the three-sided square.
5. A raised right eyebrow.
6. An accentuated left cheek.
7. An accentuated right cheek.
8. An enlarged left nostril.
9. An accentuated line between the nose and the upper lip.
10. A heavy line under the lower lip.
11. A hairless area between the lip and the beard.
12. The forked beard.
13. A transverse line across the throat.
14. Heavily accentuated, owlish eyes.
15. Two loose strands of hair falling from the apex of the forehead.

(clockwise from top left) Fig. 93. Christ Pantocrator mosaic, dome of church of Daphni, ca. 1050–1100. Fig. 94. Christ Pantocrator mosaic, in the Apse of Cefalu Cathedral, Sicily. 1148. Fig. 95. The face of Jesus, Martorana, Palermo, Sicily, ca. 1148. Fig. 96. Christ Enthroned, mosaic in the narthex of Hagai Sophia, Constantinople, late 800s. Fig. 97. Christ the Merciful, icon in mosaic, Ehemals, Staatliche Museum, Berlin, 1000s. Fig. 98. Icon at Dormitron Cathedral, Moscow, 1100s.

(left to right, top to bottom) Fig 99. Icon at St. Ambrose, Milan, 700s. Fig. 100. Mandylion of the Commenus period, 1100s. Fig. 101. Christ Enthroned, Monastery of Chilandari, Mt. Athos, Greece, 1200s. Fig. 102. Mosaic of Jesus, National Museum, Florence Italy, 1100s. Fig. 103. Pantocrator, mosaic, Holy Luke Monastery, 1100s. Fig. 104. Christ Pantocrator, fresco, from the dome of the Karanlik Monastery Church, Cappadocia, early 1100s. Fig. 105. Christ Enthroned Fresco, Church of St. Angelo in Formis, Capua, Italy, 900s. Fig. 106. Icon at St. Bartholomew's, Genoa, 1200's. Fig. 107. Bust of the Savior in the Niche of the Pallium, imediately above the Tomb of St. Peter. Vatican Grottoes, mosaic from the 700s. Fig. 108. The face of Jesus, Martorana, Palermo, Sicily, 1100s.

Fig. 109. Icon at St. Catherine's Monastery, Mt. Sinai, ca. 500–700.

Fig. 110. Sixth century face of Christ.

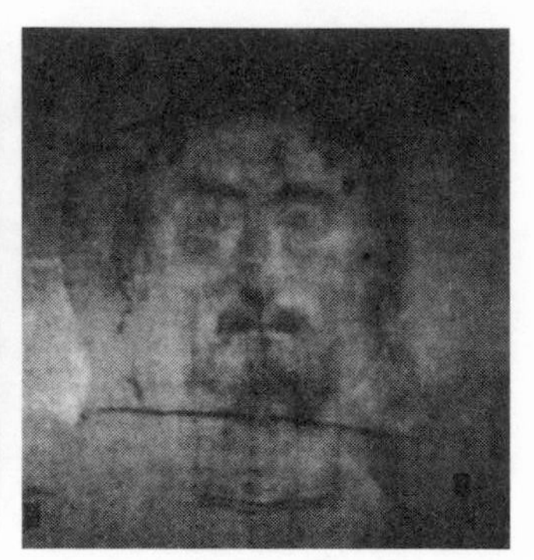

Fig. 111. Overlay of sixth century icon on Shroud facial image.

somewhat indefinite. The eyes of the Shroud image appear on the cloth to be open and staring; however the photographic negative reveals they are actually closed. It seems the early artists attempted to incorporate the Shroud image's facial features into their work by interpreting and composing them into the best representation they could. It was as if the artists were studiously attempting to follow a definitive and superior representation of Christ.

According to Wilson, it is possible to analyze statistically the frequency with which these anomalous features occur in various works of art. Wilson took a half dozen samples from the sixth, eighth, tenth, eleventh, and twelfth centuries and found between eight and fourteen of these odd features on each of them, which yields an impressive average of 80 percent incidence. Figures 93 through 110 show many portraits, still in existence today, that bear these features.

The similarities of features in artistic representations of Jesus are not limited to portraits. Dr. Alan D. Whanger of Duke University has also discovered them on Byzantine coins, particularly on Justinian II coins minted between A.D. 692 and 695 and a gold solidus of Constantine VII struck in A.D. 945. Whanger has also worked with some very early portraits of the sixth century. Using his "polarized image overlay technique," whereby one image is superimposed over another to identify points of similarity, Whanger has identified from thirty-three to more than one hundred points of congruence when these various images are matched to the Shroud face.[6] Figures 110 and 111 contain a sixth-century Byzantine icon and its appearance when overlaid on the Shroud image. Dr. Whanger has found so many points of congruence between the Byzantine icon and the Justinian II coin and the Shroud image, that he concludes the artists of the artifacts must have copied from the Shroud.[7]

When the Justinian II coin images are superimposed over the Shroud face, all three images have a transverse line in the exact same spot. This is artistically represented as a wrinkle line on the Justinian coin figures' garments, a totally unnecessary feature in and of itself. If Whanger is correct, the image on the Shroud must date far earlier than the 1350s, specifically, prior to A.D. 695, when the Justinian II coins were minted.[8]

Figs. 112, 113, 114, and 115. Note the correlations when images of the Justinian II solidus and tremissis coins minted A.D. 692–695 are laid over the Shroud's facial image.

While one could argue that the Shroud image was duplicating the coin features, and not vice versa, this would be extremely difficult to sustain. If an artist were trying to copy wrinkle lines from the coin images onto his model, he could best do this by making wrinkle lines on the model's garment. It is absurd, however, to think that he would imitate the coins' wrinkled garment lines by deciding to portray eleven feet of full-bodied frontal and dorsal images on cloth of a nude man who had been crucified, and which happens to have a transverse streak on the frontal image.

Another Shroud researcher, Dennis Mercieri, noticed similarities between the face on the Shroud image and a gold coin minted between A.D. 963 and 969, issued by Nicephorus II. Mercieri wrote, "In linking the shroud with this Nicephorus coin, I assumed the shroud inspired the coin's design. If so, then the shroud dates to A.D. 969 or before."[9] Justinian and Nicephorus were two of the very few in their times to have had access to this secluded and mysterious image of Christ on cloth. Building on this foundation established by Vignon and others, Ian Wilson was the first to assert, two decades ago, that the Shroud of Turin was actually the Mandylion, or Image of Edessa.

THE HISTORY OF THE MANDYLION

The Mandylion was the most highly revered relic in the Byzantine Empire. Numerous miracles had been attributed to the cloth. Its existence and survival were used to argue against the forces of iconoclasm, a force that destroyed countless images in the eighth and ninth centuries. The cloth was so highly regarded that

Romanus Lecapenus, a Byzantine emperor, was determined in 943 to bring the cloth to the capital of Constantinople to add to the already impressive collection of relics there. Accordingly, he sent his most able general, John Curcuas, to the city of Edessa, which had been taken from Byzantine control in 639 by the Moslems.

In what must have been one of the most unusual military missions in history, the Byzantine Army besieged Edessa and promised its emir that the city would be spared, that two hundred high-ranking Moslem prisoners would be released, that twelve thousand silver crowns would be paid, and that Edessa would be guaranteed perpetual immunity—all in exchange for the Image of Edessa. While this arrangement infuriated the Christian minority, which was tolerated in Edessa, the emir finally agreed, but only after receiving instructions to do so from his superiors in Baghdad. Before the Byzantines departed, the emperor's designee, the bishop of nearby Samosata, would accept the Mandylion only if two known copies were delivered with it. After the cloth arrived in Constantinople, the new emperor, Constantine Porphyrogenitus, commissioned the special feast-day sermon, "The Story of the Image of Edessa," which related the history of the cloth up to that time and serves today as one of our sources recounting its past.

Once in Constantinople, the Image of Edessa became known as the Mandylion. It was kept in three different locations at different times: the Church of St. Mary of Blachernae, the Hagia Sophia, and the Chapel of Pharos in the Boucoleon Palace.

Wilson believes the Mandylion, or Image of Edessa, was actually the Shroud of Turin folded into an encasement or frame so that only the head of the image was visible, with its full-length feature not becoming known until sometime after the eleventh or twelfth century. He gives numerous plausible reasons for this assertion. First, the Image of Edessa/Mandylion would have traveled from the Jerusalem area, to the eastern Anatolian steppeland of Edessa and back to Constantinople in western Turkey; these locations conform precisely to the geographical origins of the pollen that were found on the Shroud and identified by Dr. Max Frei.[10]

Second, the same descriptions and interpretations that are used for the Mandylion can be applied to the Shroud. For instance, the Mandylion has been referred to with the same terms—vague, watery, or blurry—that are used to describe the Shroud image on cloth. Upon the Mandylion's first arrival in Constantinople in 944, two sons of the reigning emperor were recorded as being disappointed upon viewing it for the first time because it seemed extremely blurred and they were unable to distinguish its features.[11] Similarly, the Mandylion was described in literary text as "an imprint," a description that also more accurately describes the Shroud. The author of the "Story of the Image of Edessa" described the image as being made by a "secretion without coloring or painter's art . . . it did not consist of earthly colors." This same also applies to the Shroud.

The vagueness of the Mandylion contributed to the various ways the Mandylion was copied. Variances are found in the length of the beard and the degree to which it forks. In some cases, the hair falls vertically to the neck, and in others it is splayed out to the sides. The eyes sometimes seem to be looking straight ahead, but other copies show them looking to the right or left.[12] It should be noted that we cannot know how many artists have copied the Mandylion directly. Some artists definitely had an opportunity to copy the image, while other artists may have copied those copies.

Moreover, the persistent claim throughout the centuries attributed to the Image of Edessa and the Mandylion that it was *acheiropoietos,* or "not made by hands," is very striking.[13] Pope Pius XI, speaking about the Shroud during the 1930s, stated that its image was "certainly not by the hand of man."[14] This statement was made some four decades before Wilson advocated his theory that the Image of Edessa, the Mandylion, and the Shroud were the same cloth. The description of the Mandylion as *acheiropoietos,* is not only similar to visual observations of the Shroud but also matches the conclusions reached by scientists after examining the cloth. For example, scientist John Heller, who conducted extensive chemical analyses of the Shroud, concluded the Shroud image "was not made by the hand of man."[15]

Wilson's hypothesis is further supported by another common feature found in portraits of Jesus dating from 540 to 940: a trellis pattern surrounding the circular halo around Jesus' face.[16] This seems to be further corroborated by the tenth century "Story of the Image of Edessa," which describes the actions of King Abgar as "fastening it to a board and embellishing it with the gold which is now to be seen."[17] This mounting and/or framing of the Mandylion apparently continued while the cloth was in Constantinople. Professor Andre Grabar, a Byzantine art scholar, has made an exhaustive study of copies of the Mandylion.[18] He has found that those copies made more than fifty years after the Mandylion's disappearance in 1204 are of the suspended type showing the cloth limp and hanging free. Copies of the Mandylion before then, however, show the cloth seemingly taut, with a fringe and frequently with a trellis pattern. Such a fringe or trellis pattern would not only add decoration around the portrait, but would also help keep the full-length Shroud hidden inside its frame.

An even stronger indication that the Shroud and Mandylion are the same object is evident from the coloring of the Mandylion's copies. The consistent color of the background of these copies is ivory white, the natural color of linen, just like the Shroud's original background. Close microscopic examination under white light reveals the true color of the Shroud body image fibers to be yellow. However, if one stands back a few feet—as one must do to distinguish the outlines of the image—the figure clearly takes on a sepia brown appearance. The color images of the Mandylion copies, too, range "from a sepia monochrome to a rust-brown monochrome, slightly deeper but otherwise virtually identical to the coloring of the image on the Shroud."[19] Furthermore, copies of the Mandylion contain the same anomalous facial features, or Vignon markings, discussed earlier. Wilson cites examples where as many as thirteen of these features have appeared on copies of the Mandylion.

The Image of Edessa/Mandylion was not necessarily a small cloth, at least according to several references to it. One early author describing the Image of Edessa actually used the word *sindon,*[20] which means "shroud" and is the word used in the Gospels to describe Jesus' burial garment.[21] The original of the "Latin Abgar

(clockwise from top left) Fig. 116. The Mandylion was usually depicted in landscape rather than portrait aspect. Fig. 117. Full-length Shroud as seen with the naked eye. Figs. 118 and 119. Mandylion copies from 1199–1200s, enclosed in a trellis frame, which is how the Mandylion was kept for part of its history according to historians.

legend" also presumed it was a cloth several yards long.[22] St. John of Damascus, writing in ca. 730, referred to the Image of Edessa as a likeness of Jesus upon a "himation," which was a full-length, oblong, outer garment worn by the Greeks over the shoulder and down to the feet.[23] Further, in the late tenth century, Leo the Deacon refers to the Image of Edessa as a *peplos,* which, too, was a full-size robe.[24]

Wilson asserts that the full-length Shroud was doubled, then doubled twice again, resulting in a cloth that is doubled in four sections. If folded in such a manner, the head appears disembodied in a landscape frame. On all but two of the copies of the Mandylion prior to its disappearance in 1204, the head was arranged on a landscape aspect as opposed to a portrait aspect, as indicated in figure 116.[25]

The portrait of a head on a landscape-shaped background is unnatural and contrary to the virtually universal artistic convention found throughout art history. Besides being visually unappealing, it is a wasteful use of artistic space. This may be one of the most compelling reasons to support Wilson's hypothesis. For this type of arrangement to awkwardly persist through several centuries suggests that each artist who copied the Mandylion must have had a similar reason. They would not all have decided to do this for different and inexplicable motives; they must have been trying to duplicate the appearance of the original that they were using for a model.

Wilson's theory is further endorsed by the unique use of the word τετραδιπλον, which describes the Image of Edessa/Mandylion. This description occurs soon after the cloth is rediscovered after five centuries of absence. The word τετραδιπλον is first found in a sixth-century text,[26] then again as ρακος τετραδιπλον in a tenth-century text written not long after the cloth's arrival in Constantinople.[27] According to Cambridge University Professor G. W. H. Lampe, editor of the *Lexicon of Patristic Greek,* in all of literature the word τετραδιπλον is used *only* in connection with descriptions of the Image of Edessa/Mandylion. As such, this usage could hardly be said to have been an idle turn of phrase. This word is a compound of two very ordinary Greek words τετρα, meaning "four," and διπλον, meaning "doubled," hence "doubled in four." The uniqueness of this term in descriptions of the Mandylion suggests that the author was trying to characterize what he may have been fortunate to observe—the way a full-length cloth was folded within its frame in a doubled-in-four manner. Otherwise, usage of this term in connection with the Mandylion has no apparent meaning.

The doubled-in-four configuration is confirmed by scientific studies. For example, a doubled-in-four folding pattern is supported by Dr. Max Frei's pollen sampling of the Shroud, which determined that the facial area contained more pollen exposure than other parts of the cloth.[28] More important, Shroud scientist John Jackson examined the locations where Wilson's proposed folding would have left evidence on the cloth and found that such a fold pattern probably remains on the Shroud today.[29] A doubled-in-four fold pattern would leave seven fold marks on the cloth. Examining X-ray, reflectance, and raking light (grazing angle illumination) photographs of the Shroud, Jackson identified four probable fold marks from Wilson's fold pattern. Tentative fold marks identified at two other locations on the cloth were consistent with Wilson's doubling-in-four method. The remaining site where a fold mark should be located cannot be detected because of fire marks, water stains, and patches at this position. Perhaps additional scientific examination of the Shroud will reveal more information on this matter, but, in the words of Wilson, "certainly there can no longer be claimed to be any absence of fold marks consistent with the Image of Edessa/Shroud identification hypothesis."[30]

The Image of Edessa/Mandylion has been lost to history. If the Shroud and the Image of Edessa/Mandylion are, in fact, one and the same, the known history of the Image of Edessa/Mandylion would complete almost the entire missing history of the Shroud of Turin.

EARLY HISTORY OF THE MANDYLION OR IMAGE OF EDESSA

We have numerous written accounts detailing the early years of the sacred cloth known as the Image of Edessa in relation to the early but brief Christianity of Edessa and the semilegendary Abgar story.[31]

The first account of the Abgar story that we have is from Eusebius, who wrote his famous *History of the Church* about A.D. 325. Eusebius states that his account is based on his own translation of Edessan archives from the Syriac into Greek. There it is stated that King Abgar of Edessa, who ruled from A.D. 13 to 50, suffered greatly from an incurable disease and had heard of the many miracles that Jesus had performed. Abgar sent a messenger with a letter to Jesus, inviting him to come to Edessa to cure him. Jesus declined but promised by correspondence to send one of his disciples to cure him after his mission on earth was complete.

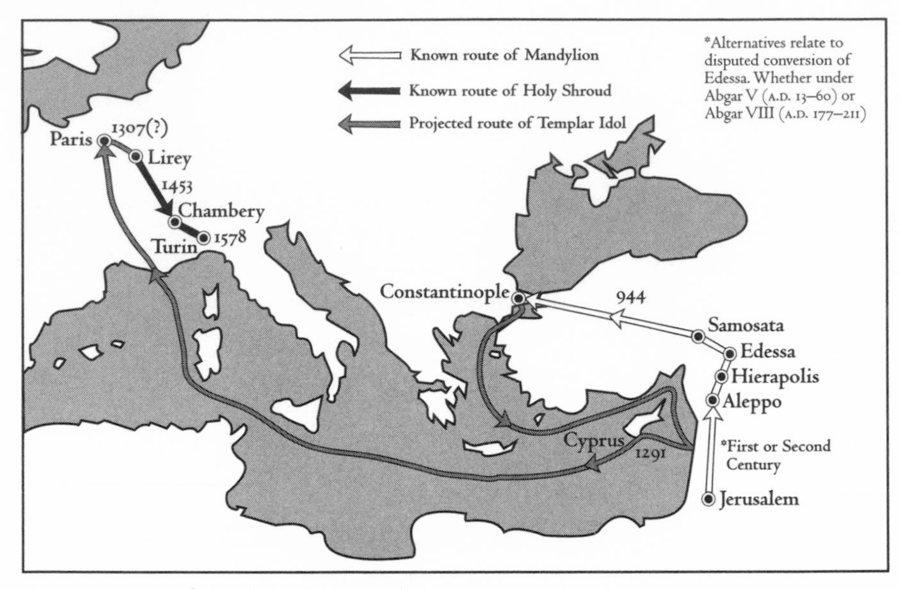

Fig. 120. Map of the Shroud's various locations.

More details of this story emerged when whole caravan loads of ancient Syriac manuscripts were recovered and brought out of the desert of lower Egypt from the Nitrian Monastery in the 1840s. Many of these manuscripts also recount the Abgar story written in the Syriac language,[32] the language from which Eusebius translated his account. These manuscripts, along with the "Story of the Image of Edessa," provide an extensive account of the Abgar story involving the early evangelization of Edessa and the Image of Edessa.

All of these accounts agree that following Jesus' departure from earth, the disciples sent the evangelist Thaddaeus (Addai, in the Syriac language) into Edessa. The more reliable versions point out that he was "one of the seventy" described in Luke 10:1. While in Edessa, Thaddaeus healed many people in the name of Christ. At this time he was staying with a man named Tobias, whose father was also named Tobias and who was a Jew originally from Palestine.[33] From the number of tombs, apparently a substantial Jewish community resided in Edessa at that time. When Abgar heard of the wonders that Thaddaeus was performing, he was reminded of Jesus' promise, and Thaddaeus was summoned to the King. As recounted in the tenth-century account of the "Story of the Image of Edessa," Thaddaeus brought with him the Mandylion and, before entering the throne room to see Abgar, "placed it on his forehead like a sign."

> Abgar saw him coming from a distance, and thought he saw a light shining from his face which no eye could stand, which the portrait Thaddaeus was wearing produced.
>
> Abgar was dumbfounded by the unbearable glow of the brightness,

> and, as though forgetting the ailments he had and the long paralysis of his legs, he at once got up from his bed and compelled himself to run. In making his paralyzed limbs go to meet Thaddaeus, he felt the same feeling, though in a different way, as those who saw that face flashing with lightning on Mount Tabor.
>
> And so, receiving the likeness from the apostle . . . immediately he felt . . . his leprosy cleansed and gone . . . Having been instructed then by the apostle more clearly of the doctrine of truth . . . he asked about the likeness portrayed on the linen cloth. For when he had carefully inspected it, he saw that it did not consist of earthly colors, and he was astounded at its power. . . .[34]

Those accounts go on to state that, upon the request of Abgar, Thaddaeus stayed and told of all the things that Jesus had said and done and of his mission on earth, thus giving the first sermon at Edessa. Most of the accounts also state that Thaddaeus stayed after this first sermon, helping to establish the first church at Edessa and the beginnings of Christianity in this area. During this time, he was assisted by Aggai, who formerly made the silks and headdresses for the pagan kings. Upon the death of Thaddaeus, Aggai then became the head of the church in Edessa. Under the reign of Abgar V and that of his son, Ma'nu V, who ruled from A.D. 50 to 57, Christianity enjoyed the freedom to develop in a formerly pagan country. However, when Ma'nu VI ascended to the throne, he chose to return to the pagan practices of old and ordered Aggai to "make me a headdress of gold, as you did for my fathers in former times." Aggai refused: "I will not give up the ministry of Christ, which was committed to me by the disciple of Christ and make a headdress of wickedness."[35] Upon Aggai's refusal, the King's men burst into the church while Aggai was preaching, broke both of his legs, and killed him. The young Christian community, imperiled, seems to have disappeared from Edessa. During this time, the Edessa Image also disappeared from all recorded history, not to appear again for hundreds of years. Denizens of Edessa between that time and the sixth century were completely unaware that the Image of Edessa/Mandylion was in their city. A typical example concerns the pilgrim Egeria, a woman of boundless energy who visited Edessa around the year 384 as part of her tour of Christian holy places.[36] She describes her numerous findings in minute detail but makes no mention at all of an image-bearing cloth. As historian Sir Steven Runciman describes her:

> She was a sightseer of a thoroughness unrivaled even by the modern American; and, had so interesting a relic then existed, she would certainly have referred to it.[37]

The same would be true for St. Ephraim, the so-called harp of the Syrian church, who lived in Edessa in the late fourth century. In all the reams of ecclesiastical verse he wrote, not a mention is made of the Mandylion. Nor was it mentioned by the monk-author of the "Chronicle of Joshua the Stylite," written at Edessa in about 507. The same is true for Jacob of Serug, also a most prolific writer of Edessa, who died in about 521.

The rediscovery of the Image of Edessa in the sixth century shed valuable light on the cloth's disappearance centuries earlier. According to the "Story of the Image of Edessa," the cloth was found in a space above the city's western gate that was carefully bricked over to conceal its contents. A brick-red tile imprinted with the face of Christ and a small lamp were found hidden with the cloth. This tile turned out to be another historical object, known subsequently as the Keramion, which was kept in Hierapolis (the sister city of Edessa) before also being transferred to Constantinople in 969. According to Wilson, this Keramion was one of many stone or clay heads of gods and gorgons commonly displayed over gateways in the Parthian Empire (examples can still be seen at Parthian Hatra).

Wilson believes the Keramion was almost certainly made from Christ's likeness on the Mandylion and displayed above the city gate until the reign of Ma'nu VI. When Ma'nu VI returned to paganism and began persecuting Christians, the Keramion was removed from view and concealed in the brickwork. Evidently, someone also hid the Mandylion there for safekeeping since persecution of Christians and destruction of their relics and vestiges were most likely occurring. This choice of a hiding place proved fortuitous, for it not only saved the Christian relics from destruction by the pagan ruler, but it also provided a hermetically sealed environment for the Mandylion for the next five centuries.

This location turned out to be important for another reason: Edessa suffered severe floods in 201, 303, 413, and 525 that caused extensive damage to the city's palaces and churches. Fortunately, the land rises steeply at what was Edessa's western gate. The Mandylion, sealed above the gate in one of the city's highest points, remained safe. The flood of 525 was by far the worst that Edessa experienced. In it, one third of the population died, and public buildings, palaces, churches, and much of the city wall were destroyed. As a result, the wall and its outworks had to be rebuilt. Wilson speculates that the Image of Edessa was rediscovered during this period of rebuilding. The cloth was not mentioned by Jacob of Serug before he died in 521, nor by Joshua the Stylite in 507, but it was mentioned by Evagrius, who described it in connection with the attack on the city by Chosroes the Persian in 544.

At the time of the flood, the Monophysites and the Orthodox were engaged in controversy. The Monophysites, who were dominant in Edessa at the time, contended that Christ could not be represented in icons or pictures because he was more spiritual than human, while the Orthodox argued that since Christ had both a human and divine nature, his human appearance could be represented in art. With the Monophysites in control, the finding of the Image of Edessa most likely was not a cause for public pronouncement. It may have been several years after the image was found before the cloth's existence was made public. At any rate, after 544, when Evagrius described the important role of the Mandylion in the defense of the city, the image was prominently mentioned.[38]

Evagrius relates how Chosroes built a huge mound of timber higher than the city wall that was to be moved next to the wall to serve as a platform from which his army could attack the city. The Edessans, in return, tunneled under the wall with the intention of setting the mound on fire from below before it could be moved forward to the city wall. Evagrius then goes on:

> The mine was completed; but they [the Edessans] failed in attempting to fire the wood, because the fire, having no exit whence it could obtain a supply of air, was unable to take hold of it. In this state of utter perplexity they brought out the divinely made image *not made by the hands of man*, which Christ our God sent to King Abgar when he desired to see him. Accordingly, having introduced this sacred likeness into the mine and washed it over with water, they sprinkled some upon the timber . . . the timber immediately caught the flame, and being in an instant reduced to cinders, communicated with that above, and the fire spread in all directions.[39] (italics added.)

It is evident by Evagrius' account that the Mandylion was then known again in Edessa.

With the rediscovery and known historical presence of the Edessan Image in the sixth century, the opportunity for close examination became possible. From then on, the Image of Edessa/Mandylion was described as *archeiropoietas* or "not made by hands," and imprinted on the cloth. The reputation of the Image of Edessa continued to grow until it became the most valued possession of the city. Thereafter, copies of the Mandylion and the likeness of Christ were made by artists throughout the Christian world. It was during the sixth century, with the emergence of the Image of Edessa, that the traditional likeness of Christ began to appear in art.[40]

While Wilson made a brilliant, well-reasoned theory twenty years ago to account for the Shroud's history from the 1350s back to the first century, I disagree with one aspect of it. Wilson speculates that the Shroud would have caused the disciples serious problems, because as a burial garment it was considered unclean, and its image would have violated the Second Commandment. He believes that sending the cloth to nearby Edessa, which was outside the borders of the Roman empire, whose king had shown an interest in Jesus, and where images were acceptable, would have solved their problem. He believes the one step remaining was to transform the burial garment into a portrait, because the gravecloth of an executed convicted criminal killed under the most degrading circumstances would be repugnant to anyone, and it was a king to whom the cloth was being sent. Wilson believes the prime candidate to carry out the transformation would be Aggai, the maker of headdresses and silks. The gold trellis work evident on the earliest copies of the Mandylion is typical of the trellis-style embellishments found on headdresses and costumes of Parthian monarchs. Under these circumstances, Thaddaeus and Aggai would be the last to know for many centuries that the image was a full burial shroud with Jesus' body imprinted on it.

I have always thought, as have many others, that the weakest point to the Shroud's authenticity is that if an image of Jesus existed on his burial garment following his resurrection, it would have been mentioned orally and in writing, not only in the Gospels, but in many other sources. But even had an image been visible the disciples would not have violated the law Moses brought down from the mountain: "You shall not make for yourself a graven image, or any likeness of anything that is in heaven above, or that is in the earth beneath," for two important and obvious reasons. First, the disciples would not have been guilty of making anything, for Jesus would have left it himself during his resurrection. Second, the prohibition was against making a graven image or an idol. Certainly no people worshiped the image itself. They never even mentioned it.

Actually, no record of any kind shows there was an image of Jesus on the cloth at this time. Not one book in the entire New Testament mentions an image left behind by Jesus on his burial garment. Other sources, such as the apocryphal Gospel of the Hebrews written in the second century, mention Jesus' burial garment as having been given to the servant of the priest (or variantly, Peter), but do not mention an image. In another tradition, it was acquired by Mary Magdalene. In a fourth century account, the apostle St. Nino reported the common belief during her youth in Jerusalem that St. Luke had acquired the burial garment from Pilate's wife and then hid it.[41] The historical accuracy of these last three sources is less important than the point that no image is mentioned in these accounts.

If the Shroud that Jesus had been buried in contained a recognizable image at the time, one of the many disciples or apostles would have mentioned or recorded this fact. It would have also been found somewhere in an oral tradition or other source. After all, any person who simply observed the image would not be in violation of the Second Commandment. Today, we are very interested in the possibility that the historical Jesus Christ may have left his image on his burial garment at the time of his resurrection. We are very interested in this possibility twenty centuries after his life on Earth, and our interest does not violate the Second Commandment, nor do we worship the image. Had there been an image, the disciples, apostles, and ordinary people of that day whose lives were contemporaneous to Jesus would also have been interested. In addition, the image would have provided a memento of Jesus to them.

The first account of the Abgar story, written by Eusebius ca. A.D. 325, does not mention an image of Jesus. Eusebius' account was based on Edessan archives written in the previous century. The tradition of Christ's letter to Abgar, however, clearly survived during this period. The first accounts of an "image not made by human hands" do not appear until the sixth century. Thereafter, several accounts and versions appear of Christ's miraculous imprint on the cloth with, perhaps, the best-known version being written after the cloth's tenth-century arrival in Constantinople. With the rediscovery of the image of Christ not made by human hands and miraculously impressed by Jesus, those who wrote such accounts from the sixth century on naturally gave the cloth and its image a first-century history. In fact, these are the only versions that can or do give the image a first-century origin.[42] Similarly, the only extant paintings depicting Abgar receiving the Image of Edessa are from the sixth century. No such paintings are known from an earlier period. The nonexistence of the image in the early centuries best explains the complete lack of a record or tradition of an image at this time.

Moreover, the yellow body image on the Shroud is made up of "chemically altered cellulose consisting of structures formed by dehydration, oxidation, and conjugation products of the linen itself."[43] The same spectral reflectance curves for the body image and background can be produced by laboratory simulations using controlled accelerated aging processes. In fact, the changes in cellulose that are known to be the result of aging are these same dehydrative and oxidative processes. Not only does the aging process produce the same spectral reflectance curves found on the Shroud, the Shroud's chemistry is also similar to the chemistry that causes yellowing of linen with age. The fact that the image is visible "tells us that the body image is due to a more advanced decomposition process than the normal aging rate of the background linen itself."[44] In other words, the body image is a result of an aging process that develops over time. Something caused the body image to age faster than the less yellowed background of the cloth. Since the image appears to have developed as the result of an aging process, the image that we see today—or would have seen centuries after it was first encoded—would probably not have been present at this very early stage.

Upon reflection, the lack of an image on the Shroud in its early history actually gives it more authenticity, for it offers the best explanation of why the Gospels do not mention an image on Jesus' burial cloth. If all the disciples or apostles had was a burial garment, it could easily have made its way to Edessa. It was still a burial cloth, and to Jews or anyone else, there would have been no reason to keep it or, subsequently, to mention it in the Gospels. They may have been reluctant to destroy it, and it very well could have made its way over to a people who spoke virtually the same language, had an early interest in Christianity, and stood astride a major east-west caravan route. Indeed, while the second- to fourth-century accounts of Jesus' burial garment are somewhat contradictory, they do maintain or reflect some type of memory of its existence—but without an image—after the time of the Gospels.

It could very well have been hidden away in Edessa at the time Wilson states and for the very same reasons (all Christian relics, symbols, or vestiges of any kind were at risk). It would be over fourteen feet long and would still have to be folded no matter where it was kept. When it is folded naturally, or doubled in four, the face is the only part showing. Perhaps, when it was rediscovered with an image of Christ, a frame was then put on it, keeping it in the same fold pattern in which it was found. It could even have been given its fold pattern at the time the image was discovered, since it was a burial garment and would easily be recognized as such in that part of the world. This could have happened in the sixth century or a century or so previously.[45]

Since the worst of several floods occurred in Edessa in 525, and the accounts of an image of Christ in Edessa "not made by the hands of man" also begins in the sixth century, this appears to be the most logical time that the cloth and its image were rediscovered. Possibly, the cloth was rediscovered earlier, but without a need to use it in order to help save the town, or without exposure to other people, it did not acquire its publicity and fame until then.

There is one other less likely scenario that should be considered. It is also possible that the Shroud had been somewhat exposed to the elements from the time it left Jerusalem until A.D. 57, and by the end of Abgar's reign, or during the reigns of Ma'nu V and Ma'nu VI, it was beginning to or had developed a more visible image than would have been seen in Jerusalem around A.D. 30. It could have been around A.D. 50 to 57 that the image was first observed, then framed, by Aggai, since it was a burial cloth and naturally folded to a visible facial area. (Even if there had been some kind of an image on the cloth while it was in Jerusalem, the disciples still may not have seen it. They may never have had the opportunity or inclination to spread and suspend the fourteen-foot-long cloth horizontally and stand six to ten feet away from the cloth to see the image with the naked eye.)

ANOTHER SHROUD ROUTE TO EDESSA

Obviously, the previous scenario is not the only way in which the cloth could have reached Edessa. Historian and archaeologist William Meacham writes: "By A.D. 66 the Judeo-Christians had migrated east of the Jordan, and thereafter little is known of them apart from their increasing isolation from the early church and their heretical tendencies. If the Shroud had been taken from Jerusalem by this group, its obscurity in the early centuries would be understandable."[46]

The exact role the Image of Edessa may have played prior to the sixth century really cannot be known for certain at this point. What we do know is that there was a real Abgar V who ruled in Edessa from A.D. 13 to 50. There was a real Image of Edessa, and all accounts written while the cloth was present described it as "not made by hands." The city of Edessa was only 350 miles from Jerusalem and even closer to one of the first Gentile Christian communities at Antioch. Edessa was located on the major trade routes of the day, and its people spoke a language almost identical to the Aramaic of the Palestinian Jews. All of these facts make the accounts of Abgar's hearing of Jesus and the early evangelization of Edessa, however brief, a very real possibility.

Part of the problem of a lack of any other early references, according to Warren Carroll, is that,

> Almost all the records of early Christianity (before 300) that have come down to us derive from the Graeco-Roman, not the Parthian world; and their total volume is small. Therefore, we are unlikely to find among them historical reports on a city in an alien realm, its people speaking an alien tongue, who had the faith only briefly, and then lost it for a long time. When the first Christian history properly speaking is written by Eusebius, the early conversion of Edessa is featured.[47]

A lack of references in this early period could also be due to the burning of the great libraries of Constantinople by the crusaders in 1204. There is no telling what information pertaining to the Mandylion may have been contained there. In addition, the great libraries of the important early Christian center of Alexandria also suffered major destruction in 391 under Theodosius I and in 642 during its fall to the Arabs. Perhaps more will be uncovered by historians and archaeologists, but much enlightening information is probably lost forever.

In light of the reputation that the Image of Edessa had attained, it was quite understandable that the Christian population was reluctant to part with the cloth in 944. Whether the Byzantines were justified in taking the cloth to Constantinople is debatable, but, in retrospect, this action may have been critical for the Mandylion's survival. Two centuries after the Shroud was taken, Turkish Moslems seized and sacked the city of Edessa in a manner it had never known in its almost fifteen-hundred-year existence. In describing how thoroughly the city was looted, J.B. Chabot states:

> . . . for a whole year they went about the town digging, searching secret places, foundations and roofs. They found many treasures hidden from the earliest times of the fathers and elders, and many [treasures] of which the citizens knew nothing. . . .[48]

The Christian population was dispersed, and every feature of Christian civilization was wiped out. If the cloth had not been removed to Constantinople, it most assuredly would not have survived.

THE FULL-LENGTH FEATURE OF THE SHROUD BECOMES KNOWN

In the eleventh and twelfth centuries, the power of the Moslems was growing. In an effort to find allies to help protect themselves against this growing power, the Byzantine emperors actively courted prominent visitors from the West. During such visits the guests, in a complete departure from past practices, were sometimes shown the treasures and most cherished relics of the empire. During this same period, the first indications of a full-length figure on the cloth come to light. One such indication comes from an interpolation recorded sometime before 1130 of an original eighth-century sermon by Pope Stephen III, in which he refers to the Mandylion:

> For the very same mediator between God and men [Christ], that he might in every way satisfy the king [Abgar], stretched his whole body on a cloth, white as snow, on which the glorious image of the Lord's face and the length of his whole body was so divinely transformed that it was sufficient for those who could not see the Lord bodily in the flesh, to see the transfiguration made on the cloth.[49]

Another account is found in a Vatican Library Codex, which also dates to the twelfth century. Its version of Christ's letter to Abgar reads:

> If indeed you desire to look bodily upon my face, I send you a cloth on which know that the image not only of my face, but of my whole body had been divinely transformed.[50]

Another example can be found from the history of the church written by the English monk Ordericus Vitalis in 1130, wherein he states:

> Abgar reigned as toparch of Edessa. To him the Lord Jesus sent . . . a most precious cloth with which he wiped the sweat from his face, and on which shone the Savior's features miraculously reproduced. This displayed to those who gazed on it the likeness and proportions of the body of the Lord.[51]

This last reference may have derived from the first reference mentioned. None of these references should be taken at strict face value. The idea that Jesus impressed his naked, crucified body on the cloth while alive is preposterous. Furthermore, the first reference is clearly an interpolation of a sermon given three centuries before. However clumsy these references are in their attempts to explain the history of the Mandylion, they testify to the knowledge at this time of a full-length figure of Jesus on the Mandylion cloth.

Another indication can be found in the appearance for the first time of new Lamentation scenes in art that begin at this same time.[52] Earlier, whenever Jesus' entombment was depicted he was invariably shown as wrapped mummy-style, wound in swathing bands rather than in a large, shroud-like sheet. Without any

(clockwise from top right) Fig. 121. Artist's copy of the Edessa cloth, tenth century, from a fresco at the Church of St. John at Sakli, Cappadocia. Fig. 122. Hungarian Pray Manuscript, 1192–1195. Fig. 123. Liturgical cloth from the 1200s, whose scenes probably began in the 1100s. Both dates are well before the Shroud first appears in Europe in the 1350s.

explanation, artists in the twelfth century began depicting scenes of Jesus' entombment with his body enveloped in a shroud. Several of these examples show, for the first time ever, his hands crossed over his loins. These examples are consistent in their portrayal of the right hand crossed over the left,[53] which is exactly as the hands appear on the Shroud cloth image. An excellent illustration of this is the Hungarian Pray manuscript, ca. 1192 to 1195, which, like the Shroud, uncharacteristically depicts Jesus completely naked (fig. 122).

Other similar full-length representations of Christ can be found on *epitaphioi,* which are liturgical cloths. The earliest examples survive from the thirteenth century, but, according to Wilson, their similarity to the Lamentation scenes, their consistent feature of Christ's body laid out in death and with his hands crossed, right over left at the loins, suggests a twelfth-century origin. A good example of such a portrayal on a liturgical cloth can be found in the Museum of the Serbian Orthodox Church, Belgrade (fig. 123). This artistry existed a full century before the Shroud was ever displayed in France in 1357.

Another indication of the full-length feature being known at this time can be found from an umbella, which was probably an ornamental tapestry canopy sent to Pope Celestine II from Byzantium in the twelfth century. While the canopy was subsequently destroyed, a record of its appearance has been preserved by the archivist Jacopo Grimaldi. His drawings reveal that at the centerpiece of the umbella was a representation of Christ in exactly the manner of the Shroud.[54]

A very interesting indication that the Shroud's full-length feature was known at this time may also be found from observing the fold marks, as well as a band of discoloration and small tack marks present on the Shroud.

As discussed earlier, John Jackson first noticed these fold marks during the STURP examination in 1978, and has spent parts of his overall Shroud study since then reconstructing the precise manner in which the cloth was folded. Within the doubled-in-four fold arrangement can also be found four closely spaced parallel lines

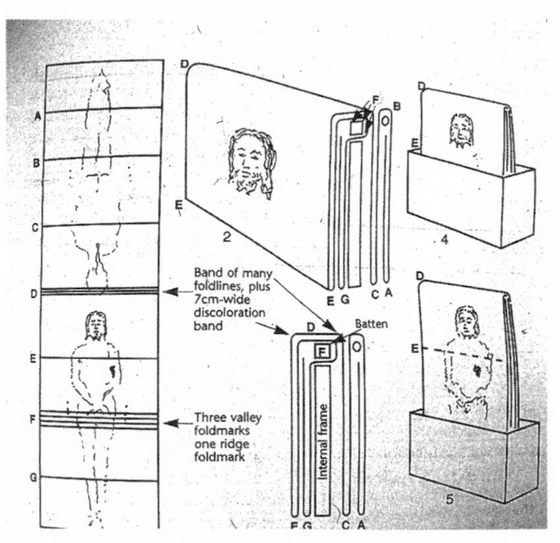

Fig. 124. The fold marks on the Shroud suggest the manner in which the Mandylion was displayed.

beginning at, and immediately below, the crossed hands. Curiously, Jackson found that these and the doubled-in-four fold marks all lined up if the Shroud wound around a square block of wood running the length of the cloth while folded inside its container as illustrated in figure 124.

However, the explanation for keeping the Shroud folded in such an odd arrangement was unknown until a reason was suggested by Shroud scholar Heinrich Pfeiffer, S.J., in 1993. If the Shroud is tacked to a frame above and behind these small blocks, the cloth can then be pulled straight up vertically and the image will "stand" or arise, but will stop at the level of the hands.[55] The similarity to Robert de Clari's description of Jesus' shroud, which stood up straight or raised itself so that the figure of our Lord could be seen, is very apparent.

Jackson even finds a band of discoloration on the Shroud of Turin that would have been wrapped around the block of wood under this fold arrangement. In addition, small round rust marks consistent with the locations where tacks would have been placed to secure the cloth to the backing boards are also present on the Shroud.[56]

During the late eleventh and twelfth centuries, the Extreme Humility or Utmost Humiliation (Man of Pity or Man of Sorrows in the West) type of paintings begin, a typical representation of which is illustrated in fig. 125.[57] In some of these representations, the man is rising out of a box that is obviously too small to contain the rest of Jesus' body, but could easily contain the rest of a folded cloth.[58] The similarity of this painting to the Shroud image, to Robert de Clari's description of Jesus' shroud, and to the earlier descriptions and full-length images just mentioned, all of which also begin at this time, is obvious. Jackson clearly thinks that the Shroud in this fold-and-display arrangement was the prototype for the Extreme Humility representations, and not vice versa, since the Shroud is not a painting and contains full-length frontal and dorsal images.

A few high-ranking Byzantine officials may have known of the Mandylion's full-length feature as early as its first arrival in Constantinople in 944. This appears

Fig. 125. Man of Pity, ca. 1300, made in Constantinople. Presently at St. Coce in Gerusalem, Rome.

from a text on the Mandylion attributed to Gregory, a high official of the Hagia Sophia Cathedral in Constantinople. This text is listed in the Vatican Library as Cod. Vat. Graec. 511, pp. 143–50b, and appears to date from the time the Mandylion was first received at Constantinople. The amount of detail describing this event, especially the ceremony and those present, strongly implies the author was an eyewitness.

Discovered by Gino Zaninotto, the text of this document was presented in a book by Werner Bulst.[59] The most interesting part of the text is its reference to the image itself, revealing that its full-length feature must have been observed. According to Bulst:

> After having related the legend of Abgar, Gregory describes and explains the image in a manner for which up to now we have found no parallel. . . . He speaks of the "side" [πλευρα] and of the "blood and water found there" [ʼαιμα και ʼυδορ εκει], and the wound in the side. He must have seen the cloth up close and thus we may suppose that he belonged to the imperial delegation responsible in 943 for the reception and verification of the Edessa image.[60]

Continuing in a paragraph on how the cloth was mounted in its frame, Bulst states: "Gregory's attestation of the side wound confirms anew the identification of the Edessa image and the Turin Shroud. That the wound with its blood and water is not mentioned elsewhere is easily explained. The image was not shown in public."[61]

While it was in Constantinople, the cloth itself was never publicly exhibited. This was consistent with the Byzantine mind-set expressed by the inscription on a twelfth-century Byzantine communion veil:

> If no Israelite might look directly on the countenance of Moses when he came down from the mountain where he had seen God, how shall I look upon Thy revered Body unveiled, how regard it?[62]

The Byzantines, accustomed to the most sacred parts of the liturgy being performed out of sight of the congregation, were raised on the classical tales of people being blinded after looking at gods with unveiled eyes. While anonymous monk-artists were allowed private expositions in order to paint copies, the Mandylion was recorded as being involved in a public exposition on only two occasions, and on those occasions it was carried in a casket and was not removed for public viewing. The same type of treatment is found while the cloth was kept in Edessa, with only the archbishop being allowed an occasional, momentary glimpse of the cloth.[63]

Great protective powers had been attributed to the Image of Edessa and the Mandylion, both in Edessa and in Constantinople. It was said to have played a role in the curing of Abgar, and its rediscovery five centuries later was credited with saving Edessa and defeating Chosroes, the invader. It was the primary example cited by those arguing against iconoclasm,[64] a force that destroyed countless images in the eighth and ninth centuries. Indeed, such protective powers and its enormous reputation were reasons the Byzantines desired the cloth. With the soldiers from the Fourth Crusade inside and outside the walls of Constantinople, and the city facing the greatest threat in its existence, it was not surprising that the Shroud was publicly displayed in 1204 for the first time that scholars know of. It was displayed at the Blachernae Palace, a rallying point for the citizens of Constantinople in times of distress.

SHROUD LOCATIONS BETWEEN 1204 AND THE 1350s

The Image of Edessa/Mandylion was a known historical object until it disappeared in 1204, never to be heard from or seen again. It disappeared from Constantinople during the invasion of the Fourth Crusade, which was composed of and commanded primarily by the French. A cloth then appeared in France the next century, matching the descriptions of a full-length figure of Christ in Constantinople and numerous descriptions and depictions of the Mandylion/Image of Edessa. In this section, we will consider what happened to the Shroud between 1204 and the mid-1350s, when it was displayed in Lirey, France, as the burial garment of Jesus.

The most popular explanation is that the cloth fell into the hands of the Knights Templar.[65] This was a group formed in about 1118 for the purpose of protecting pilgrims traveling to and from the Holy Land. It was one of several great orders that arose out of the Crusades (the Knights Hospitalers and the Teutonic Knights were others). Because the Knights Templar fought with such valor in the Second and subsequent Crusades, they established a reputation as one of the bravest and fiercest of all groups fighting in the campaigns and attracted men of the noblest blood to their ranks. The Knights Templar became one of the most powerful groups in Europe and established a series of virtually impregnable fortresses across Europe and the Near East, which became recognized as useful storehouses for national treasures and valuables of all kinds. They appointed their own bishops and, as they held monastic privilege, were responsible only to the pope.[66]

As leading money lenders in Europe, the Knights Templar served as bankers to several kings and popes and became one of the principal sources of finance for the Fourth Crusade. From 1187 to 1291 the Templars operated from a port in what is now northern Israel, called Acre. It seems plausible that the Templars came into possession of the Shroud because they were easily in the best position to act as guardians, traders, and pawnbrokers for the flourishing trade of relics that accompanied the Fourth Crusade. Also, their heavily guarded series of fortresses would ensure the secrecy of the whereabouts of such an object. Further, the group's wealth and power would eliminate the need to find another home for, or to sell, or to borrow against the cloth, and there would be no problem of inheritance, all of which were problems that faced subsequent private owners of the shroud.

The Knights Templar were a very dedicated group. Along with having a reputation of impeccability in all their business transactions, its members took vows of chastity. The ceremonies by which initiates were admitted to the order were so secretive that they were sworn upon pain of death never to reveal the details of their initiation ceremony. We do know, however, that during this ceremony, initiates were given a "momentary glimpse of the supreme vision of God attainable on earth."[67]

What was this image of the supreme vision of God that was attainable on earth? Described as an idol or head, the "vision" was normally viewed only by the Grand Master and the inner circle of the Templars. Ian Wilson cites several accounts to support his theory that it was the Mandylion/Shroud, the Holy Image of Christ before which the Templars prostrated themselves in adoration. Wilson, and others,[68] think that, like the Mandylion/Shroud, copies of it were made for the various Templar chapters scattered throughout the region.

This is further suggested by several accounts that speak of its being displayed at a ceremony taking place just after the Feast of Saints Peter and Paul on June 29.[69] The next feast that would have taken place in medieval times would have been the feast dedicated to the Holy Face, celebrated two days later on July 1.

Some contemporaneous accounts described the Templar idol head as "terrifying." Wilson points out that the most holy images of Christ often engendered fear in medieval times. In the twelfth century, for example, Pope Alexander III ordered that Gregory the Great's Acheropita image in the Sancta Sanctorum chapel be veiled because it caused a dangerous, even life-threatening trembling. Similarly, in the Grail legend it was the image of Christ that caused the knight Galahad to tremble. Might not the face on cloth of Christ after he has been beaten and bruised also appear frightening?

The Templars were said to have omitted the words of consecration at the special Templar Mass because they believed Christ was present in a far more powerful way than in the normal Mass, via his image.[70]

The strongest evidence supporting the theory that copies of the Mandylion had been distributed to Knights Templar chapters comes from the discovery in 1951 in a Templar ruin in Templecombe, England, of a panel painting (fig. 126) that matches many of the descriptions of the Templar idol: "bearded male head," "life-size," "painting on a plaque," "disembodied," "with a grizzled beard like a Templar."[71] This painting, with a distinct medieval style, is very similar to Byzantine copies of the Mandylion. The discovery of the Templar painting dispels any notion that the Templar idol may have been some form of bust. The existence of multiple copies of the original is indicated not only by the variety of descriptions—some in gold cases, silver cases, wooden panels, et cetera—but also by the information from a Minorite friar that in England alone there were four: one in the sacristy of the Temple in London, one at "Bristleham," another at Temple Bruern in Lincolnshire, and another at a place beyond the Humber River.

From 1187 to 1291, the Templars operated from Acre, site of their main treasury. When Acre fell, the treasury was moved to Cyprus for a short stay. In 1306, the treasury was moved from Cyprus by sea to Marseilles, then to Paris, where it

Fig. 126. Painting of late thirteenth or early fourteenth century workmanship found at site of former Knights Templar in Templecombe, England.

Fig. 127. Burning at the stake of Templar Order's Grand Masters Jacques de Molay and Geoffrey de Charny.

was settled at new Templar headquarters, at the Villeneuve du Temple, a huge fortress opposite the Louvre. At that time the wealth and power of the Templars rivaled that of the king, and as their power and wealth became more apparent, so did accusations that they were idolatrous and decadent. Also at that time, rumors began circulating in Europe of an idol that the Templars worshiped at their headquarters. This idol was described as like ". . . an old piece of skin, as though all embalmed and like polished cloth."[72] This description, in actuality, very much fits that of the Shroud cloth. It was further described as "pale and discolored," a description made of the Mandylion and true of the Shroud.

The combination of wealth, power, and charges of idolatry was to have disastrous effects for the Knights Templar. In 1307, King Phillip IV of France made a surprise sweep against the Templars and arrested, tortured, and extracted "confessions" from its members. When the king's men attacked the Paris headquarters of the Templars, they met fierce resistance. Only afterward were they able to search for the Templar idol, but when they did, it was not to be found. Templar goods were also seized and inventoried in England, with the encouragement of Phillip. Again, nothing was found. It is very easy to conjecture that either during the attack on the main headquarters in Paris or sometime preceding it (the Templars were certainly aware that charges against them had been floating around), the Templars removed the Shroud, cutting away its frame or the casket in which it was folded, and left it with someone for safekeeping. Jacques de Molay, the Templars' Grand Master, was arrested, confined, and tortured. He was eventually burned at the stake along with one of the other principal masters of the order, the master of Normandy, a town located only fifty miles from Paris. The name of the master of Normandy—Geoffrey de Charnay.[73]

Whether this Geoffrey de Charnay is related to the Geoffrey de Charny that owned the Shroud when it was first exhibited is unknown. (The slight difference in the spelling of their last names may easily be discounted as there was little standardization in spelling in medieval French.) To have two quite prominent people with nearly identical first and last names in medieval France without being related is somewhat unusual. Evidently, as a high-ranking Grand Master in the Knights Templar, Geoffrey de Charnay was considered very noble and powerful. In light of his vow of chastity, he was unlikely to have married or produced children. Before his death, in 1314, all of his lands and possessions would have been confiscated by Phillip IV. The younger Geoffrey begins to appear prominently in the French military records in 1337. Before his death in 1356, he would establish a reputation as one of the noblest and bravest men of his era. While the older Geoffrey died at the hands of a French King, the younger Geoffrey died at the side of French King John II (the Good), while saving his life. The younger Geoffrey had all the markings of a self-made man, or a man struggling to regain some lost family honor. If the younger Geoffrey acquired the Shroud directly or indirectly from the older Geoffrey, it would certainly explain his reticence to reveal how he acquired it.

The connection between the Shroud and Geoffrey de Charnay could also have come about in other ways. At the time Constantinople was sacked, the Shroud was kept at the palace of the Blachernae (Blakerna), which was taken by Henry of Hainault, who became Emperor of Constantinople a year later by succeeding his brother, Baudouin. (This began the reign of the so-called "Latin Emperors" in Constantinople until 1261.) After the attack of 1204, the Byzantine Empire was only a shadow of itself. During the "Latin Empire," rulers were extremely poor, and regencies between rulers were common. In 1207, the Shroud was thought to be on a list of relics in Constantinople.[74] Although such a reference is far less clear and certain than the earlier references in 1201 and 1204, it is possible that the Shroud stayed in Constantinople with the Latin Emperors. From 1216 to 1219, in a regency between emperorships, Nargeaud de Toucey served as regent; from 1228 to 1237, Nargeaud's son, Phillippe de Toucey, served as regent for the emperorship. The first wife of Geoffrey de Charny was Jeanne de Toucey, whose uncle, Guillaume de Toucey, was appointed Canon of the church at Lirey at the request of Geoffrey de Charny. The advocates of this theory suggest that Geoffrey's connection with the Shroud could have come through his wife's family.[75]

Another way in which the Shroud could have come to France from Constantinople is mentioned by Hungarian Oxford scholar Dr. Eugen Csocsan de Varallja.[76] This route would have occurred primarily through the efforts of Hungarian-born Empress Mary-Margaret, the youthful wife of the much older Byzantine Emperor Isaac II Angelus, who was blinded and overthrown twice, and who died in the tragic events of the crusaders' attack upon Constantinople. Within weeks of the emperor's death, Mary-Margaret married Boniface de Montferrat, who led the Fourth Crusade, and they moved to Thessalonica. There Mary-Margaret founded the Church of the Archeiropoietas (Church of the Image of Edessa), which is known today as the Ancient Friday Church.[77] The church may have been founded to house the Image of Edessa/Shroud, which was brought from Constantinople. Three years after Boniface died, Mary-Margaret wed Nicholas de Saint-Omer, a son of the titular prince of Galilee and a kinsman of Geoffrey de Saint-Omer, who was one of the two founders of the Order of the Knights Templar.[78] Mary-Margaret and Nicholas's son William also became involved with the Order of Knights Templar.

Whether the Shroud arrived in France from Constantinople through the efforts of Geoffrey de Charnay of Normandy, the de Toucey family, the Empress Mary-Margaret, or some other route (three other theories have been expressed by various authors[79]) is really not crucial. What is important is that any of these theories could explain how the Shroud traveled from Constantinople to France. The distances are not far, and there were countless contacts between the two regions that could account for the cloth traveling from Constantinople to France. If Geoffrey de Charny's ancestors received the Shroud through the Templar connection, or through the efforts of a regency (which probably would have entailed some dishonesty or trickery), an explanation may never have been provided to Geoffrey— or if it had,

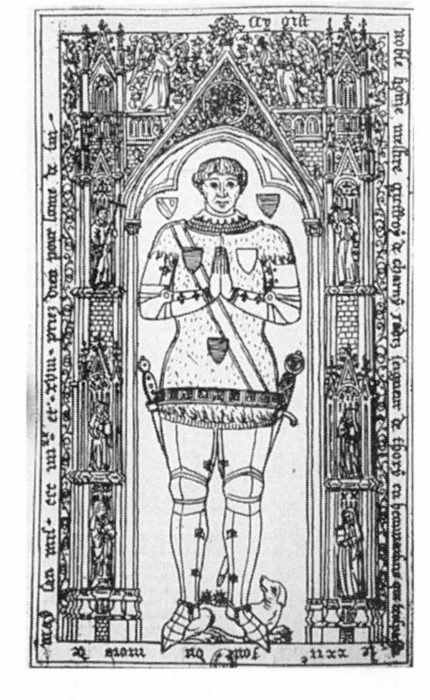

(above left) Fig. 128. Copy of the second draft of D'Arcis Memorandum, Folio 137.
(above right) Fig. 129. Pilgrim's medallion with full-length frontal and dorsal figures of the Shroud represented. The shields contain the arms of the family of Geoffrey de Charny and his wife, Jeanne de Vergy. The roundel in the center represents the empty tomb. From a damaged amulet found in the Seine in 1855 and presently in the Musée de Cluny.
(right) Fig. 130. Geoffrey II de Charny.

he may have been reluctant to divulge it. At any rate, how he came to possess the Shroud was never revealed in documented history. At the Battle of Poitiers in 1356, the numerically superior French forces were unexpectedly defeated, and Geoffrey de Charny was killed in battle.

It is often thought that the first exposition of the Shroud was not held until after Geoffrey's death. However, a pilgrim badge bearing his shield, along with his wife's coat of arms, suggests otherwise. According to the rules of heraldry, the presence of his shield would be allowed only if he was alive.

About thirty-five years after the death of Geoffrey, his family wanted to exhibit the Shroud again. At this time, a document known as the "d'Arcis Memorandum" surfaced and has since been frequently cited by detractors as proof that the Shroud is a painting. All subsequent documents and opinions asserting the Shroud is a painting revolve around this memorandum.[80] It was seemingly written by Bishop Pierre d'Arcis of Troyes to Pope Clement VII and dated to late 1389 by the learned French scholar Ulysse Chevalier. Troyes is a town in France, located twelve miles from Lirey, where the Shroud was exhibited in 1389 by the wife and son of Geoffrey de Charny. According to d'Arcis, approximately thirty-four years earlier, at the time of the previous exhibit, the then Bishop Henri de Poitiers had discovered "after diligent inquiry and examination how the said cloth had been cunningly painted, the truth being attested by the artist who had painted it. . . ."[81] This account was supposedly sent to the Avignon pope Clement VII, the first of the "antipopes" of the Great Western Schism, who was recognized as pope in France but not in Rome.[82]

The rest of the memorandum is a running diatribe accusing the dean of the modest wooden church in Lirey of the worst kind of greed and avarice. D'Arcis was apparently quite upset at the fact that the exhibitors had not asked him, the local bishop, for his permission to exhibit the Shroud, and, further, had ignored his command to discontinue the Shroud exhibition, which was attracting pilgrims not only from Troyes but from throughout the entire area.

Historian Daniel Scavone believes that d'Arcis may have had a hidden agenda:[83] he had no documents from Henri de Poitiers and gave no name of an artist. D'Arcis himself had been accused of wanting the cloth for his own benefit; he denies in the memo that he wanted the cloth for his own gain. But Scavone points out that in 1389, the nave of the unfinished Cathedral in Troyes collapsed. Records show that 1389 was the only year that expenses were greater than income. The collapse of the nave was so significant that no major effort could be made for sixty years to complete the cathedral. As Scavone says, "Recall, too, that the Memorandum alludes again and again to the avarice of the Lirey canons: Bishop d'Arcis could be telling more about himself than about the canons."

Even most Shroud scholars are unaware that a manipulation of this memorandum occurred as Ulysse Chevalier—a renowned critic of the Shroud—presented it in the early part of the twentieth century.[84] There are two handwritten copies of the memorandom attributed to d'Arcis, "Folio 137" and "Folio 138." Folio 138 is a first draft with some parts crossed out, underlinings, and some very violent expressions canceled; it is unsigned and undated and the addressee is not even shown. Chevalier made an orderly and careful transcription of this and published it in his very influential book. But Chevalier affixed the heading from Folio 137, "The Truth about the Cloth of Lirey, which was and now is being exhibited and about which I *intend* to write to our Lord the Pope in the following manner and as briefly as possible," onto Folio 138 (italics added). Thus, the document seen in Chevalier's book does not even exist, for it is actually a combination of two documents! Folio 137, the second draft, is neater than the first, and with its proper heading removed and affixed to the earlier messy draft, Chevalier gives the impression that Folio 137 was sent to Clement VII. Chevalier declares, "This Memorandum must have reached the pope at the end of 1389."[85] However, Folio 137 was neither dated nor signed.[86] Moreover, it was merely addressed to a scribe for editing, but there is no evidence that it was ever sent to the scribe.[87]

Similarly, there is no proof that the letter was ever sent to the pope. All that we have are the two copies of the rough draft. There is no copy of such a letter in the Vatican Archives; the Troyes diocesan records; in the works of Nicolaus Camuzat (the historian for the diocese); the Bibliothèque Nationale of Paris, where the copies of the rough draft were found; or anywhere else.[88] Moreover, in numerous subsequent and extemporaneous documents from Pope Clement VII concerning the Shroud and its exhibition, the pope never said a word about such a previous communication from the bishop.

Even if we assume this rough draft was changed into a letter and sent, numerous other problems arise with its contents, beginning with its reference to the earlier exhibition and inquiry. D'Arcis's letter, putatively written in 1389, states that the previous exhibition and inquiry took place about thirty-four years earlier. This would put the year for such an inquiry at 1355. Documents show that on May 28, 1356, Bishop Henri de Poitiers, the man who is supposed to have discovered the great Shroud fraud, presided at the consecration of the Lirey church, which was the very church that had acquired and exhibited the Shroud. At this consecration, Bishop de Poiters gave his unqualified and lavish blessings to the church, praising its canons and founder. If a forgery episode had occurred the year before, it seems inconceivable that the Bishop would have praised the canons and founder as he did on this occasion.[89]

Furthermore, there is no record whatsoever of any such inquiry ever having taken place. As successor to the same office held by Henri de Poitiers, d'Arcis would easily have had access to such records, had there been any. In his letter, he offers not one single piece of supportive evidence, which he also undoubtedly would have, had there been any. Nor is there any record in the work of Nicholas Camuzat that such an investigation ever took place.[90] Even if one allows for the possibility that the exhibition and inquiry did not occur in 1355, the fundamental problem remains that no documentation or record of any such investigation can be found.

Further complicating the veracity of d'Arcis's claims is the fact that they are based on several layers of compounded hearsay. The first layer of hearsay is the supposed statement or confession attributed to the unnamed artist that he painted the image. The next layer (or layers) of hearsay is the source from which d'Arcis received this same statement. D'Arcis never reveals his source, never vouches for the truthfulness of his thirty-some-year-old information. He may be relating mere gossip or rumor. Or he may be inventing "evidence" to serve his own agenda.

Finally, the medical and photographic evidence demonstrates in great detail that the image on the Shroud is that of an actual human body—something that could not be accomplished in a painting. This is perhaps the greatest argument against the veracity of d'Arcis's claims.

POLLEN ANALYSIS CONFIRMS SHROUD'S MIDDLE EASTERN HISTORY

Available scientific evidence strongly supports the contention that the Shroud of Turin was in Palestine and in Turkey sometime before the 1350s when its presence in Europe was first documented. Of the fifty-eight pollen grains from plant species that were found on the Shroud by Dr. Frei (discussed earlier), only seventeen of these—less than one third, can be found in France or Italy.[91] That seventeen of these pollen grains are grown in France or Italy should come as no surprise, since the Shroud has spent its last 645 years in these countries. What is surprising is that only a minority of pollens are native to Western Europe; the majority are native to the Middle East, including Turkey and Israel. It is interesting that the proportion of pollen grains from plants in Edessa (eighteen) and Constantinople (thirteen) is very similar to that of western European plants (seventeen). One would expect such a proportional distribution if the Shroud had a strong history associated with the Mandylion and the Image of Edessa, as postulated by Wilson and others.

Although the Shroud has spent hundreds of years in Europe, it was infrequently displayed, and was normally kept inside a closed container. Although the Mandylion spent hundreds of years in both Edessa and Constantinople, it, too, was rarely exposed to these climates. The principal exposure of the cloth to the Turkish climate would have been when it was first brought to Edessa, when it was transported west to Constantinople, and when it was displayed during times that the cities' rulers felt its exposure might help protect them from disaster or invasion, such as the attack on Edessa by Chosroes in 544 and the sacking of Constantinople in 1204.

The large number of pollen grains from plants grown in Jerusalem and the Middle East that are found on the Shroud cannot be the result of foreign contamination by winds while the Shroud was in Europe. First of all, the vast majority of the Middle Eastern species identified by Dr. Frei are from low-lying herbs and shrubs.[92] According to archaeologist Paul Maloney, "The pollens which do travel far are more likely those from trees which raise their pollen sources to the wind rather than low-lying shrubs which lie relatively protected from such winds."[93] In addition, many of these Middle Eastern species found by Dr. Frei are insect pollinated, so they would not be expected in a wind-distributed assemblage.[94]

Even completely ignoring these reasons, foreign contamination could never account for the number and variety of pollen types found on the Shroud. The geographical area between the Middle East and western Europe is interrupted by several Mediterranean basins and high mountain ranges. As a consequence, the Mediterranean wind system between these two areas is countervailing. A pollen grain from Jerusalem would have to travel through more than 2,500 kilometers of these complex and countervailing winds to reach France or Italy.

Even if a pollen grain had survived the wind system between the two regions, it would have had to land on the Shroud on one of the few days throughout its many centuries in Europe that the cloth was exposed to the outdoors. Yet under this scenario, if the Shroud was exposed to the outdoors and the pollen grain from the Middle East landed on it from wind blowing in France or Italy, the same air would easily contain many more pollens from France or Italy. Thus, such a foreign contamination theory could never account for the fact that the vast majority of pollens on the Shroud are native to the Middle East. Furthermore, the pollens on the Shroud are from plants that bloom in different seasons of the year. Not only would the same incredible accident described earlier have had to happen, it would have had to happen repeatedly.

As you can see, a historically documented, plausible provenance of the Shroud of Turin from first-century Jerusalem to present-day Turin is not hard to put together. Critics who denounce the Shroud as a fraud have not only been unable to agree on a method of forgery—they have also never agreed on a plausible, documentable place or "artist" of a forged Shroud. We have already seen that it would be impossible to forge the Shroud naturally or artistically, even with today's technology—much less during medieval times.

However, if the Shroud does have a first-century provenance and is the burial cloth of the historical Jesus Christ, then we could consider another historical explanation to account for its various features. However, before we do let us examine the one piece of "evidence" all the Shroud critics cling to—the radiocarbon dating test of 1988. Let us now turn to the one thing critics say they have been able to prove. In a way, they have proven something—that some are unable to deal with the answers the evidence points to, and that they are willing to resort to sabotage to hide from that evidence.

CHAPTER EIGHT

SCIENTIFIC CHALLENGES TO THE CARBON DATING OF THE SHROUD

If the world knew of the opportunity that was present and then lost in the 1980s, it could then begin to comprehend the tragedy that occurred in the process of radiocarbon dating the Shroud of Turin. An opportunity existed to perform an entire range of scientific testing on an object that literally defies scientific knowledge and understanding and that contains a wealth of evidence with implications for everyone. This range of scientific testing was designed to investigate a variety of issues concerning authenticity, conservation, image-formation, age, origin, and history of the Shroud of Turin. Such tests would have been performed on the cloth as a whole, samples would have been removed, and additional data would have been acquired and brought back to laboratories around the world for further analysis, as was done in 1978. In addition, the entire range of scientific testing and sampling would have been performed by the most knowledgeable and qualified group of scientists in the world. Rarely, if ever, has an opportunity presented itself for the advancement of humankind's understanding and scientific knowledge, on a subject so important to people throughout the world than did this opportunity.

The tragic fact is that only a small fraction of the planned scientific tests was actually performed, and this was performed under controversial circumstances with controversial results. Radiocarbon dating, though perceived as definitive in terms of the cloth's age, can also be the most misleading of all the tests if not performed correctly. It was not performed correctly.

This test was filled with errors for myriad reasons. Some of the underlying causes were lack of desire to advance scientific knowledge, and the desire to actually prevent the full-fledged scientific investigation and testing from taking place on the Shroud of Turin. Such motives belonged to some who had little or no knowledge or experience with the Shroud of Turin, and were used to exclude the group of scientists who had the most extensive knowledge of and experience with the Shroud.

CARBON 14 FORMATION AND MISINFORMATION

Before we review the radiocarbon dating controversy in connection with the Shroud, you should understand the fundamental principals of radiocarbon dating. Three isotopes of carbon are normally found in carbon-containing materials: carbon-12 (C-12), carbon-13 (C-13), and carbon-14 (C-14). C-12 accounts for 98.9 percent of naturally occurring carbon. C-13 accounts for the other 1.1 percent. C-14 is present only in trace amounts. C-13 and C-12 are stable isotopes that were formed when the planet's other atoms were formed. Practically all of the earth's carbon in organic and inorganic materials consists of these two isotopes. Any C-14, or radiocarbon, that was formed along with the planet disappeared long ago because this isotope is radioactively unstable and decays.

However, new, minute amounts of C-14 are continuously formed during collisions of cosmic rays with nitrogen-14 (N-14) atoms in the atmosphere. N-14 is not unusual: Air is about 78 percent nitrogen and 99.63 percent of all the nitrogen on earth is N-14. This newly formed C-14 is also unstable and disappears naturally. The amount of C-14 on earth remains nearly the same because new C-14 is created in the atmosphere at almost the same rate that older C-14 is decaying on the earth's surface. Thus, carbon-14 is said to be in balance.

However, this balance is infinitesimal with C-14 being approximately one part in a trillion of the overall carbon content (1/1,000,000,000,000). This very tiny amount of C-14 formed in the atmosphere, along with the much larger amounts of C-13 and C-12, is taken up in atmospheric carbon dioxide by photosynthesizing plants and is, thereby, spread throughout the biosphere, thus allowing all living things to have a similar ratio of C-14 to C-12. Since these carbon isotopes have the same chemical behavior, this ratio is maintained while the organism lives. However, upon its death, the C-14 disappears according to its radioactive half-life, which is approximately 5,730 years. By measuring the C-14 to C-12 ratio, scientists can calculate the date of the organism's death.

These are the general principles upon which radiocarbon dating operates. (It does not include modifications to the calibration curve for fluctuations in the radioactive level of the atmosphere or for different levels of atmospheric carbon dioxide in the oceans, neither of which directly concerns the radiocarbon dating of the Shroud.) Since the fraction of C-14 to C-12 is so infinitesimal, and since this measured ratio is the basis for calculating the organism's age, any error in measuring or counting the C-14 isotope could alter the date, perhaps significantly. A correct date can be calculated *if and only if* the very tiny trace amount of measured C-14 from the object accumulated there by the above natural process. If the measured C-14 got on the object any other way, the interpretation of the date will be incorrect.

Errors in radiocarbon dating have been quite numerous. Just a few notable mistakes are:

- dating of *living* snail shells to be twenty-six thousand years old
- dating a newly killed seal to be thirteen hundred years old
- dating one-year-old leaves as four hundred years old
- dating twenty-six-thousand-year-old-mammoth fur as fifty-six hundred years old
- dating a Viking horn to the future year of 2006
- bone tools from caribou ribs at an Alaskan site dated approximately twenty-seven thousand years old, yet a sample taken from the innermost portion of the bone dated to 1,350 years[1]

I could give you twenty more examples just off the top of my head. My point is even most scientists are unaware of how error-prone carbon dating can be. This dating measures the ratio of C-14 to C-12, so if there are any errors in measuring the small amount of C-14 or in measuring the C-12, the date will be incorrect. Even if the two isotopes are correctly measured, the isotopes that were measured must be original, and must belong only to the object from which they were taken. This dating process is not absolute and is subject to enormous error.

The reasons for the errors in the above examples were varied. In the last example, it was found that the outer portions of the bone had exchanged carbon with the air and ground water. In the example of the leaves, it was found that they had absorbed carbon from the atmosphere from the burning of oil containing old hydrocarbon. For some of the other examples, the reasons range from volcanic activity to incorporation of atomic bomb carbon. The reasons for and the extent of contamination were never fully understood for some of the remaining. Sometimes, the best and only method of evaluating the extent or effects of contamination on an object or site is to observe the divergence of the radiocarbon date from the site's historically datable context. For these and other reasons, many radiocarbon dates have been rejected by archaeologists and geologists as being anomalous or in conflict with other C-14 dates or more reliable data.[2] Good scientists do not rely on carbon dating in isolation when there is other evidence available to help confirm an accurate date.

Quotes from the most elementary textbooks show that contamination can cause errors in dating. "Carbon from other sources may easily be trapped in porous materials . . . The archaeologist is the only person who is in a position to know of these contaminating potentials."[3] "[C]ontamination of the sample may take place . . . and removal of the contaminant from the pore spaces and fissures is almost impossible."[4] Excavated samples are "liable to absorb humic matter from the solutions that pass through them (resulting in) contamination by carbon compounds of an age younger than its own . . . there is also the possibly of exchange of carbon isotopes under such conditions . . . That there are other risks of contamination and other pitfalls involved in this method is obvious enough."[5] The possibility of contamination should be exhaustively investigated and pretreatment measures should be designed accordingly whenever any critical radiocarbon dating is being attempted. Unfortunately, even with specialized pretreatment, contamination cannot always be detected and, if detected or identified, cannot always be eliminated.

LATEST SCIENTIFIC FINDINGS

The d'Arcis Memorandum that allegeded the Shroud was a medieval painting and the 1988 C-14 testing that also alleges the cloth dates to the medieval ages have many unfortunate similarities. Both were quite emphatic and perceived as definite. D'Arcis's memorandum was written with great intensity and its assertion definitively prevailed for centuries in the debate whether the Shroud was the authentic burial garment of Jesus. As shown in figure 131, the Shroud's date was smugly announced with an exclamation point. On the day of the announcement, Dr. Edward Hall of the Oxford carbon dating laboratory dismissed the Shroud during a televised interview as "a load of rubbish."[6] While carbon dating certainly is not definitive, it is perceived as such.

Both the d'Arcis Memorandum and the Shroud C-14 testing were also researched and prepared in a very poor manner. We saw in the last chapter how d'Arcis lacked any record at all from any of the several offices or people to support any part of his allegations. We will see in this and the next chapter how the directors of the carbon dating laboratories repeatedly failed to properly research, prepare, or even consider knowledgeable advice and information pertaining to the history and dating of the Shroud. Furthermore, the conduct or the motives of d'Arcis and some of the directors of the carbon dating laboratories can be legitimately questioned. Not only did d'Arcis make very serious allegations of fraud based, at best, on mere rumor, he repeatedly accused the dean of the modest wooden church in Lirey of the worst kind of avarice, while the physical and financial condition of

Fig. 131. Edward Hall, director of the Oxford University carbon dating laboratory, and Michael Tite of the British Museum as they announce the Shroud's 1988 carbon dating results.

his cathedral was in jeopardy. We shall also see in this, and especially the next chapter, repeated conduct on the part of the directors of some of the carbon dating laboratories that exemplify unprofessional motives and behavior.

Comprehensive scientific testing performed on the Shroud and its samples have conclusively proven that the contents of the d'Arcis Memorandum were impossible. The comprehensive future tesing of the Shroud recommended at the end of this chapter and chapter 10 may also completely disprove the isolated results of the unprofessional 1988 carbon dating performed on the cloth. Recent scienific experiments with carbon dating of first-century linen cloth are already casting enormous doubts on those results. Recent examination of the Shroud's radiocarbon samples, as well as other recent scientific research, is also calling into question the accuracy and location of the carbon dating sample, and the procedures utilized in its removal and dating.

In light of the medical, scientific, archaeological, and historical evidence indicating that the Shroud was the burial garment of Jesus Christ, the question naturally rises, could something have happened to the cloth, as in numerous examples previously, that caused it to be contaminated with additional C-14? The answer to that question is not only yes, but there are many substantive, scientific explanations for how this could have occurred that have only recently been published. One event to consider is the unique image encoding event that occurred to the Shroud. Another possible influence could have been the fire of 1532. Other events in the Shroud's history could also have contaminated the cloth. The Shroud is known to have been repaired on several occasions in the past, and these repairs did factor into the radiocarbon dating of the cloth. A number of different contaminants are also known to have accumulated on the Shroud during its history that could have altered its radiocarbon date in ways that were not fully anticipated by the carbon dating laboratories.

Radiation

The first area of scientific research that we shall examine may constitute the most important challenge to and refutation of the carbon dating of the Shroud. These scientific findings also indicate the key areas of future testing that must be performed on the Shroud to confirm whether it was irradiated with a particular form of radiation, whether new C-14 was created within it, to distinguish the original C-14 from the additional C-14, and to calculate the cloth's true age.

In the same issue of *Nature* in which the carbon dating report of the Shroud appeared, this scientific journal also published a letter by Thomas Phillips of the High Energy Physics Laboratory at Harvard University.[7] Phillips, also an IBM scholar, stated that if the body of the man in the Shroud gave off radiation during the image-encoding process it could have radiated neutrons, "which would have irradiated the Shroud and changed some of the nuclei to different isotopes by neutron capture. In particular, some C-14 could have been generated from C-13."[8] This same process could also form newly created C-14 from nitrogen. This newly created C-14 would make the Shroud appear much younger that it actually is. When asked by a journalist if such a process could have caused an incorrect dating of the Shroud in 1988, Michael Tite, who coordinated the carbon dating of the Shroud for the British Museum, commented: "It is certainly possible if one gave the Shroud a large dose of neutrons to produce C-14 from the nitrogen in the linen cloth."[9] Robert Hedges, one of the scientists who participated in the carbon dating of the Shroud at the Oxford laboratory, also acknowledged to the journalist that a "sufficient level of neutrons from radiation on the Shroud would invalidate the radiocarbon date which we obtained."[10] In fact it was Dr. Hedges who pointed out that the amount of neutron flux required to cause a 1,300-year difference in age was not nearly as much as first suggested by Phillips.[11] Because the amount of C-14 in the C-14 to C-12 ratio is so minuscule (one part in a trillion), if a neutron flux activated only an extra 18 percent of C-14 compared to that present naturally in the linen, it would cause a cloth from the first century to appear to be only 650 years old.[12]

In 1998, at a conference of experts in Turin held concurrently with the exhibit of the Shroud, Dr. Jean-Baptiste Rinaudo of France presented a paper that proposed how the images on the Shroud were formed by radiation and how this method would have affected the cloth's radiocarbon date.[13] We will review the image-forming aspect of his paper in far more detail in chapter 10, but for now we will note that neutrons, uniformly distributed throughout the body, would be released from its surface under his method.

Neutrons are very penetrating. While some would collide with the countless atoms in the cloth and bounce off in another direction, many would pass through the linen and the blood, bouncing off of or even penetrating the limestone surroundings, with some ricocheting back onto the cloth. After many bounces, a neutron would lose most of its energy but would then be in a position to alter the nucleus of another atom into a newly created C-14 nucleus. This could easily happen in one of two ways. Flax, from which linen is made, grows in soil that is dependent on nitrogen, so N-14 would be found within the linen itself. N-14 in the air, on the surface of the cellulose, or within the flax itself could absorb a neutron causing it to emit a proton. The remaining nucleus becomes a carbon-14 (C-14) nucleus. This reaction is illustrated below.

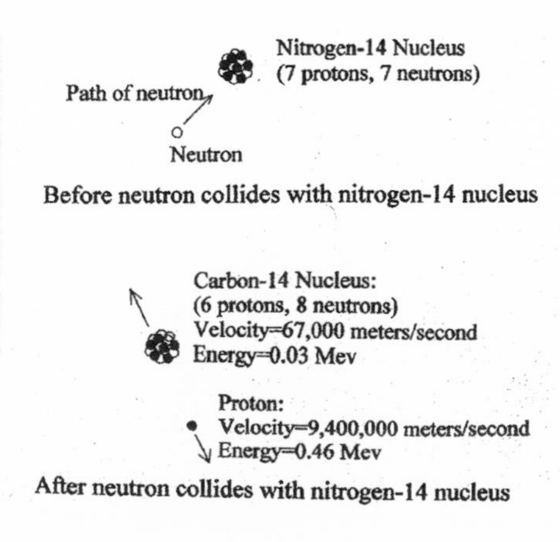

Fig. 132. Carbon-14 created from nitrogen.

When this reaction occurs with N-14 within the flax itself, or within the air permeating or passing through the porous fibers, these newly created C-14 atoms also become part of the flax. This reaction also produces energy, allowing the newly created C-14 to penetrate approximately 0.3 microns farther into the cellulose and break its chemical bonds.

The other way in which these neutrons could create new, additional C-14 is if the neutrons collided with the nuclei of C-13 isotopes or atoms within the cellulose of the Shroud. The absorptions of an additional neutron would convert a C-13 nucleus into a C-14 nucleus. This newly created C-14 would also become part of the flax itself.

When an object or sample is given to laboratories to be carbon dated, age is based on the assumption that all of the C-14 that is measured accrued in the object naturally, without any additional or newly created C-14 getting into the sample. If additional C-14 formed and remained in the cellulose of the Shroud either from N-14 or C-13, and then was measured, the age ascribed to it would be correspondingly incorrect. Rinaudo proved this by radiocarbon dating a 3400 B.C. Egyptian linen cloth before and after irradiating it with neutrons in the quantity first suggested by Phillips. The radiocarbon dating before neutron irradiation was in agreement with the mummy cloth's known historical age, but after irradiation, its age was shifted forty-six thousand years toward the future, five hundred centuries forward in time.[14]

We will review further the formation of the images on the Shroud in later chapters, but for now, note that many Shroud experts think that low-energy radiation, such as protons, alpha particles, long-wave X rays, or ultraviolet rays, was the key component of its cause. Under Rinaudo's method, an equal number of protons (at 1.135 million electron volts [MeV]) are released simultaneously with the neutrons. It is well known scientifically that air and linen attenuate protons, having an energy of about 1.1 MeV. These protons do not travel more than about 3 cm (1.18 inches) in air and 30 microns (two or three fibrils) in linen. These properties and this amount of proton energy will be shown to be ideally suited to produce the body images found on the Shroud.

Moreover, in order to create similar conditions of observation or comparison between his experimental samples and the centuries-old Shroud, Rinaudo artificially aged his samples by subjecting them to elevated temperatures for a short time for ten hours at 150°C.[15] He found that, when his sample was irradiated with the above energy at a very low intensity, the linen cloth remained white. However, after it was artificially aged, it took on a very similar appearance to that of the Shroud body image.[16] This would be consistent with the development of the Shroud's body images over time.

When Rinaudo computed the number of protons required to obtain an image like that on the Shroud, he also observed the corresponding number of neutrons necessarily present. Based on his earlier radiocarbon experiments with the Egyptian mummy linen cloth, he calculated the age change that would result from this corresponding number of neutrons. Significantly, he found the result to be an age shift of thirteen centuries.[17] This form of radiation *alone* could explain why the measured radiocarbon date for the Shroud was contrary to all the previous scientific, medical, archaeological, and historical evidence that the Shroud wrapped the body of the historical Jesus Christ in the first century. In chapter 10, we will examine a couple of image-forming methods that would give off this type of radiation and would account for such an age change.

At the same international conference at which Dr. Rinaudo spoke in 1998, a paper containing the results of experiments by three Italian scientists headed by Mario Moroni was also presented.[18] Moroni took six mummy samples that were approximately 2,110 years old and exposed all but one of them to various conditions or treatments before radiocarbon dating them again. Two samples were not irradiated, but exposed to fire conditions to simulate what the Shroud incurred during the actual fire of 1532. One sample was only irradiated. The last two samples were irradiated and then exposed to the simulated fire conditions. Like Rinaudo's radiation, and that discussed by Phillips, Tite, and Hedges, this radiation consists of a neutron flux.

The results of these experiments were very interesting: The two fire simulation experiments showed small radiocarbon age changes (discussed in appendix F). The irradiation-only sample showed an age change to 360 years younger—after it was pretreated and dissolved so extensively that a mere 10 percent of it remained. However, the two samples that were irradiated and then heated in the fire simulation models showed significant age changes—1,120 to 1,390 years younger. These samples received pretreatment processes in the standard-to-harsh range, but these were not nearly so extreme as those applied to the irradiation-only sample. These results are significant for a number of reasons.

It does not appear that the heat from the fire simulation experiments created significant amounts of additional C-14. However, in the samples first irradiated and then heated, the newly created C-14 from the neutron irradiation became bound to the cloth—with significant amounts surviving the pretreatment cleaning process.

The irradiation-only sample was cleaned and rinsed with acids and other solutions so extensively that 90 percent of it was destroyed—this could have removed most of the newly created C-14 as well. The C-14 isotopes created from N-14 in the air surrounding the cloth and on its surface would be the first to be removed by the pretreatment cleaning since they would penetrate only .3 microns into the fibers. Much of the C-14 created from the N-14 and C-13 within the flax could also have been eliminated from this particular sample. Since this sample was not heated, and was pretreated and dated a short time after it was irradiated, some or all of its additional C-14 may not have become bound, so it could have been washed away.

Almost all elements have atomic structures that require their atoms to be bound to another atom. This is true for carbon; when found in nature, it is always bound. When newly created C-14 is formed by neutron irradiation, it is extremely unstable and will bind at its first opportunity. During the pretreatment cleaning process, various chemicals and rinses are applied consecutively to the cloth sample. During this time, many of the pores of the sample's cellulose could open up. If water enters the pores, the carbon will attach to the oxygen within, become a gas (carbon monoxide or carbon dioxide), and bubble away.

Each chemical attack in the pretreatment process reduces the quantity of available material. Since only 10 percent of the original irradiation-only sample remained, it was obviously exposed to repeated rinsings and chemical attacks. However, not even this extensive pretreatment could remove all of the additional C-14 created from the neutron irradiation. In fact, the Isotrace Radiocarbon Laboratory in Toronto, which pretreated and carbon dated these particular samples, advanced the idea that for strongly irradiated samples, pretreatment was not effective in removing *all* of the C-14 atoms produced during irradiation by the collision of neutrons with nitrogen atoms.[19]

The above pretreatment cleaning techniques employed by the Isotrace Laboratory on the irradiation-only sample in Moroni's experiments were also much harsher than those used by the laboratories when they carbon dated the Shroud in 1988. In Moroni's last two experiments, two samples originally carbon dated to 160 B.C. were irradiated and subsequently placed in fire-simulation models. Both were then radiocarbon dated again—to the Middle Ages. One of these samples was dated 1,120 years younger. Its pretreatment cleaning technique left 79 percent of the cloth sample to be carbon dated. For the other sample, the age changed to 1,390 years younger. Its pretreatment cleaning left more than half of the sample to be carbon dated. These pretreatment cleaning methods better approximate the extent of the cleaning methods that were applied to the Shroud samples in 1988. At that time, those labs announced that the Shroud of Turin—which all other evidence consistently points to as originating from the first century A.D.—carbon dated from 1260 to 1390 A.D.

When these high-energy neutrons bounce throughout the cloth, they also break chemical bonds within the cellulose. This provides an opportunity for the newly created C-14 atoms to bind to and become part of the linen over time, and survive the pretreatment cleaning process better. This can occur in different ways. The breaking of the chemical bonds causes other atoms within the broken molecular structure of the cellulose to become unbound. These atoms (oxygen, hydrogen, and carbon) also need to bind to other atoms and, like the newly created C-14, will actively seek to bind to other unbound atoms. The newly created C-14 and the oxygen and other carbon atoms from the broken bonds will bind to each other within the molecular structure of the cellulose. For this reason, the longer period of time that passes, the more embedded the C-14 can get into the molecular structure of the cellulose. Heat would naturally speed this process as well as cause the binding to be more extensive.

The fact that the irradiation-only cloth was 360 years younger, while the cloth samples that were irradiated and heated in the fire models dated eleven to thirteen centuries younger, could possibly be explained by three different reasons, or a combination of them. The irradiation-only sample was pretreated so extensively that only 10 percent of it remained. If it had been pretreated similarly to the last two examples, Rinaudo's sample, and the Shroud, it, too, may have appeared much younger. Neither did this sample age for many, many years, nor was it artificially aged before it was dated. Moreover, unlike the last two examples and the Shroud, it was not subsequently heated in a fire or fire-simulation model. Moroni's results are very similar to Dr. Rinaudo's results. He irradiated his sample and aged it artificially by baking it at 150°C for ten hours. His pretreatment cleaning method was also similar to the Shroud's and to Moroni's last two samples.[20] When he then radiocarbon dated his sample and calculated the dose of radiation that would have created the image, he, too, got a shift in age of thirteen centuries.

Both groups of these scientists may very well be simulating what happened to the Shroud. All the previous and subsequent evidence in this book consistently indicates that the Shroud is from the first century and that some form of radiation was involved in the creation of its image. Only after many centuries was it pretreated (and carbon-dated) for the first time. The passing of many centuries would also have allowed a great deal of time for the newly created, unbound C-14 atoms to single-or double-bind with the unbound chemicals from the broken bonds caused by the neutron radiation. Moreover, the Shroud was also involved in a fire in 1532 and was clearly exposed to its heat for an unknown period of time. This would have speeded up the above processes and more permanently or extensively bound the newly created C-14 to the molecular structure of the linen. When the Shroud's three samples were pretreated, they, too, were treated in a manner similar to Rinaudo's sample and Moroni's two samples, which were originally irradiated with neutrons and then heated. *All six* cloth samples were then dated to a period between 1,120 to 1,390 years younger than their ascribed historical dates.

Location of the Sample

Recent analysis has confirmed that the samples taken from the Shroud for carbon dating were *not* representative of the non-image samples that comprise the bulk of the rest of the cloth.[21] This was scientifically determined only when access to the radiocarbon samples was acquired by STURP chemist Alan Adler. Dr. Adler compared fifteen threads from the radiocarbon sample with nineteen assorted fibers from the non-image, water stain, scorch, image, backing cloth, and serum-coated locations on the Shroud. These samples were examined by Fourier Transform Infrared (FTIR) microspectrophotometry and were also studied by scanning electron microprobe.

FTIR spectral patterns of these samples clearly indicate that the radiocarbon samples contain a *different* chemical composition. These differences were further confirmed to Dr. Adler and his associates by peak frequency analysis using computer software that generates spectral data. The scanning electron microprobe data showed gross enrichment of the inorganic mineral elements in the radiocarbon samples, even when compared to water-stained samples taken from the bulk of the Shroud. In an article published in *Archaeological Chemistry,* Dr. Adler stated, "In fact, the radiocarbon fibers appear to be an exaggerated composite of the water stain and scorch fibers thus confirming the physical location of the suspect radio sample site and demonstrating that it is not typical of the non-image sections of the main cloth."[22] Dr. Adler also compared the radiocarbon samples to image fibers taken over an area covering approximately five feet of the body image. He found much more variation in the chemical composition of the radiocarbon samples taken from an area that covered, perhaps, a couple of inches, than existed over approximately five feet of body image on the Shroud.[23] While referring to the differences in the chemical composition of the radiocarbon sample with those taken from various locations throughout the cloth, Adler warned, "This calls into question the accuracy of the radiocarbon date."[24]

Obviously, the most significant event that occurred to the Shroud in its history was the encoding of its images. As we have already seen, but will see further in chapter 10, there are a number of reasons most Shroud scientists and experts think that some form of radiation caused the images on the Shroud. According to cloth drape configurations studied by Dr. Jackson and other STURP scientists, the radiocarbon site would have been positioned over the man's legs at the time of the image-encoding event, when the ends of the cloth were folded onto the body.[25]

Based on the Shroud's burn holes and scorch marks from the fire of 1532, scientists can also ascertain the exact forty-eight-fold pattern of the cloth at the time of the fire. When scientists refolded a shroud cloth model according to this known pattern, they observed that the radiocarbon site was just a few thin layers below the cloth's outermost surface and quite near the part of the cloth that was burned by the silver lining of the container.[26] Figure 134, taken by STURP in 1978, is a close-up of where the radiocarbon sample was removed in 1988. The part of the cloth immediately above and to the right of the white spot (caused by the removal of the Raes textile sample in 1973) and along the seam is where the radiocarbon sample was removed. The close-up shows that this location has a darker, dirtier appearance than the adjoining area of the cloth below. Furthermore, figure 135 shows that the radiocarbon site and its adjoining area both have a much darker or dirtier appearance than the rest of the cloth.

Figures 134 and 135 also help illustrate that at least four to five known repairs have been necessary in this area. The rectangular piece missing from the very corner of the Shroud necessitated that the remaining part of the cloth be sewn down at this location. Most likely, the cloth was neatly cut off and mended at this site after having gradually torn away from years of exhibiting the cloth horizontally. (Another even rectangular piece is missing from the other far end of the cloth on the dorsal side, where it was also attached while being displayed horizontally.) There may have been several mendings or repairs before the present one as the cloth gradually tore away from these two corners. A seam, visible in both photos, runs the entire length of the cloth; many experts think this was added to the cloth to support it while being displayed horizontally, perhaps after the cloth tore away at the corners. This seam runs directly adjacent to the location where the radiocarbon sample was taken. Following the removal of the adjacent Raes sample in 1973, it was also necessary to sew the remaining part of the Shroud onto the backing cloth. (The same procedure was again required in 1988 when the radiocarbon sample was removed.) In 1534, the Order of the Poor Clare nuns performed extensive repairs all over the cloth, including adding more than sixteen patches, as well as a backing

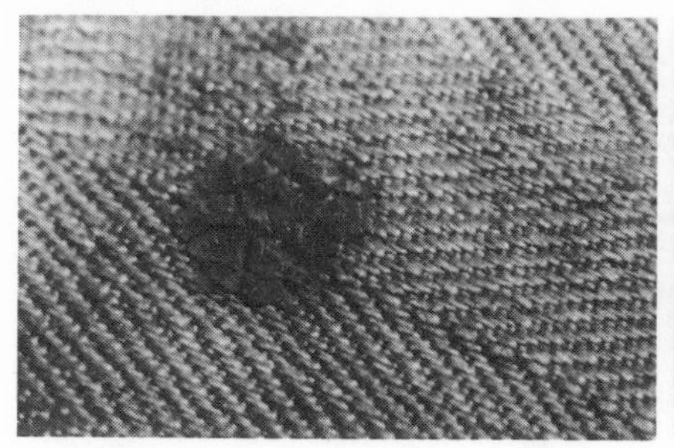

Fig. 133. A large deposit of wax on the Shroud.

Fig. 134. The darkened site from which the radiocarbon sample was taken, as seen with the visible eyes.

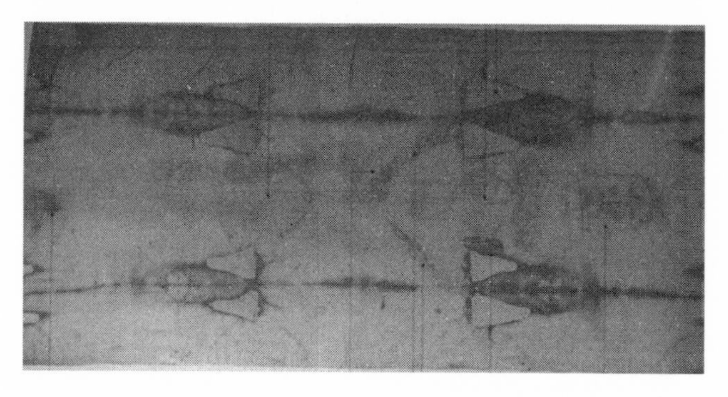

Fig 135. Even from this full-length view, the darkened area next to the rectangular-shaped backing cloth in the upper left-hand corner can bee seen with the naked eye. The darkened area is the location where the 1988 carbon dating sample was removed.

cloth, all of which remain to this day. It is quite possible that they also performed repairs or mending in the areas just mentioned.

Contamination

New scientific research related to the carbon dating of the Shroud has also been conducted at the University of Missouri, St. Louis, by Dr. James Chickos, a chemist. He has found that tallow, which candle wax is made from, can become chemically bound to cellulose by a process known as transesterification.[27] Tallow contains a long carbon chain, which, attached to the glucose molecular structure of the linen, will naturally add carbons and make the cloth appear much younger. This process is illustrated in figure 136.

Wax has been identified on samples removed from the Shroud by STURP scientists Alan Adler and John Heller, and also by Walter McCrone.[28] Wax can be seen with the naked eye in places on the Shroud itself (fig. 133). The sample from which Dr. Adler positively identified wax came from the edge of one of the relatively small, round holes found in a pattern separate from the fire marks of 1532. This series of small holes (discussed in the last chapter) was first observed in a painting of the Shroud from 1516. Many people think the most likely explanation for this pattern of round burn holes could be cinders falling onto the folded Shroud from

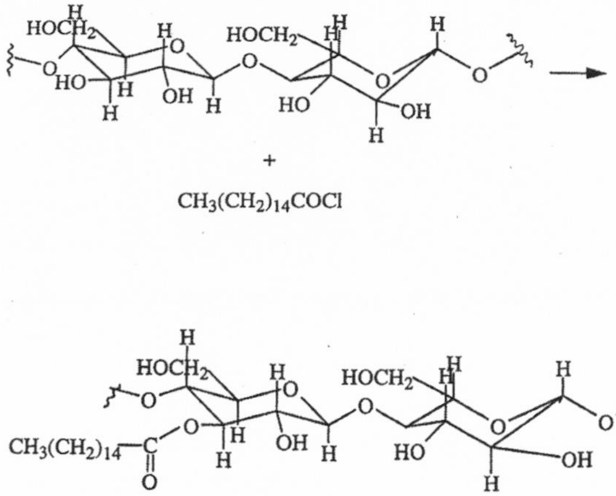

Fig. 136. The chemistry of tallow contamination.

+

$CH_3(CH_2)_{14}COCl$

a censer, an incense container swung during a religious ceremony.

Interestingly, wax is an excellent substance with which to stiffen cloth in order to give it strength or support. This would help keep the cloth from fraying or tearing at the edges of these burn holes, and could very well have been applied intentionally. The area on the Shroud from which the radiocarbon sample was removed had at some point torn away and was repaired. Starch, like wax, could also stiffen and support the cloth and has also been identified on a sample taken immediately adjacent to the radiocarbon site.[29]

Tallow or candle wax that is chemically bound to the molecular structure of the linen is certainly not visible to the naked eye, yet even a small amount could affect the results of a sample by many centuries, especially if wax was applied in the cloth's recent past. Not only would the chemically bound tallow be invisible, but a great deal of it would survive the normal pretreatment cleaning processes used on samples prior to their dating. After Dr. Chickos applied both acids and bases to linen with which tallow had chemically bound, significant amounts of it still remained on the cloth.

Each of the above experiments and findings easily justify another carbon dating of and future research on the Shroud; however, there are many other reasons to justify another dating. In this and succeeding sections, we will begin to scrutinize the unacceptable samples that were taken during the radiocarbon-dating process and the unscientific manner in which this process was conducted.

OTHER ARGUMENTS AGAINST THE 1998 CARBON-DATING RESULTS

The arguments above only begin to hint at the bad science involved in the 1988 testing. As you will see, everything was done wrong, from the choosing of the sample, to the extent of pretreatment, to the manner of reporting, and so on. The scientists fought for years over what protocols to use, and then ignored them. It is a wonder these test results have been taken seriously for so long.

Single Sample, Taken from Worst Location on Cloth

Of all the locations on the entire Shroud cloth from which to choose a sample for carbon dating, a worse location could not have been selected than the one chosen in 1988. Many of the reasons why were not known to those participating in the sampling nor apparent to the naked eye. For instance, starch was found on a thread from the Raes sample that is located immediately adjacent to the radiocarbon site. Starch could have been used to stiffen cloth and aid in any of the known repairs in this area. In 1982, without permission from the cloth's custodians or knowledge at the time that the thread contained starch, STURP sent a thread from the Raes sample to be radiocarbon dated. Interestingly, one end of the thread dated to A.D. 200 while the opposite end—containing starch—dated to A.D. 1000.[30] The molecular structure of starch is very similar to that of cellulose. Like wax, it, too, could chemically bind to cellulose, not be detected with the naked eye, and not be removed by the standard pretreatment processes. It could also alter the thread's radiocarbon date by many centuries.

This test was performed at the University of California nuclear accelerator facility using a linear accelerator mass spectrometer technique, which was one of the newly developed methods for dating small samples.[31] This was not a dedicated laboratory that regularly carbon dated samples, but the results from both ends demand attention. If one end of the Shroud thread dated to A.D. 200, the error range of the dating could place the Shroud in the first century A.D. The presence of starch could explain why the other end of the thread dated to A.D. 1000. Since starch was located near the 1988 Shroud sample site, this and all future sampling locations on the Shroud should be investigated.

Ultraviolet fluorescent photographs and reflected light imagery, which allowed details to be seen that were not visible with the naked eye, also revealed that the radiocarbon site was at the edge of a water stain and in the midst of a scorch mark.[32] This observation was first made by Vernon Miller, chief photographer of the 1978 Shroud exhibition. After examining and noting the differences between the radiocarbon samples and those from the rest of the cloth, Dr. Adler confirmed these findings.[33] While flowing over cloth, water picks up extraneous material, most of which would be deposited at the edges of the stain. After observing the above photographs, Miller stated that this location was "probably as highly marked as any place on the Shroud."[34]

Under these and other circumstances, archaeologist William Meacham cautioned before the carbon dating of the Shroud that:

> Penetration of the pore spaces may have occurred and a water-soluble, carbon-bearing solute deposited therein. The bound water of hydration may have been penetrated by other sustances, lipids and proteins may have been deposited among the fibrils. . . . If this catalogue of contamination possibilities seems overly pessimistic, one must bear in mind the various substances recorded to have been in contact with the relic since 1356—oils, wax, soap, paints, ointments,

> open wounds, saliva, sweat and smoke. Preservatives, starch and image-inhalers may also have been applied. Earlier, the cloth may have been sealed in a city wall for several centuries with a votive oil lamp and the relic may have been attached to a wooden frame for additional centuries absorbing decay products from the wood cellulose. Some of the penetrating organic substances may through time have degraded to low-order residues not detectable as specific contaminants and shielded by the cellulose substrate.[35]

As mentioned previously, several textile additions are also present in the area of the radiocarbon sample. As a point of fact, foreign threads were found on the sample that was removed from the Shroud for the 1988 carbon dating by the technician Giovanni Riggi. Riggi stated:

> . . . I was given permission to cut about eight square centimeters from the cloth in the same area where in 1973 a sample was taken by Professor Raes. This was eventually reduced to about seven square centimeters *due to contamination of the cloth with threads of different origins which even in small quantities could cause variations in the dating due to their being of later addition.*"[36] (italics added)

Professor Riggi implied that he was able to detect and eliminate all such foreign fibers. However, this is not necessarily so. When the Oxford laboratory received its sample of the Shroud, fine yellow strands of cotton were found in the sample. Peter South of Precision Processes Textiles in England, which assisted Oxford University in preparating its sample for the carbon dating, noted, "The cotton is a fine, dark yellow strand, possibly of Egyptian origin and quite old. It may have been used for *repairs* at some time in the past, or simply became bound in when the linen fabric was woven in" (italics added).[37] This is also consistent with similar cotton findings of unknown age and origin on the 1973 Raes sample.[38]

At minimum, noted STURP scientist Alan Adler in regard to the radiocarbon sample and its dating, "Only a single sample was taken in the lower corner of the main cloth of the frontal image below the so-called sidestrip from the selvage edge in an obviously water-stained area. . . . The selvage edge was trimmed off before portions of the sample were divided among the participating laboratories. Whether such an obviously contaminated sample is truly representative of the rest of the cloth is clearly questionable and the accuracy of the reported date is clearly doubtful."[39]

After consulting with art and textile experts, Dr. Adler has learned that medieval restorers were extremely skilled in repairing textiles. In particular, he found that medieval restorers could easily repair cloth in a way that was not visible to the naked eye. Dr. Adler has expressed concern whether the area surrounding and including the carbon dating sample might have been one entire repair piece.[40] Dr. Stuart Fleming, formerly of the Oxford Laboratory for Archaeology and now of the University of Pennsylvania, stated that a restorer "could certainly have re-woven a damaged edge to a standard not visible to the naked eye."[41]

Interestingly, Dr. Adler found that not only was the chemical composition of the radiocarbon samples different from the rest of the Shroud fibers, the samples were "saltier" than the water-stain fibers taken from the rest of the Shroud. This indicated to him that the water flow at the radiocarbon site was prevented from continuing by a "bounded edge" or the side strip, indicating that the side strip or seam was already on the Shroud at the time this water flowed in 1532.[42]

Numerous paintings of exhibitions from medieval times onward always show the Shroud displayed sideways and held or suspended at the corners just above the side seam or strip. It is much more practical and easier to display the cloth horizontally than vertically. Many authors have suggested that the purpose of the seam was to lend support to this area of the cloth and facilitate hanging it like this. When held or displayed thus, maximum stress is placed at the corners, so tearing inward from here could be expected.

According to historical accounts, certain noteworthy people were given pieces of the Shroud. As Adler pointed out, it would be logical to assume these came from the corners of the Shroud that are now missing near the side strip. Adler thinks that, since the side strip appears to have been added before 1532, the de Charny family could have added it as part of its repairs when they displayed the Shroud. Adler has stated, "Maybe the radiocarbon sample is simply rewoven material from the time of this repair."[43] Furthermore. Dr. Rinaudo observes that if the Shroud body image was formed by radiation, it would have developed gradually over a period of many years. Since the Shroud was known to have been first exhibited in Europe in the 1350s, he notes, "This means that it was already centuries old, an observation which is contradictory to the C-14 datation results."[44]

When an error occurs in carbon dating a sample, it is usually because of the addition of foreign contamination in some manner. This was the problem for most of the carbon dating errors listed in the beginning of the chapter. Rarely do carbon dating errors come from faulty equipment or improper measurement of the isotopic content within the sample itself. Among the duties of a carbon dating laboratory, identification and removal of contamination from the sample may be the most critical. This is all the more important with the new methods of carbon dating small samples. Since the samples dated by these methods are so much smaller, if contamination is present, it constitutes a much greater proportion of the sample, thus its influence on the dating can be much greater. For example, if starch or candle wax had become chemically bound and undetected over a two- to three-inch area on the Shroud, it would not cause great problems if the conventional method of carbon dating was used—because that method required the sample to be about the size of a man's handkerchief. Yet if this two- to three-inch area was the location where the Shroud's radiocarbon samples were taken, either contaminant could alter the radiocarbon date by many centuries.

Commenting the day after the results from the 1988 carbon tests were officially announced, Dr. Harry Gove, who helped develop one of the new methods to carbon date small samples, stated:

> The piece that was removed from the Shroud was divided among the three labs, and that piece came from one specific spot in the Shroud. If there were some reason why the carbon-14 content in that particular piece was contaminated, it's inaccurate. All of the labs used the same cleaning technique, and if there's some kind of contaminant that was not taken care of, it would give the same answer to all three labs, and all three would be wrong.[45]

Unfortunately, not all forms of contamination are known at the time a sample is chosen or are found when the sample is treated at the carbon dating laboratory prior to its dating. When carbon dating a sample, certainly a sample as significant as the Shroud of Turin, exhaustive investigation of the possibilities of contamination should be undertaken beforehand and pretreatment strategies devised accordingly. Any contamination added to a cloth or other sample after its manufacture would make the cloth appear younger or older than it actually is. Contamination of an object that causes an erroneous date occurs far more frequently than the public realizes. Prof. W. Wolfli, director of the Zurich laboratory that dated one of the samples removed from the Shroud in 1988, has stated in a jointly authored paper,

> The existence of significant indeterminant errors can never be excluded from any age determination. No method is immune from giving grossly incorrect datings when there are non-apparent problems with the samples originating in the field. The results illustrated [in this paper] show that this situation occurs frequently.[46]

Porous materials are some of the most difficult from which to remove contaminants. Many of the examples of very aberrant radiocarbon dates cited at the beginning of this chapter consisted of porous, cellulosic material such as wood, plants, bone, and charcoal. Archaeologist William Meacham has cited a number of other instances where intrusive materials have filled pore spaces in such a manner that they were unable to be extracted.[47] Cellulose in linen is porous—quite capable of admitting contaminants. These would not necessarily be removed by the standard pretreatment methods used on the Shroud samples by the carbon dating labs in 1988.

In light of these characteristics, knowing the history of each part of the cloth was important so that the best location could have been chosen for sampling. For example, the Shroud's configuration at the time it lay over the body during the image-encoding event was important. Similarly, knowing its fold configuration at the time of the fire and each location's proximity to the source of the scorches would have helped. The same can be said for the locations of the cloth's water stains, repairs, and carbon-bearing contaminants. The Shroud clearly required a great deal more knowledge and preparation than it obviously received when its sample was removed and carbon dated in 1988.

A revealing experiment occurred not long after the new methods for dating small samples became established. In order to establish that the new, small-sample dating techniques were reliable with cloth and capable of giving an accurate radiocarbon date on the Shroud of Turin,[48] six carbon-dating laboratories participated in a comparison of the two new methods. The test was coordinated by the British Museum. Laboratories received the first two cloth samples in 1983 and a third cloth sample in 1984. The results were announced at an international radiocarbon conference held in Trondheim, Norway, in 1985. The report, published in 1986, focused most heavily on the results from the first and third samples. One of the laboratories, the Zurich lab, was approximately one thousand years too young on the first sample and approximately one thousand years too old on the third sample. The report stated, "This must have been due to contamination introduced by a new method of pre-treatment" used by the laboratory.[49] It was not explained how the contamination could make one sample seem much older and another much younger, but the report does point to the obvious pitfalls that can result from contamination and from an inadequate pretreatment method.

At the time of this intercomparison test and at the time the Shroud was carbon dated in 1988, the new methods of carbon dating small samples did not have a lot of experience carbon dating cloth. However, one of the new methods, the small proportional gas counter method, was similar to the conventional method of radiocarbon dating. As it turned out, the other method—the accelerator mass spectrometry (AMS) method—was used for the 1988 carbon dating of the Shroud. An expert in the field of radiocarbon dating noted in 1988 that, "[the] cloth has undergone years of testing with the conventional dating technology. On the other hand cloth does not seem to be an item which has commonly been dated by the accelerator method."[50]

A further indication of how little forethought and preparation actually went into the sampling and dating of the Shroud can be seen in the number of samples removed from it. In 1984, STURP advised in writing that three samples should be taken from three different locations on the Shroud. In 1986, Shroud expert and archaeologist William Meacham recommended in an article and at a conference in Turin designed to determine such matters that four samples should be removed from four different locations on the Shroud. At this same conference, STURP chemist Alan Adler also advised that samples be removed from various parts of the cloth.

Unfortunately, only one piece of cloth was removed from the Shroud when it was carbon dated. All three laboratories' samples came from this single piece of cloth. This had the effect of compounding the contamination problems. If several samples had been taken from different parts of the Shroud, a contaminated area could have stood out. As it is, if the one tested sample was indeed contaminated, all we have proof of is that all three laboratories had contaminated samples.

SCIENTIFIC PROTOCOLS ESTABLISHED BUT NOT FOLLOWED

Objections to the 1988 results don't end there. After taking note of the carelessness with which the sample was chosen, it should come as no surprise that the unscientific protocol that was followed and the manner in which the sampling and dating were performed also justify a new round of carbon dating.

The first so-called attempt to carbon date a sample from the Shroud actually occurred in 1977 when Walter McCrone visited Gilbert Raes, who possessed the Shroud samples from his 1973 textile examination. Dr. McCrone proposed testing portions of these samples by first placing them on a photographic plate and, as described by Dr. Gove, "[t]hen, after some time, the plate could be developed and the tracks made by the electrons emitted where carbon-14 decayed could be counted and the sample dated."[51] At the time, this method was new and controversial. New methods, subsequently used by the carbon dating laboratories on small samples from the Shroud, were several years from being developed and operational. Fortunately Raes did not give McCrone a sample but contacted two other Belgian scientists, who were in charge of a nearby carbon-dating facility. These scientists pointed out that McCrone's idea was scientifically nonsensical on several counts and declared it unworkable.[52]

In February 1979, Harry Gove, of the University of Rochester, wrote to the Archbishop of Turin on behalf of his laboratory and the Brookhaven National Laboratory to offer to carbon date the Raes sample.[53] These two laboratories each had recently developed new but different methods for carbon dating small samples. The person to whom he sent the letter apparently did not deliver it for many years, but when it finally did arrive, the Archbishop of Turin's scientific advisor, Luigi Gonella, did not consider it since he wanted a sample to go directly from the cloth to the testing labs.

As far as the Archbishop of Turin's office was concerned, the first formal proposal to carbon date the Shroud was received from STURP in 1984, and recommended that at least three samples be removed from three different areas of the cloth, in addition to taking a sample from the Holland backing cloth as a control.[54] As we saw, Archaeologist William Meacham published a similar recommendation and repeated this suggestion, as did STURP chemist Alan Adler, at a workshop held in Turin in 1986 to establish protocol for carbon dating the Shroud.

In October 1978, during a conference at the end of the exhibition, the Archbishop of Turin, Cardinal Anastasio Ballestrero, announced that, when the scientific community could decide the best way to determine the age of the cloth, he would consider asking the then owner, former King Umberto II of Italy, to allow samples to be removed for that purpose. As a result of this statement, efforts to have the Shroud dated and to conduct further scientific examination of the cloth began in earnest in the scientific community.

In 1979, STURP formed a C-14 committee and contacted scientists at carbon dating laboratories throughout the world, asking them if they would be interested in joining STURP's effort to carbon date the Shroud of Turin. Laboratories were just beginning to develop the capabilities of carbon dating small samples by either of the two new methods at this time. Five laboratories initially joined STURP's effort: the Arizona, Brookhaven, and Rochester laboratories in the United States, and Oxford and Harwell in the United Kingdom. Thereafter, the laboratory in Zurich, Switzerland, and the British Museum joined the effort.[55] At this juncture, it was extremely critical that not only qualified, but true scientists with a quest for truth and knowledge, be recruited into the effort. At minimum, good scientists should put that quest above nonscientific considerations, maintain their objectivity, and keep their word.

In the early 1980s, various laboratories, including some of the above, were acquiring the operational ability to carbon date small samples using either the accelerator mass spectrometry (AMS) method or the small proportional gas counter method. Naturally, STURP, the carbon dating laboratories, and many other people wondered if the new technology could be shown to have the requisite amount of certainty to carbon date the Shroud. In May 1983, the British Museum, in its role as coordinator, mailed two cloth samples to the above six laboratories, which were participating in an intercomparison test of the two new techniques as a prelude to dating the Shroud of Turin. This study was discussed earlier in reference to one laboratory being approximately one thousand years too young on the first sample and one thousand years too old on the third sample. However, here, we shall focus on a fundamental and systematic error, which all the participating laboratories and the coordinating institution committed, but which the report barely mentioned. This focus is extremely critical, for as a result of this intercomparison test, the new methods for carbon dating small samples were deemed to be reliable for dating cloth specimens taken from the Shroud of Turin.

All of the participating laboratories were originally given two samples to date; however, when *all* the laboratories that carbon dated the second sample dated it to approximately half of its historical age, the sample was simply replaced with a very similar third sample. According to the report, the second sample was replaced because its radiocarbon dates "suggested that the material was of much more recent date than expected." Yet, the report then stated, "this was probably erroneous as it turned out."[56] If this is correct, then it means that *all* of the laboratories that dated this sample came up with "probably erroneous results." Each of the five carbon dating labs that tested sample 2 dated it to half its historical age. The report does not provide an *explanation* why of the radiocarbon dates were "probably erroneous as it turned out." Since *all* the laboratories made a fundamental and systematic error with one of two original cloth samples, and since the report provides a probable explanation for the one-thousand-year errors on sample 1 and sample 3, it should have ascertained and provided an explanation for the results of sample 2. The authors of the report have a duty to investigate and publish this explanation.

These errors are significant. They deserve far more attention than the report gives them. According to the data from the intercomparison experiment (even with the replacement sample), there is a one in three chance that *all* of the laboratories using the new methods will date a cloth sample at half its historical age. *This is an entirely unacceptable result!*

Even if the chance were one in ten, it would be an unacceptable level of certainty for this type of scientific testing, which is often perceived as definitive. The results from this intercomparison experiment did *not* establish that the new methods for dating small samples were capable, with the requisite degree of certainty, of dating cloth samples from the Shroud of Turin. In the case of sample 2, five laboratories provided dates, which is two more than were used on the Shroud of Turin. Furthermore, both new methods were used. When the Shroud was dated, only the AMS method was used. Note also that the samples used in the intercomparison were not known to have any of the numerous contamination problems that were associated with the Shroud samples.

Even though sample 2 had never been carbon dated, the historical period from which it derived had been known. A sobering aspect of its dating is that, if this sample's historical period had not previously been ascribed, the coordinators may never have known that all of the laboratories were "probably erroneous" when carbon dating this sample. The Shroud's wealth of historical, archaeological, scientific, and medical evidence indicates it comes from the first century. Yet it seems the scientists behind the 1988 testing were willing to disregard all this evidence—even though a similar situation with sample 2 caused them to reject the carbon dating test results.

Why didn't the same carbon dating laboratories and their coordinator disregard the contrary results that they had obtained and request another sample from the Shroud, especially since the samples all came from the same controversial site?

Unfortunately, the reason or reasons *all five* laboratories were "probably erroneous" for one of the three samples were not investigated further (or if they were, the results were never published). Even with the report basically ignoring what occurred with sample 2 and basing its conclusions just on the results for sample 1 and sample 3 (which both contained one-thousand-year errors), its authors from the coordinating institution concluded that carbon dating the Shroud of Turin would now be possible in principle if several laboratories were involved and both new techniques for dating small samples were used. This itself is hardly a vote of confidence, and it merely places the best possible interpretation on the less-than-ideal results for only two of the three samples. If *all* five laboratories were likely inaccurate in carbon dating one of only three cloth samples under both new techniques, then this intercomparison report shows that both new methods failed this experiment with small cloth samples.

The reason for this failure does not necessarily have to do with the honesty or integrity of the participating scientists, but it does indicate they were overanxious to date the Shroud with the new techniques, which weren't yet ready for the task. These laboratories are extremely expensive to run and are heavily supported by national institutions in each country, such as the National Science Foundation in the United States and the Science Research Engineering Council in Great Britain. Naturally, being one of the laboratories that carbon dated the Shroud would look very impressive in the next round of funding applications. Besides, there would be a certain amount of prestige and fame from dating the Shroud.[57]

In 1987, Paul Maloney, an archaeologist and vice president of the Association of Scientists and Scholars International for the Shroud of Turin (ASSIST), prepared and sent a twenty-seven-page document to the Vatican that apparently carried a tacit endorsement of the Vatican nuncio to the United States, for the document was contained in the nuncio's diplomatic pouch. (In 1983, the Vatican became the new legal owner of the Shroud when former king Umberto II died, leaving the cloth to the Vatican on the condition that the relic stay in Turin.) As reported in *Science News,* Maloney had interviewed a number of people prominent throughout the world in the field of carbon dating and found they tended to share " 'a grave concern . . . that accelerator technology [the only method eventually chosen to date the Shroud] is not yet ready to do what the Church wants it to do'—largely because of the frequency of spurious readings from small samples."[58] In the document, Maloney stressed that he found "across the board there had been very little testing of linen samples with either of the new methods."[59] In addition, Garman Harbottle, inventor of one of the two new methods for carbon dating small samples (small proportional gas counter method), informed Maloney that each type of item, whether it was leather, wood, charcoal, fish, cloth et cetera, had its own peculiarities when attempting to carbon date it under the new methods, and that he wished this material (cloth) was also better understood before the Shroud's actual dating.[60]

Unfortunately, this intercomparison experiment was one of a series of disappointing efforts by the carbon dating scientists and their coordinator in a process designed to establish a scientific protocol for an additional round of extensive scientific testing of the Shroud, including its carbon dating. This process, begun by STURP, included various church or church-affiliated officials. A scientific protocol is basically an agreement among the concerned parties on how and what scientific procedures shall be conducted. As time went on, this process would only become further flawed. Several protocols were developed before one was arrived at, and then ignored. Ultimately, the worst procedures imaginable were used in the removal of a sample from the Shroud of Turin. This process would also result in the elimination of an extensive round of scientific testing of the cloth in twenty-five different areas.

In 1984, concurrent with the intercomparison experiment, STURP submitted a 177-page scientific proposal describing work packages in twenty-six areas of scientific examination of the Shroud, including the dating of the cloth by radiocarbon analysis, with the dating proposal being approved by the original five carbon dating laboratories.[61] Following the announcement of the intercomparison test results in Trondheim in 1985, the six participating laboratories, each of which had joined STURP's carbon dating effort, agreed upon a protocol—known as the Trondheim Protocol—consisting of the following:

1. The British Museum would be the coordinating institution, and "guarantor" of the tests.
2. *The "good offices" of STURP would be used to arrange for appropriate samples to be removed.*
3. The British Museum would provide two additional samples both of whose ages were known (+/– 150 years or more). All three samples, including that from the Shroud, would be unraveled and the threads cut into short lengths to render them as indistinguishable as possible.

4. *The British Museum would obtain written agreement not to reveal results to anyone other than designated officials of the museum.*
5. *The six laboratories would use whatever methods they chose for preparing the samples for measurement, but detailed description of procedure and how they calculated their mean values and their uncertainties would be carefully recorded.*
6. *Results would be sent to the Vatican and the Archbishop of Turin. A press release would be issued only when Turin and the Vatican had received the results.* (italics added)[62]

Over a year later, more than twenty individuals with expertise in Shroud matters and/or the field of carbon dating gathered at a conference held in Turin, Italy. (By this time, a seventh laboratory had been added to the group, the Centre des Faibles Radioactivités, in Gifsur-Yvette, France.) They met over the course of three days, and despite a number of scientific differences that occasionally became intense, a consensus on a protocol for carbon dating the Shroud was reached, according to the participants. This protocol—the Turin Protocol—contained the following points:

1. *A minimum amount of cloth will be removed,* which is sufficient (a) to ensure a result that is scientifically rigorous and (b) *to maximize the credibility of the enterprise to the public.*
2. The samples should be taken from an unobtrusive part of the Shroud, and *from a portion which is not likely to yield other useful information. Selection of this material to be removed and the actual removal will be the responsibility of Madame Flury-Lemberg* [an independent textile expert]. The samples should be prepared in a form, not too small, so as to allow reasonable pre-treatment processes. In addition to the Shroud samples, the British Museum will also prepare and provide two control samples for each laboratory.
3. For logistic reasons, samples for carbon dating will be taken from the Shroud *immediately prior to a series of other experiments planned by other groups.*
4. *Seven samples* containing a *total of 50 mg* of carbon will be taken from the shroud. These shroud samples will be *distributed to the seven laboratories in such a way as to ensure that the seven laboratories are not aware of the identification of their individual sample.* This distribution will be the responsibility of three certifying institutions: the Pontifical Academy of Sciences, the British Museum, and the Archbishop of Turin.
5. The taking of samples will be done so that representatives from the seven laboratories will have complete knowledge of the process.
6. At this time, a date will be chosen for *submission of experimental results from the seven laboratories to the following three analyzing institutions:* the Pontifical Academy of Sciences, the British Museum, and the Meteorological Institute of Turin, "G. Colonnetti." These institutions will keep the results in sealed envelopes until an agreed upon date, at which time they will be opened for statistical analysis.
7. *After the analysis of the experimental results by the three analyzing institutions, a meeting will be held in Turin among the three analyzing institutions and representatives of the seven laboratories to discuss the results of the statistical analysis with the objective of deciding the final result of the measurement program.*

8. The Archbishop of Turin will issue a press release concerning this Turin workshop. (italics added)[63]

As can be seen in the labs' Trondheim Protocol, STURP would arrange for the appropriate samples to be removed, which was consistent with the 1984 proposal by STURP. This only made sense, since they had a better knowledge and understanding of literally every location on the entire fourteen-foot cloth than any other group or individual. The carbon dating laboratories also wanted two additional samples to be included with the Shroud, and for all three samples to be unraveled and cut into short threads so as to make all samples unrecognizable. All of this was to ensure the public that "blind testing" occurred. The carbon dating labs also wanted the coordinator, the British Museum, to obtain written agreement not to reveal the results, and to send them to the Vatican and Archbishop of Turin, who would then release them to the press. While the labs could use whatever pretreatment methods they chose for measurement of the samples, their scientists were to record detailed descriptions of their procedure and calculations of mean values, as well as their uncertainties.

At the meeting in Turin, the labs decided not to shred the samples, but to ensure blind testing in other ways. Other items worth highlighting in the Turin Protocol are: A minimum amount of cloth would be removed, from a portion that is unlikely to yield other useful information. After the samples for carbon dating were taken, a series of other scientific experiments would be performed on the Shroud. Seven samples, containing a *total of 50 mg* of carbon, would be taken from the Shroud. According to the scientific advisor to the Archbishop of Turin, the labs this time wanted the dates to be statistically analyzed independently. Following this, the analyzing institutions and representatives of the seven laboratories would meet to discuss the statistical analysis and decide the final results.

It had taken many years to arrive at this protocol during a process that had become entangled with side issues having nothing to do with science per se, which would ultimately leave deep divisions within the scientific community and in the field of Shroud research. In some areas, this protocol had more requirements than may have been necessary, while in other areas, such as sample selection, it could have used more. However, if everyone had followed the protocol and performed their jobs properly, a professional and scientific procedure could have been adhered to, thereby providing the best likelihood of acquiring an accurate dating of the samples—the whole objective of this particular test. Moreover, the 1984 STURP plan covering an entire range of testing in twenty-five other scientific and medical areas could have taken place, which would have revealed an enormous amount of evidence on a variety of matters concerning the cause of the image, and the age, origin, history, authenticity, and conservation of the Shroud of Turin.

However, none of this occurred. What followed was a process and a procedure that were so riddled with inconsistencies and unscientific behavior that the public may find it hard to believe. Before the samples were ever removed, Turin announced that no more than three samples would be taken, which, in turn, would be given to three selected laboratories (Oxford, Arizona, and Zurich). Their reasoning—contrary to both protocols—was that in most archaeological studies only one laboratory does the dating; rarely are as many as three involved. They expressed concern over the amount of material that would have been removed for seven laboratories, and picked the corresponding three laboratories on the basis of the labs' experience in the field of archaeological dating, the internationality of the group, and the sample size required by each laboratory.[64] Since the AMS method could date even smaller samples than the small proportional counter method, the latter method was eliminated entirely.

This reduction provoked an unexpected outcry on the part of the laboratories' directors, who went to great lengths to complain but found little sympathy. All seven carbon dating laboratories signed a letter in July 1987, complaining to the Archbishop of Turin about the reduction to three laboratories that closed with the statement ". . . we would be *irresponsible* if we were not to advise you that this fundamental modification in the proposed procedures may lead to *failure*" (Italics added). In November 1987, the directors of the Zurich, Oxford, and Arizona laboratories—the three laboratories selected to carbon date the Shroud—wrote again to the Archbishop: "While we understand your desire to use a minimum amount of material from the Shroud, we believe that the increased confidence in the measurements which would result from the inclusion of more than three laboratories in the program would justify the additional expenditure of the material." Harry Gove next drafted an open letter to the Pope himself. Gove stated: "The procedure that the Cardinal of Torino is suggesting is bound to produce a result that will be *questioned in strictly scientific terms by many scientists around the world* who will be very skeptical of the arbitrarily small statistical basis . . ." (italics added). Gove actually closed his letter to the Pope by referring to the change as "an ill-advised procedure *that will not generate a reliable date but will rather give rise to world controversy, we suggest that it would be better not to date the Shroud at all*" (italics added).

Carbon dating is perceived incorrectly by the public as being close to absolute and definitive, even to the exclusion of other scientific tests. (Indeed, for many, that is precisely what has happened with the Shroud.) Yet the alarms raised by these and many other scientists indicate that they did not think this test, at that time, on this cloth could achieve the level of certainty, reliability, and definitiveness that is required, even by using an already unusually large number of three laboratories.

Actions and statements did not end there. In an earlier letter written toward the end of 1987, Gove urged the president of the Pontifical Academy to meet personally with the Pope about the laboratories' concerns. In this letter, Gove even enclosed a press release on this subject and informed the president, "We propose to resort to the press release only if our appeal to the Pope fails." Indeed, Gove and Garman Harbottle of the Brookhaven National Laboratory held a press conference in their home state of New York, but only about a dozen people showed up. At this conference, it was again suggested that it was "probably better to do nothing than to proceed with a scaled-down experiment."[65] In addition to each speaking separately to the *New York Times,* Harry Gove also arranged that *La Stampa,* a well-known newspaper in Italy, be contacted, which then published the open letter to the pope. Gove also wrote to the director of the British Museum that the Archbishop's scientific advisor "has little standing in the world scientific community," adding "he has a vested interest in the Shroud which renders him suspect. . . ." In desperation, Gove also called and wrote to both United States senators from New York, and even reluctantly called at home another member of the Pontifical Academy whom he knew, suggesting that he might intervene. With this, Gove finally terminated his efforts, conceding "It was another case of my clutching at any straw."[66]

Following Turin's indication that it was not going to increase the number of labs to seven, representatives of the three remaining laboratories met with Luigi Gonella—the Archbishop's science advisor—and Michael Tite—the coordinator of the testing from the Brittish Museum—in London in January 1988, a few months before the samples would be taken from the Shroud. According to Michael Tite, from a published letter to the editor in the journal *Nature*, the parties decided that:

a. Each laboratory will be provided with a sample from the Shroud, together with two known-age control samples.
b. The Shroud samples will be taken from a single site on the main body of the Shroud.
c. The weight of each cloth sample will be 40 mg, but will not be unraveled or shredded.
d. A *blind test procedure* will be adopted in that the three samples given to each laboratory will be labeled 1, 2, and 3 and *the laboratories will not be told which sample comes from the Shroud.*
e. The removal of the samples from the Shroud will be undertaken under the supervision of a qualified textile expert.
f. On completion of their measurements, *the laboratories will send their data* for the three samples *to the British Museum and the Institute of Meteorology "G. Colonnetti" in Turin for preliminary statistical analysis.*
g. The laboratories have agreed not to discuss their results with each other until after they have deposited their data for statistical analysis.
h. *A final discussion of the measurement data will be made at a subsequent meeting between representatives of the two above institutions and the three laboratories at which the identity of the three samples will be revealed.* The results will form a basis for both a scientific paper and *for communications to the public.* (italics added)[67]

Commenting on this new procedure, though still disappointed in the reduction to three laboratories, Harbottle communicated his concerns to archaeologist Paul Maloney around the time the sample was actually removed from the Shroud: "We have then, in my opinion, a shaky experiment, badly designed, innocent of peer-review, and having a reasonable chance of failing to produce a result convincing to everyone. . . . The experiment is to be performed by the same scientists who not three months earlier clearly and effectively objected to its terms."[68]

The Removal of the Sample and Its Dating

The "qualified textiles experts" who were present to supervise the removal of the sample from the Shroud may have been textile experts, but they apparently had little or no knowledge of the Shroud. Sadly, these men were seeing the Shroud for the very first time.[69] Reportedly, one of them pointed to the side wound and asked, "What's that large brown patch?"[70] Giovanni Riggi, the technician who actually cut the sample from the Shroud, argued with these two individuals for more than an hour over where to remove the sample from the Shroud.[71] One of these "experts" discussed the criteria for the sample cutting during a symposium held in Paris the following year. According to a translated summary of this presentation, "care was to be taken to avoid any risk of contamination, due to the 1534 mendings, to the lateral strip and to the partly burned sections."[72] Evidently, neither this care nor the argument served any productive purpose, for the sample was taken from an area that had serious problems, including the mending at the lateral strip and the fact that it came from a scorched area. Under the 1984 comprehensive proposal and an earlier protocol, the group of scientists possessing the most knowledge and experience of anyone concerning the Shroud, would have decided where the sample should be removed from the cloth. Moreover, this group had recommended removing several samples from several different locations on the Shroud.

To make matters more absurd, a piece of cloth approximately 285 mg was cut and removed from the Shroud. This is a full 75 mg more than the total amount required by all seven laboratories.[73] One of the reasons for the reduction from seven laboratories to three had been the concern for minimizing the damage to the cloth. One of the laboratories, Harwell, had been eliminated on this basis, and it had performed more radiocarbon experiments than all the other laboratories combined. So much material was removed that more than ten laboratories using the new methods could have carbon dated the Shroud. Giovanni Riggi kept almost half of this huge sample for himself! Subsequently, Riggi and one of the "qualified textile experts" would give several conflicting versions of the size and weight of the three samples given to the three laboratories and from which half they came.[74] Their final version was still at variance with the size and weights of the samples recorded at the carbon dating labs.

All the protocols emphasized the need for blind testing. The last protocol stated, "a blind test procedure will be adopted in that the three samples given to each laboratory will be labeled 1, 2, and 3 and the laboratories will not be told which sample comes from the shroud." Yet, incredibly, when the samples were given to the laboratories, they were accompanied by a certificate containing the dates of the samples.[75] The wording of the certificates began as follows:

> The containers labeled . . . 1, . . . 2, and . . .3 to be delivered to representatives of [named laboratory] contain one sample of cloth taken in our presence from the Shroud of Turin at 9:45 A.M., 21 April 1988, and two control samples from one or both the following cloths

supplied through the British Museum: *First-century cloth: eleventh-century* [cloth] (italics added).

[signed] Anastasio Ballestero
Michael Tite

In addition, Jacques Evin, a French C-14 expert who was present at the sample taking in 1988 and at the Turin workshop in 1986, published these dates in an article in *Shroud Spectrum International* in June 1988, so the labs clearly knew the dates of the control samples.

When asked in an interview at a Shroud symposium in Paris in 1989 as to why blind testing did not occur, Michael Tite stated, "We had decided it could not be a blind test because they'd been given whole pieces of the Shroud which they could immediately identify and therefore it could not be a blind test."[76] This is true. STURP's photographs at different magnifications and distances had been published for several years in numerous publications. In addition, the lineal density of the Shroud's warp and weft threads had been disclosed at the Turin workshop and was known to each of the laboratories. Furthermore, representatives from each of the three labs were present and observed the removal of the sample from the Shroud. Yet, while the Shroud sample was clearly recognizable, the labs still didn't need to be given the dates of the control samples, which could have served as an independent check on the labs' dating results. For example, none of the dates of the three samples in the intercomparison experiment had been given to the labs prior to their dating.

Even though the Shroud samples would be clearly recognized by the laboratories and the dates of the control samples were given, Michael Tite inexplicably took all three samples for each lab (the Shroud sample and the two control samples) into another room and packaged them in stainless steel containers. All of the other activities were being recorded on camera; however, there was no camera in this other room where Tite packaged the various samples. Around this time, unexpectedly, one of the textile experts provided a fourth sample, known ahead of time to be datable to the same period to which the Shroud was ultimately dated. This sample was placed in separate envelopes for each of the laboratories. The packaging of these samples also occurred off-camera in the other room. This sample, which came from the cope of St. Louis of Anjou, has a normal linen weave, "but the weave of its gold thread embroidery is in fact herringbone and on the linen apparently there is an impression from the gold that shows up as herringbone."[77]

Because this sample was not part of the protocol; was unexpected and unnecessary; was put into a separate envelope, in another room off-camera; had a similar description; and, unlike the other known controls, had a date similar to the Shroud's, a lengthy series of accusations arose that Dr. Tite had switched the Shroud samples with this fourth sample, causing the Shroud to date to medieval times. There is absolutely no evidence that any such thing occurred, and the accusation defies logic. Cardinal Ballestrero, as well as Luigi Gonella and Giovanni Riggi, were all in the off-camera side room at various times while Dr. Tite would supposedly have been making the switch. The laboratories would recognize the Shroud sample regardless of which container it was in. A switch would have required the cooperation of numerous people involved in the dating at all three laboratories. While the description of the fourth sample sounds similar to the Shroud, the laboratories were given threads from the back of the cope, which are not like those of the Shroud. This episode with the fourth sample is one more example of a very inconsistent and unorganized procedure that failed to follow any of the earlier scientific protocols.

The failure to follow scientific protocol, as well as earlier pledges, continued even after the directors of the laboratories brought the samples back to their institutions. The laboratories' own protocol, which the directors had devised in Trondheim, called for their coordinator to obtain written agreement not to reveal the carbon dating results to anyone other than designated officials of the British Museum, and it also stated that a press release would be issued only when Turin and the Vatican had received the results. The Turin protocol called for the results to be sent to three analyzing institutions, which would keep them in sealed envelopes until an agreed-upon date when they could be statistically analyzed. Only then would the laboratories meet with the analyzing institutions to discuss the statistical analysis with the objective of deciding the final result of the measurements. In the last procedure agreed upon by the three dating laboratories and the scientific advisor to the Archbishop, a similar arrangement between the laboratories and two analyzing institutions was also published. However, the ultimate actions of the scientists from the carbon dating labs would fall far short of following any of these protocols and procedures.

Leaks of the Shroud's carbon dating results first started appearing in the British press in July 1988, when *The Times* reported that David Sox implied the Shroud had been shown to be a fake.[78] Sox even had a book already printed but not released, containing the C-14 results weeks before they were announced.[79] Another British newspaper also reported in August of that year that the Shroud dated to the 1350s.[80] A month later the *Sunday Times* of London headlined an "official" medieval carbon dating result. About this same time, leaks also occurred in the wake of the *Timewatch* program, which was associated with part of Sox's research. The British Society for the Turin Shroud Newsletter stated that Sox, or an advance copy of his book, was at least indirectly responsible for the *Sunday Times* leaks and those following the *Timewatch* program.[81] We do know he was aware of the outcome of a bet that Harry Gove had with his companion, Shirley Brignall, over the results of the Shroud date, before the date was even announced. Sox's book relates that Gove bet on the Shroud being medieval and that he won the small wager.[82] Two days before Sox arrived in Zurich, Gove attended the dating at Arizona and learned the Shroud's date from their measurements, so Sox's ultimate source was either Gove or someone at the Arizona laboratory.

Furthermore, the laboratories sent the results only to the British Museum; the other analyzing institution received only the results of what the British Museum worked out.[83] Because of the numerous press leaks, no meeting between the analyzing institutions occurred, and no subsequent meeting with the carbon dating laboratories and the analyzing institutions ever occurred. While the laboratories' own protocol called for "detailed descriptions of procedure and how they calculated their mean values" and for "their uncertainties [to] be carefully recorded," the official report from these three laboratories did not contain any raw data or measurements whatsoever to indicate how they arrived at their dates. These were the same three laboratories whose directors had written to the Archbishop of Turin the year before, complaining of the reduction from seven to three laboratories and stated, "clearly, it is the reduction of the possibility of unrecognized non-statistical errors in measurements that leads to increased confidence in the final result."

A reflection on the accuracy of carbon dating, particularly by accelerator mass spectrometry, in the late 1980s can be acquired from a report that appeared in the *New Scientist* in 1989.[84] The article reported the results of a trial conducted by Britain's Science and Engineering Research Council (SERC), that compared the accuracy of thirty-eight carbon dating laboratories with artifacts of known age. "The margin of error with radiocarbon dating, an analytical method for finding out the age of ancient artifacts, may be two to three times as great as practitioners of the technique have claimed . . . . Of the thirty-eight, only seven produced results that the organizers of the trial considered to be satisfactory."[85]

One of the organizers of the trial, Murdoch Baxter, director of the Scottish Universities Research and Reactor Centre at East Kilbride, thinks that the reasons for the errors go beyond uncertainties in the accuracy of the counting of the pulses of radioactivity from the sample. "It is now clear, that other unaccounted for sources of error occur during the processing and analyzing of samples." According to the *New Scientist* report, Baxter "suspects that the most serious unforseen errors arise in the chemical pretreatment of samples." All three methods—gas proportional, liquid scintillation, and accelerator mass spectrometry—participated in the trial, with the AMS and liquid scintillation methods using the most chemical pre-treatment.

The report also contained the organizer's comments about the performance of the accelerator laboratories:

> Baxter says that accelerator mass spectrometry, used last year by a laboratory, at the University of Oxford to date the Turin Shroud, allegedly the burial shroud of Jesus Christ, came out of the survey badly. Five of the thirty-eight participating laboratories used this technique, for which samples weighing a few milligrams are acceptable. . . . Baxter says that some of the accelerator laboratories were way out when dating samples as little as 200 years old.[86]

Robert Stuckenrath, one of the earlier pioneers of carbon dating, many years ago pointed out the very real dilemma that exists when you have a problematic radiocarbon date. "The date of a sample whose provenance is in doubt is worse than useless—it is misleading."[87]

As we saw at the beginning of this chapter, carbon dating is hardly an exact or absolute test. There are innumerable ways an incorrect measurement of a sample's original C-14 content can occur, especially when the C-14 isotope was only one part to a trillion while the object was alive. In reality, while carbon dating is a useful tool in most instances, neither this nor any other single test should be the sole criterion for evaluation. Anna Hulbert, an Oxfordshire specialist in the conservation of medieval paintings, has noted, "Carbon dating, like X rays or any other analytical technique, should be regarded as one tool among many. It is chiefly useful in the dating of undisturbed archaeological material."[88] Moreover, as Middle Eastern archaeologist Eugenia Nitowski observed:

> In any form of inquiry or scientific discipline, it is the weight of evidence which must be considered conclusive. In archaeology, if there are ten lines of evidence, carbon dating being one of them, and it conflicts with the other nine, there is little hesitation to throw out the carbon date as inaccurate due to unforseen contamination.[89]

In light of the fact that there are so many substantive scientific and procedural challenges to the dating of the Shroud; that the events and images depicted, as well as the cloth's contents, could relate to a critical part of history that would be relevant to all the world; and that the carbon dating results are contrary to a comprehensive amount of medical, scientific, archaeological, and historical evidence, the owners of the Shroud have every right to and are under the highest obligation to the world to demand that new and proper carbon dating be performed on the Shroud of Turin.

FUTURE RESEARCH PRIOR TO REDATING THE SHROUD

Before another carbon dating of the Shroud occurs, preliminary research on the cloth itself should take place concerning the various substantive scientific challenges to and refutations of the dating of this linen. In the last year or two, a superb form of spectral imaging has been developed and proposed to be applied to the Shroud of Turin.[90] This technique has the ability to obtain information on wavelength intensity from each type of light, at each point in space, and from every point in an optical image—simultaneously.

When the human eye sees an object, it perceives only the colors of the visible spectrum from red to violet. Below red lies the infrared part of the spectrum. Every chemical compound has a unique infrared spectrum, and this characteristic is routinely used to determine the chemical composition of unknown materials. Previously, one of two procedures was used to obtain spectral information from an object: either obtaining a complete spectrum of only one point on the object and repeating for all points of interest, or obtaining a complete image of the object at only one wavelength in the spectrum and repeating for all spectral wavelengths of interest. However, with this new technique, an object can now be viewed under the entire visible and infrared light spectra simultaneously, allowing its corresponding composite image to become visible from all spectra simultaneously.[91] Items such as the vessels in a retina, the hemoglobin bands in the eye, and the spectral patterns of human chromosomes are routinely viewed and examined with this new spectral-imaging technique.

This new technology could scan the entire Shroud in only six hours, thereby allowing scientists to spend years analyzing all of its data. It could map the entire cloth and identify not just every fiber of every thread, but what is on every fiber. It can even identify tissue at the molecular level. By using carefully calibrated light sources, the pattern of reflected light can identify individual chemical compounds on the Shroud.

In particular, this technique may allow us to view whether any chemical bonds in the cellulose had been broken by protons, neutrons, or newly created C-14 and whether the chemicals had reattached themselves to molecules throughout the cloth. These broken bonds and chemical reattachments could confirm or corroborate the scientific challenges to the Shroud's carbon dating posed by the various experiments of Dr. Rinaudo, Mario Moroni, and their associates. Correspondingly, their absence could refute these challenges.

Similarly, this technique could allow us to view whether substances such as wax or starch had attached themselves to the molecular structure of the cellulose. As we saw, both contaminants are known to be on the Shroud with starch present, yet invisible, immediately next to the radiocarbon site. Wax was found at another location similar to the radiocarbon site. The application of this technique on remaining samples at each of the carbon dating laboratories would not necessarily tell us whether these contaminants were present on the samples that were dated and destroyed, but their presence would strongly indicate and confirm that these substances did chemically bind to the molecular structure at the radiocarbon site. The application of this nondestructive technique on the rest of the cloth could also tell us what other locations may have such foreign substances physically or chemically attached to the linen and, thus, what areas to avoid. This technique should also be applied to each sample that has or will be removed from the Shroud, so that all foreign contaminants or carbon of any kind can be identified and, if possible, removed.

Moreover, this spectral imaging technology should be applied to the areas over the eyes of the man in the Shroud. This could help reveal whether the appearance of the letters, lituus, and motif of a Pontius Pilate coin minted between A.D. 29 and A.D. 32 is due to encoding on the cellulose fibers of the linen thread. Confirmation of such a coin over the eye(s) of the man in the Shroud would far more specifically indicate the age of the burial cloth and when the images and the events depicted thereon occurred than would any carbon date.

By the same token, spectral imaging would be useful in confirming or determining whether plant images are on the Shroud. As noted in a previous chapter, numerous flower images matching pollens from the same species or genus have been identified primarily in the off-image regions of the Shroud. Furthermore, pollen and images from thorns have also been identified on the Shroud. If spectral imaging, or other scientific examination, can confirm the presence of such plant images and/or pollen, it would additionally confirm the location and time in which these events occurred—Jerusalem in the spring. Such findings, along with much other evidence, would be consistent with and help corroborate the time of the events depicted on the Shroud and the age of the burial cloth. Since this is a new area of investigation developed since the Shroud was last examined in 1978, and since these images are subtle, examination by this new technique of spectral imaging could be particularly useful for this subject.

Moreover, if the Shroud was irradiated with a flux of neutrons, this would have other measurable consequences: radioactive or unstable isotopes would have been formed.[92] More than two decades ago, STURP scientists discovered that calcium (along with strontium and iron) was distributed uniformly throughout the Shroud, probably as a result of the retting process when the cloth was originally manufactured.[93] Almost 97 percent of all calcium consists of calcium-40 (Ca-40); the other 3.1 percent consists of Ca-42, 43, 44, 46, and 48. Conspicuously absent is Ca-41, which does *not* occur naturally. However, if a neutron flux had irradiated the Shroud, it would convert some of the Ca-40 in the cloth to Ca-41. If Ca-41 were found on the Shroud, it would confirm that the cloth had been irradiated with neutrons.[94]

Since calcium has been found distributed uniformly over the Shroud, any portion of the original cloth could be examined for the presence of Ca-41. Similarly, limestone matching the same rock shelf that Jesus was buried in has also been found on the Shroud. Limestone consists mainly of calcium carbonate. If this limestone resulted from the man's burial, similar limestone particles might still be found trapped between the cloth's threads, especially on the back of the dorsal side, which would have been in contact with the limestone floor of the tomb. STURP was not allowed to remove the backing cloth in 1978; the back side of the Shroud has never been examined scientifically. In 1998, the Holland backing cloth that was sewn onto the Shroud by the Poor Clare nuns in 1534 was removed and replaced with another backing cloth, which rests against the now straight and horizontal Shroud in an atmospherically controlled container. The Shroud's entire back side and both backing cloths, especially the one from 1534, should be examined for microscopic particles of limestone. These limestone samples could be examined to see if they, too, match the same rock shelf in which Jesus was buried. Furthermore, these particles could be examined for Ca-41 because, if the Shroud was irradiated with a neutron flux, some of the Ca-40 within the limestone's calcium carbonate could have converted to this radioactive isotope. (If access could be acquired, and if enough remnants still exist, the ceiling, walls, and floor of both tombs reputed to be Jesus' could also be investigated for the presence of Ca-41.) An investigation for and of limestone particles could help confirm the location of the burial and its victim and whether a neutron flux occurred at this location after burial.

In addition, when STURP scientists made X-ray fluorescence measurements on thirteen threads that had been removed from the Raes sample, they detected small traces of chlorine.[95] If a neutron flux irradiated the Shroud, it would convert chlorine-35 (Cl-35), found naturally, to chlorine-36 (Cl-36). Like Ca-41, Cl-36 does *not* occur naturally. As stated by Thomas Phillips in the scientific journal *Nature,* "The presence of either [Ca-41 or Cl-36] would confirm that the Shroud had been irradiated with neutrons."[96]

Determining whether Ca-41 and/or Cl-36 are present on the Shroud would involve the destruction of samples from the cloth. Fortunately, there is an abundance of samples on the Shroud that could be removed for this purpose without disfiguring or damaging the cloth in any way. At eight different places over the entire length of the Shroud, there are basically two sets of patches covering various burn holes from the fire of 1532. Shroud cloth can be found behind each of these sixteen locations. Excluding charred material, from behind just some of these patches can easily be found more than enough material for these analyses. So much cloth lies behind these patches that the Shroud could even be carbon dated by the conventional method with this material.[97] In fact, chemists Larry Schwalbe and Ray Rogers of STURP stated that:

> Large amounts of material could be removed from beneath the patches. Mottern et al. presented a radiograph of one of the patched-hole areas that reveals some of the Shroud material that could be recovered; a total Shroud area of approximately 400 cm^2 lies concealed beneath all of the patch work. This is sufficient for literally hundreds of carbon-14 tests with the latest techniques. Any of this material could be removed without affecting the visual appearance of the Shroud or damaging the fabric structure. (Indeed, as a conservation measure, this charred material should be removed to prevent further irreversible soil-contamination of the cloth.)[98]

Furthermore, as we saw earlier, scientists concluded that the blood marks were on the Shroud first, and shielded the underlying cloth from the body image encoding event. If this event involved a neutron flux, it would also have affected the blood chemically. Iron, abundant in blood, will undergo nuclear reactions with neutrons. A likely product is chromium-53 (Cr-53), which is not normally found in blood. Cr-53 found in blood samples from the Shroud would also confirm that the cloth was irradiated with neutrons.

Moreover, if the Shroud was irradiated with neutrons, it could have affected the blood in another significant way. The solid part of dried blood contain mostly proteins, which typically contains about 12 percent nitrogen by weight. This is a much larger amount of nitrogen than is found in cloth. If a neutron flux irradiated the blood on the cloth, it could convert the nitrogen-14 (N-14) into C-14 on a much larger scale than it would convert in cloth. As such, the blood would carbon date to a much younger date than the cloth. In fact, it could easily date well into the future. Such a date alone would refute the 1988 radiocarbon dating of the Shroud. (Any date appreciably younger than 1350 would seriously discredit the 1988 dating since the Shroud with its body and blood images has been known in Europe since then.)

The blood from the Shroud of Turin should be examined for Cr-53 and should also be carbon dated. Performed in that order, these tests could determine if the Shroud was irradiated with neutrons and if that affected the 1988 carbon dating of the cloth. These tests could not only explain the effect, they could completely refute the earlier radiocarbon dating of the Shroud. Both tests could be performed on the same sample with a mass spectrometer (see also appendix H).

Fortunately, an adequate amount of blood for both purposes is also easily available and, unlike the other blood on the Shroud, (a) is off of the body image, and (b) provides no other useful information. This blood can be found off the man's anatomical right foot on the dorsal side of the Shroud. (The right foot on the photographic negative image—the left foot on the image as seen with the naked eye. Normally, blood is not the best candidate for carbon dating, because it does not contain a great deal of C-14. However, if the C-14 is enhanced by the large amount of N-14 found in blood, there would easily be enough C-14 present. The approval to carbon date blood from this location could even be given after the above Ca-41 and/or Cl-36 tests are completed to confirm whether the Shroud was irradiated with neutrons.) Naturally, the Shroud's owners should be reluctant to destroy any blood on it, especially in light of the possibility that it was shed by Jesus Christ. Yet, there is an enormous amount of blood on the cloth, literally from head to feet on both the front and back of the man.[99]

These tests could have much greater significance for the world. From only the tests to measure the ratios of Ca-40 to Ca-41, Cl-35 to Cl-36, and Fe-56 to Cr-53, we could calculate the original age of the Shroud! From these ratios we can determine the average amount of neutron flux required to produce three different, independent amounts in both the Shroud linen and blood, which were removed from various parts of the cloth. (If the Ca-40 to Ca-41 ratio can also be determined from the limestone particles, it could provide an additional independent measure of the amount of neutron flux that occurred after the body was placed in the burial cloth inside the tomb.) From the amount of neutron flux, scientists can determine the amounts of newly created C-14 from the known "cross-sections" or conversion rates of N-14 and C-13. This amount of newly created C-14 can then be subtracted from the C-14 found in the C-14 to C-12 ratio in the above samples to arrive at their true original C-14 to C-12 ratio.[100]

The equipment that would most likely be used to measure the Ca-40 to Ca-41, Cl-35 to Cl-36, and Fe-56 to Cr-53 ratios in the cloth, blood, and limestone particles would be an accelerator mass spectrometer or a thermal mass spectrometer. The primary focus of those using accelerator mass spectrometers has been on measuring the C-14, C-13, and C-12 isotopic amounts and ratios, at least as compared to measuring the above calcium, chlorine, and iron-to-chromium ratios. If the operators of this equipment cannot measure these calcium, chlorine, and iron-to-chromium ratios as accurately as they can the carbon ratios, then they should develop and refine such existing technology. However, it should certainly be refined and confirmed by testing that is far more detailed, certain, comprehensive, and proven to be repeatedly accurate on similar samples of cloth, blood, and limestone than was the Trondheim intercomparison experiment and report for cloth samples. Failure to develop, refine, and employ this ability, with the same type of equipment used for carbon dating, could perpetuate—and fail to correct—the most infamous error of all on the most important object ever carbon dated.

Cloth samples from behind the patches, blood from the off-image area of the right foot, and limestone particles are all that are necessary to be removed from the Shroud to determine if a neutron flux affected an accurate carbon dating of the cloth. This would allow the most minimal and unobtrusive amount of material to be taken from the Shroud that contains vital and dynamic information, and that serves no other useful purpose. Since these samples are critical in establishing whether a neutron flux irradiated the Shroud and the amount of flux, it would be the height of negligence not to also carbon date them prior to their destruction, especially since that would also allow for the amount of newly created C-14 to be independently calculated, along with the original remaining C-14, and the true age of the Shroud to be determined.

A piece of the thread that was sewn along the entire length of the Shroud, possibly to give it support, should also be removed and carbon dated to help determine when this thread was sewn onto the cloth. The sewing thread, as opposed to the cloth, would not likely have been part of the original Shroud, and would not have been present if the cloth was irradiated with neutrons.

One could argue that the C-14 to C-12 ratios obtained from the samples behind the patches might be suspect. While this is one *additional* reason carbon dating the blood sample off the right foot is critical, there is plenty of material behind the patches that is not charred that can be used. STURP thought these samples were excellent choices. The carbon dating part of their 1984 proposal recommended taking "200 mg from each of the burned areas under and around the patches" (correctly excluding a dorsal shoulder fold mark intersection) as well as from other locations. The carbon dating laboratories even looked into this prior to the 1988 dating and had concluded that this material would provide an accurate date.[101] (They only avoided this area to be on the safe side, but it would have been a much better choice than the area from which the 1988 sample was taken.)

Remember also that other portions of the Shroud are not going to be so unobtrusive as these samples, and, if the Shroud was irradiated with neutrons, would also contain additional C-14, thus presenting a misleading C-14 to C-12 ratio, as well. It should also be remembered that many other excavated samples necessarily from or near fires have repeatedly been dated successfully. (Frequently, the reason archaeological remains were originally destroyed, as well as preserved, was due to fire.) Moreover, the fire would not have affected the calcium, chlorine, and iron-to-chromium ratios of these samples. These ratios must be taken regardless. Since taking these ratios would necessarily destroy these samples, their C-14 to C-12 ratios should also be measured at the same time. If it is determined that these samples were affected by the fire, much smaller samples than these could be removed to measure their C-14 to C-12 ratios.[102]

A flux of neutrons onto the Shroud at any time would have *necessarily* been an unprecedented event. Since all of the other evidence indicates that the events depicted occurred to the historical Jesus Christ, we need to consider whether any unique events also occurred to him after his death and burial. Historical sources that we shall examine further in chapter 11 indicate that a very unique event also occurred to him after he was buried in a tomb in the first century. If the recommended research establishes that the Shroud was also irradiated with a flux of neutrons, even if scientists can't specifically measure and calculate the exact century prior to medieval times in which it occurred, its uniqueness would still strongly indicate that it, too, occurred in the first century.

This recommended research of the Shroud must take place to corroborate or refute its 1988 carbon dating. This dating has been challenged substantively in a number of ways scientifically, and has also been considered one of the most controversial, as well as amateurish, tests of all on the Shroud. The recommended research could not only test many of these substantive challenges, identify new locations for future tests, and provide a great deal of information about the cloth, it could correct one of the greatest errors and misconceptions in all of science on a subject of monumental importance to the entire world.

CHAPTER NINE

THE SCANDAL EXPOSED

The mishandling of the 1988 radiocarbon dating of the Shroud was tragic. Because of those results, millions of people believe there is scientific proof the Shroud is a fraud. We've seen this simply isn't true. But how could such a thing have happened? How could those who inherited the legacy of the diligent 1978 STURP investigation have fumbled so badly?

During the 1980s, STURP members continued their research on the data and samples taken from the Shroud, published their findings, and prepared a twenty-six-point proposal for a new round of scientific examination of the Shroud of Turin. This comprehensive examination could have built upon and expanded their earlier 1978 scientific examination of the cloth and provided valuable evidence relating to the age, origin, and history of the Shroud. It also included scientifically sound protocols for its radiocarbon dating. However, during this same period, the carbon dating laboratories, especially through the actions of their reputed leader, would not only fail to undertake preparations commensurate to the importance of their job in dating the Shroud, but also would prevent the advancement of scientific knowledge and humankind's understanding on a critical subject, thereby violating the true spirit of science.

As author David Sox, who acquired intimate contacts with representatives of the carbon dating laboratories, noted over a decade ago, "Harry Gove was the natural leader and spokesman for the six laboratories. No one attached to the six had been longer involved with the Shroud or knew more about the intricate politics concerning the relic. And none was more forthright in his opinions."[1] In 1989, Edward Hall, head of radiocarbon dating operations at Oxford, confirmed that during the process leading up to the 1988 dating of the Shroud, Gove had assumed leadership of the laboratories.[2] In 1996, Gove published his own book, admitting to acting as spokesman for the six labs and the coordinating institution, the British Museum, and described in detail his involvement, and much of theirs, in the lengthy process covering this period of about a decade.[3] Gove's book makes it absolutely clear that he assumed leadership on every possible issue that he could, and that he was a major influence not just on the laboratories but also on the final decisions that led to such controversial results. Moreover, his book revealed that Gove was a major influence in the tragic outcome that excluded STURP from any involvement in the carbon dating or the twenty-five other areas of comprehensive scientific testing of the Shroud.

Gove's book also revealed his deep-seated animosity toward STURP, the full extent of which would be revealed only by his publication seventeen years after joining STURP's effort, and eight years after the radiocarbon dating of the Shroud had taken place. Gove's true feelings and motives toward STURP and its involvement in the scientific testing of the Shroud are clearly revealed on the first page, on which he mentions them, in the very first chapter of his book. "I was determined to prevent their involvement in its [the Shroud's] carbon dating, if that were ever to come about. . . . Fortunately in this I was successful."[4] On the next page, he states, "I am happy to say that, in the end, they [STURP] played no role in its carbon dating."[5] One of the sad ironies is that STURP's participation could have prevented many amateurish aspects of the procedure and precluded the questionable sampling. While discussing the early period of 1978 to 1979, Gove further informs us of his actual "disdain for those [STURP] scientists."[6] In his 1996 book, he shows that his hypocritical and concealed attitude existed before he even made any pledges to or agreements with STURP. Immediately before he called STURP in 1979, accepting that its C-14 committee members could be present during the removal of the sample from the Shroud and during the sample's preparation and measurement, he blatantly reveals his duplicity: " . . . they [STURP] had good connections in Turin, and could be useful in obtaining a shroud sample for dating—if only they could be prevented from playing any other role."[7]

Gove would maintain the full extent of this attitude toward STURP throughout the entire process while continuing to look for an opportunity to eliminate them from any further Shroud testing. He clearly admitted in 1985, "I felt, naively as it turned out, that I now had a golden opportunity to eliminate STURP once and for all."[8] Although that particular time was premature, it clearly indicates that his constant goal was to eliminate them. Gove understood from the start that STURP had always had great success in its dealings with Luigi Gonella, the scientific advisor to the Archbishop of Turin, Cardinal Ballestrero, and with the office of the Archbishop. By no coincidence, Gove had a number of conflicts with and criticisms of Gonella. Toward the end of the carbon dating process in 1987, when one of the directors of the carbon dating labs stated to Gove that he would avoid supporting Gove again in any further conflicts with Gonella, Gove clearly admitted his real motivations. "Let me hasten to assure you that my 'hectoring' as you call it is directed toward STURP and only peripherally toward Professor Gonella to the extent he champions STURP's cause."[9]

Throughout his involvement, Gove lobbied behind the scenes to eliminate STURP from any role in the further scientific investigation of the Shroud. The principal method Gove employed was through his influence with Prof. Carlos Chagas, the president of the Pontifical Academy of Science, and his colleague Dr. Vittorio Canuto. According to Gove's book, these men followed almost every suggestion Gove made. However, rarely, if ever, do any of Gove's suggestions deal with substantive scientific matters; Gove focused on maneuvering and procedure.

A couple of examples of this can be seen in 1986, when there appeared to be two opportunities to finally acquire approval to radiocarbon date the Shroud and perform further scientific tests on it. In February 1986, Gove learned and revealed that Luigi Gonella had informed the president of STURP and Dr. Harbottle that two weeks after a Turin workshop scheduled for June 9 through 11 of that year, STURP could perform its testing in the twenty-six areas of inquiry and that samples would also be removed then for radiocarbon dating. Only eight days earlier, Gove and Gonella, along with another person, had met in New York. At that meeting Gove himself made it clear that Gonella favored STURP performing its tests for a number of reasons, the principal ones being that conservation of the image was one of the main focuses of their tests and that new techniques for testing had also become available since 1978. He stated the STURP proposal was the only one being considered in Turin and that their carbon dating proposal was identical to the Trondheim Protocol submitted by Gove and the carbon dating laboratories. Gonella and STURP had also agreed to the British Museum serving as the coordinating institution under the protocol. Gonella wanted the scientific workshop held at the Turin Polytechnic, where he held a faculty position and where a meeting of scientists from different countries would attract less media attention than if it were held in a building owned by the church. Lastly, Gonella emphasized that the Shroud was Turin's responsibility and that they believed STURP's textile experts would be the best choice for removing the sample.[10]

STURP not only had spent 120 hours examining and photographing every inch of the cloth, it also had retained textile experts to assist in further sampling and testing. In addition to investigating the cloth under a variety of scientific tests and photographs, STURP had extensively examined and tested samples from the image, non-image, burn marks, water stains, scorch areas, side-strip, and their various margins, acquiring an extensive knowledge of the complexities of the cloth. If Turin had concluded that STURP and its textile experts were the most knowledgeable and appropriate experts to select the sample, they would be correct. The British Museum and many others would be present during the removal of the sample, and, of course, the laboratories' scientists would be solely responsible for preparation and measurement after they received the sample from the British Museum.

Gonella even met with Ambassador Celli, the Vatican ambassador to the United Nations, while in New York, and Canuto reported to Gove that the ambassador thought "Gonella was really a fine fellow and he seemed convinced by everything that Gonella said."[11] Yet Gove's reaction to Gonella's meeting and the important news that the radiocarbon dating and other tests would be performed on the Shroud indicates that his concern was on trivial procedural matters and not on any substantive matters of science. The next day following his meeting with Gonella, Gove called Canuto. "What really *upset* him [Canuto], as it did me, was Gonella's *suggestion* that the *Turin* workshop should be held at the Turin Polytechnic. . . . it would give the advantage to STURP" (italics added).[12] These two were exceedingly concerned that the meeting should be held in Rome. The following day Gove made this same complaint to Chagas, who then set in motion a series of contacts and telexes over this matter of where the meeting should be held.

Consistent with the above quotation in the conversation between Canuto and Gove, that merely holding the meeting at the Turin Polytechnic "would give the advantage to STURP," Gove spoke with Michael Tite of the British Museum three days after calling Chagas. In part of this conversation Gove related to Tite "that Gonella had suggested that a STURP textile expert would be best and *that appalled me*" (italics added).[13] In this same conversation, Gove and Tite discussed using a textile expert named Mechthild Flury-Lemberg, whom neither appear to have ever met, instead of STURP.

Gove's reaction in March of 1986, after learning of the approval of the scientific testing and dating of the Shroud, is even more illustrative. He again informed Canuto of his opposition to the sample taking and, apparently, complained again about the STURP testing scheduled for two weeks following the workshop.[14] When Gove learned from Canuto through Chagas that the meeting would be held at the Turin Polytechnic, Gove indicated he "was desperately disappointed in Chagas's lack of leadership."[15] He then made an incredible statement: "If the Pontifical Academy of Sciences had to knuckle under to 'Princes of the Church' in matters of science, then I really must reconsider the involvement of our group."[16] The "group" comprised the carbon dating laboratories.

He actually considered withdrawing the laboratories when the seven-year project stood on the verge of achieving its scientific objective—dating the Shroud—because of the location of a meeting. He also refers to the location of a meeting as a "matter of *science*" (italics added). Gove continued, "I had found that dealing with *prelates*, no matter how elevated, was a *hopeless enterprise*. I was not opposed to the workshop being held in Turin. However, it must not be held in Gonella's home territory, it must be held in a place controlled by the Turin Archdiocese" (italics added).[17] A prelate is a high-ranking church dignitary, but there would certainly be far more prelates at the Turin Archdiocese than at the Polytechnic Institute. At the end of the quote, Gove revealed his real reason for being upset and disappointed in Chagas's leadership and for threatening to withdraw the laboratories, when he stated, "As far as I was concerned, Gonella's location was a STURP location."[18]

The entire focus was changed from important scientific matters of testing and dating the Shroud to insignificant and petty matters such as the location and sponsorship of a meeting. The extensive scientific examination that had been planned for the Shroud did not occur. Another two years passed before the radiocarbon sample was even removed.

These are some of the reasons that the process took nine years and produced a poorly planned, poorly executed, very limited, and quite controversial scientific testing of the Shroud in 1988. During the process, Gove consistently insisted that the Pontifical Academy be involved, for as the examples above indicate, as long as it was involved, Gove could easily influence them and negated plans for STURP to participate in any further testing of the Shroud.

Until Gove's book was published in 1996, STURP scientists were at a complete loss to explain their elimination from the testing and dating of the Shroud or to explain fully the controversy surrounding the removal and dating of the Shroud sample. Gove's book supplied many answers. From the beginning of STURP's carbon dating efforts in 1979 until the publication of Gove's book in 1996, STURP scientists were unaware of Gove's deep-seated animosity toward them and of his efforts behind the scenes to eliminate them.

One effort to disassociate STURP from the carbon dating process was to have a textile expert remove the sample from the Shroud instead of STURP. In this vein, Gove's book showed that he engaged in many conversations and efforts to have a woman from Belgium, or so-called textile experts from Italy or Brazil, make one of the most important decisions of all regarding the carbon dating of the Shroud—the selection of the locations from where the samples should be taken. Gove, Tite, and others seem not to care about the qualifications of the so-called textile experts, so long as they could replace STURP. Only a few months before the scheduled June workshop, Gove stated that, "Tite would probably suggest the names of people that might be appropriate. Chagas might know of some textile experts in Italy or Brazil. . . . "[19]

Their first choice was Madam Flury-Lemburg of Belgium, whom neither Tite nor Gove had met at the time. Her status as a textile expert was merely one criterion that should have been considered. The most important criterion for this task would have been her knowledge and understanding of the Shroud itself. Unfortunately, she had never seen the Shroud, and she indicated her complete lack of understanding of the complicating factors and history associated with the Shroud when she stated at the workshop that the Shroud "is the same from one end to the other. There is no need to take samples from various places. One could take strips from the edges of the main cloth from any place and it would be the same."[20] This critical mistake of not taking samples from more than one location was made in the presence of the two textile experts who were appointed to the job and who also had apparently never seen the Shroud. A sample was taken from the edge of the main cloth as suggested by Flury-Lemburg, but this resulted in removing a single sample from the most controversial location on the cloth.

One of the other ways to disassociate STURP from any role in carbon dating the Shroud was to prevent its members from attending the workshop. Gove admitted, "I said that unless STURP could show that it had some role to play, I personally could not justify their presence at the workshop."[21] This statement was made in spite of the fact that Gove and the carbon dating laboratories had joined STURP's carbon dating effort and had agreed that STURP could remove the sample (and be present during the lab's preparation and measurements), and that the lab's own protocol, which Gove himself had submitted, called for STURP to arrange for the samples to be removed from the Shroud.

In April of 1986, Gove wrote to Professor Chagas and "urged him again to send all of us official invitations on behalf of the Pontifical Academy of Sciences to attend and participate in the workshop as soon as possible."[22] He enclosed a list of people who he thought should be invited; however, the only names on the list were from the carbon dating community and Dr. Canuto of the Pontifical Academy of Sciences. This and other instances had caused Cardinal Archbishop Ballestreros, the custodian of the Shroud, to appeal to the Vatican secretary of state and to the Pope that the Pontifical Academy was taking matters away from him concerning the Shroud.[23]

Gonella contacted Chagas asking him to also invite Robert Dinegar, the STURP C-14 coordinating investigator; Jacques Evin, a carbon dating specialist from France; and William Meacham, an archaeologist and an expert on the Shroud of Turin. Gove volunteered to Chagas that he did not understand why the latter two men should be invited.[24] Gonella also suggested that STURP chemist Dr. Alan Adler and James Druzik, conservator with the J.P. Getty Museum, be invited to attend the meeting. Chagas volunteered to Gove that he did not think he was going to invite either of those two.[25] Chagas eventually agreed to Gonella's requests and invited these men, all of whom except Druzik attended the workshop. Meacham and Adler, whose lengthy and combined expertise of the Shroud included practically every aspect of its numerous features, not only attended but also provided valuable advice, yet this advice was ignored by those who conducted the carbon dating procedure and sampling from the Shroud. As an archaeologist with a great deal of experience in having samples retrieved on site and carbon dated, as well as having a wide range of knowledge on all facets of the Shroud, Meacham was an especially qualified participant for the workshop. His advice was not only ignored; it was ridiculed by Gove.

At the workshop, Dr. Adler repeated STURP's position that samples from various parts of the Shroud should be removed for carbon dating purposes. He discussed STURP's proposal for overall tests, emphasizing its concerns for the conservation of the image and the cloth, and repeated that, in addition to the STURP scientists, four to five textile conservation specialists would also be present (as well as many observers). As another example of the unscientific concern for the Shroud by Gove, his own book shows that when an inquiry was made to Adler as to when STURP could take on the job, Gove stated, "This comment appalled me but I remained silent. As far as I was concerned, STURP would never take on the job."[26]

In light of the Shroud's known and unknown contaminants and history, Meacham warned the lab directors at the Turin workshop, and in an article published the same year, "[t]o measure Shroud samples, one must consider every possible type of contamination and attempt to identify and counter them all, before the measurement is made and a 'radiocarbon age' assigned."[27] Meacham further warned that "an elaborate pretreatment and screening program should be conducted before the samples are measured."[28] If the laboratories had followed this advice, perhaps they could have determined whether wax or starch had chemically bound to the cellulose in the sample taken from the Shroud for dating. Chemical analysis by combustion of the linen before and after the cleaning treatment could have revealed if any wax had become chemically bound to their samples. Enzymes could have detected the presence of starch. Meacham's article, as well as previous articles, had even listed wax and starch among the known contaminants discovered on the Shroud.

Meacham further warned in the same article that, "unless there are specific conditions which warrant specialized pretreatment, most laboratories process samples with acid and alkali washes. While this standard pretreatment is usually effective in removing modern contaminants, it may not do so for intrusive materials deposited much earlier."[29]

We have seen how a radiation event could have infused C-14 into the cloth's porous cellulosic structure. If the carbon dating laboratories had measured the amout of nitrogen present in their samples, they might have helped provide a basis for determining the amount of neutron flux, if any, that irradicated the Shroud. This could have ultimately aided in determining the cloth's additional and original C-14, and its actual age, if the cloth had been irradiated in this manner. Unfortunately, the labs failed to prepare for the possibility of a neutron flux, and if they did measure the amount of nitrogen in the cloth, they did not disclose it in their report.

This is ironic, because in the journal *Nature* and in interviews published shortly after the Shroud's date was announced, Dr. Robert Hedges, Dr. Edward Hall, and Michael Tite all acknowledged that if the cloth had been exposed to a neutron flux it could have altered the radiocarbon date that was assigned to the Shroud.[30] Hedges even admits these processes were "considered by the participating laboratories,"[31] but were evidently rejected. When these carbon dating scientists have acknowledged the above possibility, they have also skeptically stated that it would have been quite a coincidence that the amount of neutron flux would have been so exact as to give the Shroud a date that approximates the time it was allegedly painted.[32] Yet these statements only show a further lack of understanding about the Shroud. If a neutron flux irradiated a cloth, the amount of carbon remaining on its samples would only partially result from the amount of radiation they received; it would primarily depend upon the extent that the carbon dating labs pretreated the samples and removed the additional C-14 from them.

Incidentally, neutrons hitting N-14 or C-13 and creating new C-14 in the Shroud would not have given the cloth a dirty or contaminated appearance. (It would have made an insignificant contribution to the yellowing of the Shroud's linen fibers wherever it occured; however, such yellowing has occurred throughout the cloth.) Yet this would have greatly contaminated the Shroud with additional C-14, significant amounts of which would have survived the standard pretreatment cleaning techniques that the labs applied to it.

William Meacham recently acknowledged that at the Turin workshop the directors of the carbon dating labs

> ridiculed the notion that contamination could account for more than 1 or 2 percent of the C-14 after the standard pretreatment. Their stance was decidedly haughty then, and now shown to be dead wrong. The truth is that there are many possible sources of error which are not fully understood, and it simply behooves us to at least look for all the possibilities that we can.[33]

Similar lack of preparation and diligence was exhibited by the same group of carbon-dating laboratories and their coordinating institution during the intercomparison experiment. This was the experiment discussed earlier that the carbon-dating laboratories and the British Museum conducted in order to validate new C-14 dating methods on cloth—and failed.

Meacham also advised that samples should be taken from as many zones on the Shroud as possible, as an archaeologist attempts to do at a field site. He argued that taking samples from various areas would give greater credibility to its carbon dating. Instead of conceding even this basic point, Gove stated that these comments "seemed remarkably inappropriate as applied to the Shroud."[34] Meacham's suggestion was, in fact, remarkably appropriate to the Shroud, especially in hindsight after what actually happened during the Shroud's sampling. Even an uneducated layperson would have to admit that the taking of only one sample, especially from such a controversial location, was unfortunate. Gove's remark shows a failure to analyze the subject from any kind of scientific perspective and indicates a complete lack of objectivity.

Yet the real reason for Gove's inaccurate and unfavorable remarks about Meacham may be found from what occurred later during the workshop. After Gove and Chagas openly stated at the workshop that STURP's larger scientific tests should not even be considered until the radiocarbon testing was completed, Meacham did not agree. He drew the analogy that an archaeologist would not abandon all of the important activities associated with an excavation or archaeological dig because of the results of one carbon date. He argued that all of the tests concerning the Shroud's authenticity, the cause of the image, its conservation, age, and origin, would have relevance regardless of what the carbon dating results were. Meacham said they should go forward. Instead of addressing these comments on their merits, Gove then stated that Meacham "should never have been allowed to play a role in the workshop."[35]

From 1986 to 1987, Gove lobbied against STURP harder than ever. During this time, the carbon dating laboratories' directors wrote letters to Chagas, and Chagas met personally with the Pope, echoing their complaints about STURP's planned tests. In the summer of 1987, Gove wrote a remarkable letter to Chagas, which contained a number of untrue and unsubstantiated comments that, of course, concerned STURP and even those in Turin. Gove himself stated:

> In the letter, I sharply criticized the way in which the Turin authorities had handled matters concerning the Shroud in the past and the way they were continuing to mishandle them. I noted that the Shroud had been subjected to a number of scientific tests of dubious value carried out in ill-conceived ways by scientists of unknown reputation. . . .
>
> I stated that almost every aspect of the STURP organization was distasteful. . . . This included their clear religious zeal, their questionable sources of support, their military mind set. . . .
>
> Now, however, the Pontifical Academy of Sciences, under Chagas, had a chance to change all this.[36]

Gove's incredible letter continued; before he finished, he would even compare STURP to the Spanish Inquisition.[37] This time he not only threatened that the carbon dating laboratories would withdraw without Chagas's continued support, he *guaranteed* their withdrawal "if STURP participates in the carbon dating enterprise in any way."[38]

In October 1987, the Archbishop of Turin officially announced that the Shroud would be carbon dated by three laboratories. Word had leaked in the summer that such a decision was going to be announced. The Archbishop's letter stated that he received instructions in late May from the Holy See on how to proceed with the radiocarbon dating of the Shroud, but that it took more time to work out a number of logistics concerning various aspects of the procedure. The Archbishop's letter also stated:

> The decisions took more time to be worked out than originally wished, owing to the situation without precedents created by a number of competing offers tied to a rather rigid proposal, and also *by (the) initiative of some participants in the workshop who stepped out of the radiocarbon field to oppose research proposals in other fields, with implications on the freedom of research of other scientists and on our own research programmes for the Shroud conservation that asked for thorough deliberations* (italics added).[39]

Gove admits in his book that the Archbishop's "thinly veiled accusation that we were attempting to prevent STURP from carrying out its scientific investigations was quite accurate."[40]

In November 1987, at a meeting in New York with Gonella, STURP learned that all of the other scientific tests on the Shroud had been canceled or not been authorized. (Other groups, with much less experience and knowledge of the Shroud than STURP, had also submitted competing proposals to test the Shroud, but they, too, failed to receive such permission.[41]) None of these extensive tests have ever been performed to this day, which stands in the way of a tremendous opportunity to advance the state of scientific knowledge on a subject that challenges both science and humankind in a fundamental manner, perhaps more than any object in history. In addition, STURP was not even allowed to influence or participate in the decision of where to remove the radiocarbon samples from the Shroud. Whether these decisions had also been made by the Holy See is not known for sure, nor whether other members of the Pontifical Academy, or other individuals, had lobbied against STURP's participation in further testing of the Shroud. What is known, according to Gove's book, is that Gove, members of the carbon dating laboratories, Chagas, and Canuto led the effort against STURP's participation, with Gove initiating and leading.

STURP's testing in the twenty-five other scientific areas could have provided complementary, confirmative, or relevant information for the radiocarbon dating of the Shroud. For example, we saw earlier how a much greater proportion of the pollens identified by the Swiss botanist Max Frei were from Palestine and Anatolia than from Europe, which reflects both on the cloth's origin and its history. The STURP scientists had hoped to acquire many more pollens that could have shed further scientific evidence on the cloth's history and origin. This information could have possibly confirmed or contradicted the radiocarbon date, but it would have related to it in some fashion. You would think that the carbon dating laboratories, which would soon consider it to be a "catastrophe" that only three labs were going to date the Shroud instead of seven, would have welcomed additional relevant scientific information pertaining to the cloth's history and origin.

In the same manner, STURP scientists would also have measured the isotopic compositions of the cellulose's oxygen and hydrogen, both of which are influenced by the climate in which the flax plant grew; temperature and humidity can cause this climatic influence.[42] It had previously been shown that oxygen and hydrogen ratios of cellulose prepared from historic and modern linens known to have originated in the Middle East and Europe were related to the climates in those regions.[43] This previous study had been undertaken to establish a database so that similar measurements could be taken from the Shroud, then compared to these data to establish its geographical origin. This research had concluded that such a determination was indeed possible.[44] Such measurements from the Shroud could have been relevant to its origin and, thus, complementary to the cloth's history and age. Such isotopic measurements should have particularly interested radiocarbon scientists, indeed any scientists interested in ascertaining knowledge and truth on an important subject.

Moreover, STURP would naturally have prepared follow-up research of the area over the man's eyes, particularly his right eye to confirm or rule out the presence of the lepton. This could have provided dramatic independent archaeological and historical evidence of an even more specific and historical date than could be found by carbon dating. To take part in a process that eliminates further testing in this area indicates that the radiocarbon scientists were more interested in other matters than they were in obtaining the true age of the Shroud.

The carbon dating scientists were eager to date the Shroud but failed to realize the important tasks that went along with dating this unique object with a new method(s). Throughout the entire lengthy process, they failed to do the study and preparation required for dating an object with such a different and varied history, even though the sources were readily available. They even ignored expert advice in this regard. They also went back on their own written word several times and worked behind the scenes to eliminate other more knowledgeable scientists from conducting far more extensive tests of the Shroud.

Historian Ian Wilson, who had numerous contacts with scientists from the carbon dating laboratories during the entire carbon dating process, and thereafter, commented upon their activities following the workshop: " . . . there was some intense politicking going on behind the scenes. Some of the radiocarbon-dating laboratories, now within a whisker of getting the go-ahead they had been waiting for, began to voice their disapproval of the idea of other scientific experiments being carried out on the Shroud at the same time as theirs, concerned that these might steal some of the thunder from their work."[45] Italian writers Orazio Petrosillo and Emanuela Marinelli, who investigated the entire carbon dating process, also stated that the carbon dating laboratories' " . . . manoeuvre had reeked of a political action intended to eliminate those [members of STURP] who use modern methods to deal with the problem in various disciplines, thus favoring access to the Shroud by only a very restricted circle of persons. . . ."[46]

Professor Gonella, the scientific advisor to the Archbishop of Turin, who was quite familiar with the events that occurred during the lengthy carbon dating process, reflected on these matters later. His translated remarks begin with comments concerning STURP's original proposal to carbon date the Shroud and the carbon dating laboratories' involvement.

> This proposal was put forward in 1984 within the context of a vast multidisciplinary research programme presented by the group which had already carried out an investigation in 1978, and by other individuals. These investigations were above all intended to understand better the physical and chemical structures of the sheet and of the image, specifically with the ultimate aim of understanding how it had been practically formed and also the possible means for conserving it. Naturally, it would have provided a background, a much wider backdrop to evaluate all the possible implications of the carbon-14 test. Originally the laboratories had agreed to this research; they were practically committed to start working on the project; then they made it decidedly clear by all means possible, that they preferred to work alone, and that they wanted the carbon-14 test to be conducted independently of any other investigation. Thus the investigations became separated.[47]

Commenting further on the carbon dating laboratories' involvement, Gonella said:

> At the beginning, when they themselves had asked us to be allowed to examine a sample of the Shroud, they had guaranteed us the utmost seriousness and completeness in the analysis, as well as promising to collaborate with the custodian of the Shroud, the Archbishop of Turin, and with his scientific consultant, the undersigned. Seized however by a feverish desire for celebrity, they began to renege on their promises: no further interdisciplinary investigations; just the carbon-14 test. They even badgered Rome, bringing pressure to bear so that Turin would have to accept their conditions. Through the intervention of Professor Chagas, then president of the Pontifical Academy of Sciences, they set aside the undersigned so that they could do whatever they wanted.[48]

Gonella also stated that:

> Scientifically I would have been happier and have my mind at ease if the dating operation had been carried out in the context of comprehensive, wide-ranging, and thorough chemical and physical investigation of the Shroud as had been originally planned. The carbon-14 laboratories preferred to work independently and they did not wish to collaborate with other scientists, something which, from the point of view of scientific methodology, left me greatly puzzled and certainly not satisfied.[49]

Gonella's comments included his statements as a scientist.

> It is a question of general principles of scientific methodology. At every congress and in every scientific discussion it is said that science must be interdisciplinary and multidisciplinary. Therefore it is with astonishment, with the regrets of a scientist, that I can say that there was once the real possibility of carrying through a vast multidisciplinary project but that unfortunately a section of scientific researchers opted to act completely alone.[50]

It is difficult to say why the carbon dating laboratories acted the way that they did. It would be easy to assert that their motivations were selfish and that they wanted all the attention centered on them and their new carbon dating methods. Yet, I think this would be too simplistic. I think the best explanations for the laboratories' behavior is that they greatly underestimated the task they had undertaken, were not willing to ascertain or put forth the extra efforts that were required, placed far too much reliance upon the new method's application with cloth, and, to some extent, may have wanted all the attention centered on them and their new methods. Their actions derive from the central facts that the vast majority of the scientific, medical, archaeological, and historical evidence concerning the Shroud was unknown to them, and that they excluded or ignored those scientists and archaeologists who possessed much of it. This lack of knowledge about the Shroud contributed to their failure to appreciate the scope of the task they had undertaken or to put forth the effort that was required for its undertaking.

Giovanni Riggi actually removed the sample in the presence of Gonella and others. However, note that these two men had worked for years with STURP and wanted its expertise and knowledge present. STURP would clearly have been able to influence the locations and number of samples selected. Not only would they have influenced these questions long before the samples were removed, their scientists would have stepped forward at the time of the actual removal had an incorrect location been chosen. You can see an example of this in 1978 in a confrontation between Dr. Max Frei and one of the STURP scientists: They literally stood face to face, confronting each other over the question of whether Frei would be allowed to put sticky tape on the face of the man in the Shroud and then peel it off. STURP won a ruling from the authorities present, thereby preventing possible harm to the facial image.

While others certainly shared the blame for the amateurish, controversial, and even tragic events surrounding the carbon dating of the Shroud, distinctions can be made among the participants. The carbon dating laboratories had the primary responsibility for pretreating the sample and for screening and eliminating any contamination. Their failure to heed the very pertinent advice in this regard is their sole responsibility. The carbon dating labs also had the responsibility to record and reveal their raw data. Not only did the labs fail to reveal their raw data, they did not measure and record all of the relevant raw data that they should have. Furthermore, their activities to eliminate tests in other areas that could have yielded relevant scientific information regarding the sample's age, origin, and history also bore directly upon their obligation to assess the age of the cloth accurately. Moreover, the carbon dating laboratories had made written and oral agreements concerning these and other matters but failed to honor them.

It is difficult to say why the laboratories and the coordinator acted as they did. Much of the information about their activities was not made public until Gove's book was published in 1996. Gove's book basically covers the period from the Shroud exhibit and scientific examination in 1978 through the carbon dating announcement in 1988. However, in the more than twenty years that he has been involved with or known STURP, Gove has never given any valid scientific reasons why he was obsessed with preventing their involvement in further testing the Shroud.

From the very first page that he discusses STURP in his book, and continually thereafter, he refers to them as "true believers in the shroud's authenticity."[51] Yet nowhere in any of the numerous scientific articles that STURP has published have they ever made anything close to such a statement. The reader should review this scientific evidence not only to verify the last statement, but also to ascertain the tone and objective scientific findings of the STURP scientists. Gove himself tacitly admits this when commenting upon the release of STURP's first scientific articles in 1980: "None of these articles were definitive in terms of establishing one way or the other whether the Shroud image was similar to a photograph or not. They mostly concluded that the evidence was against its being a painting. On the other hand, they didn't really present any indication of what else it might be."[52] This hardly sounds like scientists who were "true believers in the Shroud's authenticity" or who had lost their scientific objectivity or integrity. I have been studying this and other literature, as well as extensively interviewing members of STURP, and many other Shroud experts, for almost twenty years. This group of scientists stands out among various Shroud experts in many ways, one of which is that most of them refuse to offer even a private opinion as to whether they think the Shroud is authentic. Even when such private opinions can be ascertained, they cover a range of conclusions from the Shroud's not being the burial garment of the historical Jesus, to uncertainty, to the conclusion that it is Jesus' authentic burial garment. This simply mirrors the public's opinions, however, it is actually contrary to the opinions of most Shroud experts who have studied the scientific, medical, archaeological, and historical evidence pertaining to the Shroud. Most of those experts who are familiar with the total evidence from all the relevant fields, conclude that the Shroud is authentic. In fact, STURP members have often been criticized for not expressing their personal opinions on this subject, and for only analyzing or evaluating the evidence in absolute scientific terms.

David Sox's 1989 book, *The Shroud Unmasked,* clearly stated that STURP scientist John Heller reported "that in his interviews only three STURP team members . . . believe that it is probably the authentic burial cloth of Jesus of Nazareth; 'The rest of us have to say that we do not know.'"[53]

In any group of thirty people—whether they be scientists, lawyers, farmers, or truck drivers—there will usually be individuals with personalities, egos, or attitudes that can be annoying. STURP is no exception in this instance. However, as scientists, this group and its members should be evaluated on the quality of their scientific work, which has been excellent. If scientists' individual traits or qualities are going to be considered, they should be evaluated only if they affect the quality or quantity of their scientific work or the work of others. Throughout Gove's exposure to STURP's scientific work, he appears to have avoided discussing or analyzing it in scientific terms. This can be seen in his own book, when as early as the spring of 1979, STURP scientists gathered from around the country, along with other interested parties, including Gove, to review the critical findings from their scientific examination. Gove describes some of what was presented:

> The two sessions after lunch, chaired respectively by Jumper and Jackson, were devoted to microphotography, X-radiography, X-ray

> fluorescence, infrared measurements, analysis of particulate matter picked up on sticky tapes and ultraviolet and visible light spectroscopy.
>
> Sam Pellicori showed some spectacular color microphotographs, including a known burn mark from the 1532 fire that went all the way through the fibers, and evidence that the image itself was superficial. In the "blood" areas there are clumps of reddish-yellow material in the crevices. The "blood" penetrates right through the cloth. Under ultraviolet light neither the "blood" stains nor the image fluoresced but scorches, water stains, candle wax and background did.[54]

Even at this early date, it was known that the Shroud image is superficial as well as three-dimensional. It also contains the high resolution and detail of a photographic image; and appears to contain actual blood, the stains of which differ completely from the body image. Information is beginning to be acquired about the scorch marks and water stains. Candle wax is known to be on the Shroud at this early date. In addition, the historical allegation that the Shroud was a painting appeared to be false. Yet instead of reacting to this new and interesting evidence as a scientist, Gove's reaction is fixated on STURP, not on science. After Gove's description (above) of these initial findings, he stated that STURP "seemed to me to be a group of kids playing with expensive toys, hoping they would reveal some ultimate truth. . . . I summarize it to show the kind of information STURP was collecting and how little it mainly signified."[55]

Gove next related how Dr. Jackson discussed the possibility that ancient coins may have been placed over the crucified man's eyes when he was laid in the Shroud. Jackson stressed that STURP members from the Jet Propulsion Laboratory would be doing image enhancement work on the eye areas when some of their current work with the Jupiter program was completed. He stressed the importance of this work, because if such coin(s) could be identified, it could relate to the age in which the image was encoded. You would think that a scientist like Gove, who specializes in dating objects, would, at least, be interested in this discussion. However, here are Gove's remarks that immediately follow Jackson's comments: "Since neither the photographs nor, as we have seen, the three-dimensional analysis provided sufficient evidence to settle the question of whether the man had a navel, it seemed unlikely one would be able to establish the presence of coins, much less to identify them."[56]

The closest thing to a substantive scientific complaint in Gove's entire book comes when he mentions that ultraviolet fluorescent photographs taken by STURP may have been harsh or intrusive on the Shroud. This complaint was based on Gonella's slide presentation of the 1978 scientific examination of the Shroud. However, these ultraviolet fluorescent photographs were all taken within the guidelines of the American Society for Testing and Materials (A.S.T.M.) and did not cause any harm to the cloth. In fact, if Gove and the members of the carbon dating laboratories that dated the Shroud had looked at these ultraviolet fluorescent photographs, they would have seen that the radiocarbon site was in the midst of scorch marks and water stains. Their presence at the radiocarbon site was first observed and documented by STURP photographer Vernon Miller by examining the photographs taken under ultraviolet light. These photographs of the Shroud were valuable scientifically in a number of ways, and would have helped the laboratories' directors and the coordinator, who knew months in advance where the sample was to be removed. Instead, they sat and watched the sample be removed from the worst location on the entire cloth without saying a word.

Not only are Gove's criticisms unscientific, they also seem childish. In three different places in his book, he complained that the radiocarbon dating experiments were mentioned sixth in STURP's twenty-six experiments. Gove paraphrased his colleague Garman Harbottle's remarks to him, with which Gove concurs: " . . . STURP diluted the relative importance of the carbon 14 experiments by imbedding it as number six out of twenty-six in the proposal of 1984. He said that very much disappointed him."[57] This is silly. The importance of a test is ultimately determined by how it is conducted and what its results are. Furthermore, STURP did not number the tests according to their perceived importance, but concentrated on making an extensive list of important tests and desired for all to be performed. STURP had always considered radiocarbon dating the Shroud to be of paramount importance. In this regard, one of STURP's founding members had even contacted Willard Libby, the inventor of radiocarbon dating, as early as 1976.[58]

It is easy to conclude that either Gove and Harbottle were acting childishly over what number the carbon dating test was listed in a proposal containing many tests, or, perhaps, they knew of the additional testing that would be performed regarding the coins, pollen, image, and other isotopic measurements, each of which related to the age, origin, and history of the Shroud, and were upset that the carbon tests would not get all the attention.

During the 1979 conference, Gove also complained that three members of the STURP team had crosses around their necks. Gove called this an "ostentatious display" in his book, and it upset him so much that he stated, "I suppose if I had not been on my best behaviour I would have baldly asked them the reason for this Christian ornamentation, but I refrained."[59] There were approximately thirty members of STURP; for three of them to be wearing crosses is hardly unusual. Probably 10 percent or more of the general population wears crosses or other religious ornamentation. This just indicates that STURP represented a rough cross section of the population. One of the STURP scientists from the Los Alamos National Laboratory happened to also be a minister. Does he, or anyone else, have to give an explanation for this "ostentatious display"? This particular scientist does not happen to think the Shroud is Jesus' burial garment. A person can still think that and wear a cross, or be a Christian or a member of another religion, or not subscribe to any religion. Under Gove's criteria, a Christian, Muslim, Buddhist, Jew, atheist, et cetera, would each be suspect of having prior beliefs on the subject and, therefore, not be sufficiently objective to investigate it. By these criteria, the subject might never be investigated by anyone.

Gove must not have realized that throughout the Shroud's history, the most influential and incorrect critics who formulated and reinforced the view that it was a painting were Christian clerics. Bishop d'Arcis and his infamous memorandum began the widespread public misconception that the Shroud was a painting. Canon Ulysse Chevalier's writings and pronouncements, following the startling findings from the first photographs ever taken of the Shroud around the turn of the century, helped reestablish this historical misconception, as did the subsequent arguments of Reverend Herbert Thurston. In addition, Reverend David Sox, a close collaborator of Gove's, who long ago concluded the Shroud was a forgery, is an Anglican priest.

Moreover, Gove's attitude was hypocritical. His book, as well as Sox's, showed that Doug Donahue, codirector of the University of Arizona radiocarbon dating laboratory, was hoping, along with his wife, that the Shroud would date to the first century and prove to be authentic. In fact, Donahue was clearly disappointed in the result, according to Gove.[60] Yet Gove never mentioned once that this disqualified him from participating in carbon dating the Shroud, nor should it. Moreover, Shirley Brignall, an administrative assistant at the University of Rochester and Gove's constant traveling companion from 1985 on (he dedicated his book to her), had an abiding belief that the Shroud was authentic. Gove stated, "She has told me that, even now, her heart still tells her it is Christ's shroud."[61] Gove had her attend every kind of meeting or function regarding carbon dating of the Shroud, starting in 1985. He put Brignall, who is not a scientist, on the invitation list to the Turin workshop, which she did attend, but did not invite any of the thirty STURP scientists. He even wanted her to attend the Shroud's dating at Arizona, but Donahue would not allow it. That she obviously believed strongly that the Shroud was Christ's burial garment did not disqualify her in Gove's mind from participating in everything that Gove did. Gove's attitude toward STURP was nothing less than blatant hypocrisy.

Gove's obsession toward STURP manifested itself in his deceitfulness, resentment, hypocrisy, and unscientific behavior. Throughout his book, Gove seems to brag and gloat about his conduct and attitudes, which began with STURP's initial research and which never had any scientific or other valid basis.

Explanation of Gove's conduct might come partly from the fact that the subject of the Shroud can be a little frightening. The subject relates not only to the question of whether it was Jesus' burial garment, but also to the crucifixion and resurrection of the historical Jesus Christ. The implications of these issues are staggering, and sooner or later, one is confronted by them in one's research or investigation of the Shroud. When these issues and implications arise, they are not necessarily expected or welcomed. These issues relate to such fundamental matters and can have such permanent (even eternal) consequences that to have to deal with them in an unexpected manner can be overwhelming. As a result, one can lose one's objectivity on this subject and act emotionally or irrationally. I have seen this happen to scientists, as well as to people of all backgrounds, with this subject. My initial reaction to these implications was very similar. Fortunately, such a reaction is usually only temporary, and once the individual accepts that he/she may be dealing with these issues, objectivity usually returns.

However, some people never accept that these issues are raised as a result of an investigation of the Shroud. Perhaps, this might explain why Gove was so obsessed with preventing STURP from playing any future role in the testing of the Shroud. It was STURP that provided most of the scientific evidence from the Shroud that related to and corroborated the accounts of the crucifixion and resurrection of the historical Jesus Christ. Throughout Gove's book, he constantly raised the question of why the Shroud should even be dated. Moreover, he never addressed the merits of any of STURP's extensive scientific evidence or findings, yet, unlike other members of the carbon dating laboratories, Gove had extensive exposure to STURP's findings from the beginning. Interestingly, he closed his book with the following emphatic statement concerning his result: "Meanwhile, the Shroud still lies in its silver casket. . . . No further scientific forays on [this] most famous artifact have been authorized. The mystery of how its image was produced remains just that—a mystery. *Maybe that is as it should be.*" (Emphasis in original)

All of the above actions and quotes appear to be those of a man who wants to avoid the evidence acquired to date, as well as to avoid acquiring any new evidence on the subject. They do not appear to be the actions or words of a true scientist regarding an important subject.

The actions of Gove and the carbon dating laboratories were somewhat reminiscent of the infamous actions of the French Academy of Sciences at the turn of the century. The academy was considered to be the most prestigious group of scientists in its day, yet its actions in regard to the Shroud were contrary to the spirit of science. They gave little or no recognition to the first scientific investigation of the Shroud conducted by Vignon and Delage, and they rejected attempts to investigate it further. True scientists—objective and impartial—would have studied the initial scientific findings of their day and encouraged further scientific investigation of a subject that posed such a challenge to humanity, as well as to science. Yet the inaction of the French Academy may be preferable to the inaction or inadequate actions of the carbon dating community because the academy did not undertake any correlating responsibilities for its commitments, unlike members of the carbon dating community. The inactions and actions of the carbon dating community, unfortunately, resulted in numerous unnecessary controversies. Sometimes controversy can be a good or welcome event; however, rarely are unnecessary controversies welcome. Unfortunately, these numerous controversies may have contributed to a misleading test result—worse than no result at all. And, while a member of the French Academy may have deceitfully omitted or censored part of Delage's report, the actions of Gove were also deceitful and hypocritical, yet covered a much longer time period. Moreover, while the French Academy of Sciences may have immediately declined to attempt any further investigation of a subject or object that may have enormous importance for all humankind, it did not seek to prevent further extensive examination of the Shroud by other qualified scientists, as did the directors of several carbon dating laboratories.

Until now, not many have been aware that twenty-five areas of comprehensive testing of the Shroud were eliminated, or of the underhanded way members of the carbon dating community helped to bring this about. Yet, the public at large has the right that these and other tests take place. Few objects in history have the potential to affect people throughout the world in the way this could.

No one should be afraid of these tests or their results. Scientists should research and approach this subject as they would research medicine or outer space or any other subject that could benefit and enrich humankind. They should approach the subject objectively, studying the contaminants, qualities, and history of the cloth, as well as considering other scientific evidence, especially if it relates to their scientific tests and objectives. They should be honorable and be willing to work with other knowledgeable scientists in a multidisciplinary effort to benefit or educate the public at large.

Think of all the information that was acquired from the only comprehensive scientific examination of the Shroud in 1978 and from the data and samples that were brought back for further analyses. For having never seen the cloth before, STURP performed an intelligent and thorough investigation of the Shroud and its components over twenty years ago. Yet scientific technology, imaging techniques, medical knowledge, and so many other ways of examining the cloth have also greatly improved since then. Applied Spectral Imaging, which we discussed in the last chapter, is just one example of these new methods. Not only has technology improved, but since our knowledge of the Shroud is so much greater, we can better evaluate and focus our areas of investigation.

The underhanded efforts and unjustified elimination of the twenty-five areas of scientific tests on the Shroud could be a blessing in disguise. They could serve as an impetus for extensive testing in the near future—which could reveal even more information than would have been learned in 1988—and the tests and their results will receive far more attention than they ever would have back then.

In light of all the evidence that resulted from the first scientific examination of the Shroud in 1978—that twenty-five areas of further scientific testing of the cloth were tragically eliminated; that many substantive scientific and procedural challenges exist to the 1988 dating; that many areas of testing remain that relate to the age, origin, and history of the cloth and when its images were encoded; that the images on the Shroud are unique and contain such a wealth of information, which is discovered and revealed as technology advances and develops; that scientific, medical, and imaging technology have noticeably advanced so that much more information can now be acquired from an examination; that this subject poses a great challenge for the knowledge and understanding of science and humanity; that it is a subject that could fundamentally affect everyone in the world in this and future generations—it is obvious that future comprehensive scientific testing of the Shroud must take place.

CHAPTER TEN

THE CAUSE OF THE IMAGES ON THE SHROUD

As we have seen from the previous chapters, the Shroud of Turin has an exhaustive number of original and incomparable features. As science and technology have advanced, the quality of the resulting Shroud images has also improved, unlike any image ever known. The Shroud of Turin seems to be the most extraordinarily encoded image in all the world. The question of how these images were encoded is clearly one of the most important questions of all concerning the cloth. Let us examine the pertinent scientific, medical, archaeological, and historical evidence pertaining to the cause of the unique images of a crucified man on the Shroud.

The scientific evidence found on the cloth itself gives a good deal of information about its images, in particular that they appear to have resulted from some type of radiation. The first indications of this can be found by comparing the body images with the scorch marks found on the Shroud. Both are stable in water, neither impeding its flow on the cloth nor being altered by it. Both are also stable to further heating and do not change color up to temperatures and times that would produce equivalent scorches.[1] Neither the scorch marks nor the body images have faded with time, and neither were caused by foreign materials or particulates. Also, neither the chemical components of the scorch marks nor the body image are soluble in acetic acid, redox, or organic solvents.[2]

The translucent fibers from the lightly scorched areas on the Shroud closely resemble the Shroud's body image fibers.[3] The body image fibers are also similar to a light scorch chemistry in their microscopically corroded appearance and their lower tensile strength.[4] This similarity between the body images and the light scorches can be further observed with ultraviolet visible and reflectance spectra. The relative spectral reflectance curves for the light scorches were also very similar to those of the body images on the Shroud.[5] The spectrophotometric examination further reported that the light scorch areas of the Shroud reduced the background fluorescence, as did the body image.[6] With controlled timing and heat, superficial scorches (similar to the Shroud body images) can also be produced on cloth without affecting its gross mechanical properties.[7]

Another indication of the cause of the body image formations can be found by analyzing the cloth's coloration. The Shroud, like all linen, has yellowed or darkened with age. As we saw earlier, when a Shroud fibril is cut in half, its original white color can be seen on the inside. However, the body images have yellowed faster and darker than the background or non-image areas on the Shroud. The body image fibrils are more chemically degraded than the background fibers; their cellulose has been chemically altered and consists of structures formed by dehydration, oxidation, and conjugation products of the linen itself.[8] Normal darkening and degradation of cellulose (the main component of linen) occurs through gradual oxidation and loss of water—by exposure to heat and light.[9]

Scientists who have studied the full-length Shroud and its samples note that "the body image is due to a more *advanced* decomposition process than the normal aging rate of the background linen itself" (italics added).[10] Scientists have also established that the application of either light or heat to cellulose will artificially darken it in what amounts to a rapid simulation of the aging process.[11] It should be noted that laboratory simulations by scientists using such controlled, accelerated aging processes produce the same overall properties, as well as spectral reflectance curves, as the body image and background areas on the Shroud.[12] The body images on the Shroud appear to have been exposed to more light or heat than the rest of the cloth and to have aged, decomposed, or degraded at a faster rate than the background.

For these and many other reasons, most scientists, as well as other Shroud experts, have concluded that some form of light or heat (or radiation) caused the images on the Shroud.[13] However, the absence of pyrolitic compounds or products expected from high-temperature cellulose degradation indicates that the image-forming process took place at a fairly low temperature. This type of low-temperature radiation would not leave any residue on the cloth, as is the case with the Shroud's body images. Neither would it leave any directionality across the width or length of the image. In addition, radiation is an agent that can operate on skin, hair, coins, or flowers, and can uniformly encode the fibers on a cloth.

It is extremely difficult to imagine how the subtle shades of light and dark on the Shroud's body images could possibly have been obtained without using light or radiation. These body images are not saturated or diffused. The edges of the man's body at the sides, top, and bottom break off sharply. Furthermore, the agent, acting at a distance, barely penetrated the cloth. As one noted scientist who has studied the Shroud for more than two decades observed, "An agent acting at a distance with decreasing intensity is, almost by definition, radiation. The limitation of the cloth darkening to the outermost surface pointed to a non-penetrating, non-diffusing agent, like radiant energy. . . ."[14]

A vertical beam or beams of light or radiation also best explains how the Shroud's body image was encoded through space in a straight line from the body to the cloth. STURP scientist John Heller stated, "It is as if every pore and every hair of the body contained a microminiature laser."[15] This vertical directionality of the Shroud body image has only been accounted for by methods involving radiation. As scientist Luigi Gonella explained, "Whatever the mechanism might be, it must be such to yield effects as if it were a burst of collimated radiant energy."[16]

We also saw earlier how the various shades of light and dark on the cloth's frontal image directly correlate to the various distances that they were from the underlying body. This ratio exists throughout the length of the body image, even in places where the cloth could not have been touching the body, resulting in an image that contains precisely encoded, three-dimensional information. Such a precisely encoded correlation over such a distance could seemingly only be achieved by radiation. The Shroud's highly resolved image is also difficult to imagine unless light or radiation coming from the body is somehow directed onto the cloth.

All of the evidence points to a very unique occurrence that caused the images on the Shroud, something that could never have been created by the technology of the medieval ages (or even by the technology of today). Only through simulation have today's scientists been able to come close to the Shroud's three-dimensionality, vertical directionality, and finely resolved and highly focused image; their simulation achieved by a mechanism in which light was attenuated in a liquid, then traveled in a vertical, straight-line direction from the plaster reference face while it was being focused in a camera.[17]

Dr. Giles Carter, Professor Emeritus, Eastern Michigan University, has conducted years of experiments with X rays. He has noted that cloth samples placed in an X ray beam and exposed to low-energy, long-wave X rays for different periods of time will produce superficial, straw-yellow discoloration like that found on the Shroud body images.[18] He also noted that these same types of X rays are easily absorbed in air. Because of this absorption or attenuating quality, Carter stated that X rays given off by the body would also convey three-dimensional information onto the cloth.[19]

Dr. Carter first suggested in 1984 that the finger bones are visible on the photographic negative images of the man in the Shroud. In addition, he noted that the bones extending into the hand, over the wrist, could also be visible, helping to explain why the man's fingers appeared so long. Since then, other scientists and physicians have confirmed the identification of these finger and hand bones.[20] Carter stated that these "images may be due at least in part to x-rays emanating from the bones in the body."[21]

Scientists and physicians have identified other possible internal skeletal features on the man in the Shroud. Dr. Jackson has noted that part of the skull at the forehead may be visible on the man. Surgeon Alan Whanger, utilizing his modified Polarized Image Overlay Technique with the Shroud's negative and positive images, has also identified features from the skull, as have Dr. Carter and Dr. August Accetta.[22] Dr.

Accetta, a physician, has also conducted experiments concerning radiation-imaging of skeletal and other bodily features. Dr. Jackson and Dr. Accetta have further identified faint images of the curved and inverted thumb under the man's left palm.[23] Carter, Whanger, and Accetta have stated that images of the man's teeth could be partially visible, especially on the right side of the man's mouth.[24]

Dr. Carter also first stated that, "Part of the backbone may be visible on the dorsal image . . ." of the man in the Shroud.[25] This identification has also been confirmed by Dr. Whanger.[26] Recently, I enlisted the services of Dr. Joseph Gerard and Dr. Cheri Ellis, who, in their profession as chiropractic physicians, make and view more X-ray images of the spinal column than almost any other profession. After studying quality photographic negatives of the dorsal area, they were able to specifically identify numerous vertebrae in the neck and backbone (and even a few pedicles of the vertebrae with disc spaces prevalent).

All these skeletal features lie near the surfaces of the frontal or dorsal sides of the man in the Shroud. All are encoded correctly, and none were visible for hundreds of years—until the development of modern technology. The existence of just some of these features shows not only that the radiation came from the body, but that it resembled or had qualities analogous to X rays.

Since we know the Shroud contained a body, the fact that both the frontal and dorsal images are contained on the inside of it is an indication that the body wrapped within was the source of radiation. We also saw that the reason a truly proportional three-dimensional image resulted was because the lightness or darkness of the image on the cloth correlated to the distance that it was from the body. Since the various degrees of the body image's lightness are all contained *on* the Shroud's surface, and the cloth itself *received* this information indicating the corresponding distances between it and the body below, the light had to have come from the body. In fact, all the numerous body image features, that are encoded over the entire lengths and widths of both the frontal and dorsal body images, indicate that the radiation emanated throughout the length and width of the body. Having studied the various features of the Shroud body image that have been discovered since its first extensive scientific examination in 1978, physicist Kitty Little wrote:

> It was already known that the image was inside the Shroud and not on the outside. With this further examination it became certain that the source of the illumination that had formed the image came from within—that is, from the body—and that whatever caused it had a range of about four centimetres . . . [with] the "illumination" coming from the body as a whole.[27]

That the light or radiation came from the body is apparent not only from the body image surface features, but also from the various internal or skeletal features. Normally, magnification or enlargement of the body's bones, ligaments, and skin occurs when X rays are made. That is because the rays leave an external tube before hitting a part of a person's body and being recorded on film. The degree of magnification will vary with the degree of distance. For the short distances that existed between the Shroud cloth and the underlying body, extensive magnification would have clearly been present if the source of radiation came from outside the body. However, the Shroud's body image clearly lacks any such magnification. According to Gerard and Ellis, this is one further indication that the source of light came from under the cloth or film—from the body itself.[28]

PROCESSES THAT ACCOUNT FOR THE SHROUD'S BASIC BODY IMAGE FEATURES

Every single method that has ever been attempted has failed to duplicate all the various features found on the Shroud. Not only have artists and proponents of naturalistic methods failed in their attempts to duplicate the Shroud's images, but so have scientists from among the most prestigious institutions in the world. Scientists have attempted these methods, modified them, and created methods of their own. Yet all have failed.

In light of this, we should consider the advice of John Jackson, who has made many of the most important discoveries concerning the Shroud, has attempted most of the above methods, has caused more scientists throughout the world to study this subject, and has published more scientific papers on the Shroud than any scientist alive:

> Therefore, perhaps the time has come to ask if we ought to start thinking about the Shroud in categories quite different from those that have been considered in the past. In particular, perhaps we need to be more flexible in our scientific approach and consider hypotheses that might not be found readily in conventional modern science, for it is conceivable that the Shroud image presents, if you will, some type of "new physics" that ultimately requires an extension or even revision of current concepts.[29]

In light of these various—and repeated—failures, to consider what could be called an "unconventional method" would not be illogical. Moreover, if the unconventional method could account for or explain all the various features found on the Shroud image, its consideration would be mandatory. We will consider three methods that account for or explain the various features found on the Shroud, and (in the words of Dr. Jackson, who developed one of these methods) all "will be developed and argued strictly from consideration of image properties and no support will be derived from extraneous speculations."[30]

Another point should be noted independently. There is an exhaustive amount of evidence that cannot occur naturally or be forged, indicating the Shroud is the burial garment of the historical Jesus Christ. If the Shroud wrapped the body of the historical Jesus Christ and contains an unprecedented image of him, then an unconventional method could also be independently considered from a historical viewpoint since, according to several historical sources, a unique event occurred to him after he was buried in a shroud in a tomb in Jerusalem.

We will examine the only three methods that have ever been proposed that account for the numerous characteristics of the Shroud's images. We will see how one of them accounts for every single feature of the Shroud's unique images, and the unusual qualities of the cloth itself, is completely consistent with and corroborated by the historical sources, and can be verified by future testing of the Shroud.

Protonic Model of Image Formation

We alluded to the first of these methods in chapter 8. This method, the Protonic Model of Image Formation, has been developed over the last several years by Dr. Jean-Baptiste Rinaudo, of the Department of Biophysics, Institute of Biology, at the Center of Nuclear Medical Research in Montpellier, France. Starting with this scientific dilemma and the above indications that the body gave off some form of radiation, Dr. Rinaudo theorized that, if the body gave off vertically polarized gamma rays from its surface, protons and neutrons would be released from the nuclei of deuterium atoms, uniformly distributed over the body. Deuterium is an isotope of hydrogen; its nucleus consists of one proton and one neutron. When a sufficiently energetic gamma ray collides with a deuterium nucleus, the latter will break apart, with its protons going in one direction and its neutrons going in the opposite direction. The protons and neutrons would thus travel in straight-line directions, with half going straight up and half traveling straight down. The protons would cause the body images on the frontal and dorsal sides of the Shroud, while the neutrons would alter the C-14 content throughout the cloth.[31]

This reaction occurs at maximum efficiency when the gamma ray has an energy of 4.5 MeV. Breaking the nucleus apart would consume 2.23 MeV of its energy, with its remaining energy being given to the proton at 1.135 MeV and the neutron at 1.135 MeV. Unlike neutrons, protons are not penetrating particles. Once they reached the cloth, they would not travel farther than two to three fibers, thus producing a superficial image. Protons are also easily attenuated or absorbed in air. As they traveled, their energy would subside or dissipate. The maximum range of protons in air would be three to four centimeters, or less than two inches.[32] Air and linen will attenuate or absorb protons well at 1.135 MeV. Accordingly, those parts of the body farthest away from the cloth would encode the fewest number of superficial fibrils, and those parts of the body closest to the cloth would encode the largest number of superficial fibrils. As a result, the protons would leave a superficial, three-dimensionally correlated image on the cloth and would have traveled in a vertical direction from the body to the Shroud.

The lengths and widths of the front and back of the body would be precisely encoded negatively and with equal intensities. Even parts of the body that could not have possibly been touching the Shroud, such as the sides of the nose, would become encoded. However, under this method there would be a blank space around the groin region as found on the Shroud. Because the arms have been intentionally folded with the left hand placed on top of the right hand, there is a gap of more than 4 centimeters (about 1.6 inches) from the cloth to the inner part of the legs below.

Dr. Rinaudo and his associates performed numerous experiments irradiating white linen cloth with proton beams of various energies with a particle accelerator at the Grenoble Nuclear Studies Center in France. Linen naturally fluoresces under ultraviolet lighting, as does the Shroud's background, or off-image, areas. However, when their experimental linen was irradiated with proton beams with energies of 1.4 MeV or less, the cloth's natural fluorescence disappeared, as is the case with the Shroud's body image. They were also able to duplicate the microchemistry results of dehydratively oxidized, degraded cellulose, as is also found with the Shroud's body image.[33]

The resulting superficial coloration on the test linen was straw-yellow, just like that produced on the Shroud. Its fibers and threads lacked any cementation or added pigments or other materials. Where body image fibers crossed, underlying fibers were protected and remained white, as found on the Shroud; in addition, the inner part of the straw-yellow image fibers remained white, like the image fibers on the Shroud.[34]

Furthermore, Rinaudo's straw-yellow color also resulted from conjugated carbonyl groups within the molecular structure of the cellulose, as is also found with the image fibers on the Shroud.[35] This means that many of the carbon atoms are double-bonded with other atoms. These groups absorb light and reflect it as the straw-yellow color that is visible to us. Carbon occurs naturally in cellulose, but only as single bonded atoms. In order to become double bonded, something must break the bonds within these groups, causing them to reattach and reunite in other arrangements. Protons would certainly comprise one of the most effective candidates to accomplish this, since it does not penetrate and its damage would be concentrated at the topmost two to three fibrils of the cloth.

Dr. Rinaudo has also duplicated another novel feature with his Protonic Model of Image Formation and experiments. In order to duplicate the background color of the Shroud, he followed the technique of previous scientists. He artificially aged his experimental cloth by heating it at 150°C for ten hours. However, when he then combined this technique with his other image-encoding experiments, he achieved some very interesting results. He exposed the experimental cloth to a lower intensity of radiation, which left the exposed portion white without any discoloration. However, after artificial aging, the sample produced a superficial, straw-yellow color like that on the Shroud.[36] This duplication of coloring was further confirmed by experts familiar with the Shroud. Rinaudo stated, "these experiments suggested that aging might be an *essential process* to unveil an initially concealed image into a visible one" (italics added).[37] Dr.

Rinaudo's statement is consistent with the historical evidence that the Shroud's image developed over time.

The Protonic Model is also the first image-forming method to offer an explanation for the Shroud's 1988 carbon dating results. Under this model, the protons and neutrons from the deuterium atom go off in opposite vertical directions. Unlike the proton, the neutron is very penetrating. Some would collide with atoms in the cloth and bounce in another direction, while many would pass through the linen and blood, bouncing off of or even penetrating the limestone surroundings, with some ricocheting back onto the cloth. After the neutron lost most of its energy, it would be in position to alter the nucleus of nitrogen-14 or carbon-13 into a newly created C-14 nucleus as previously discussed in chapter 8. Thus, in addition to duplicating or accounting for all the Shroud's basic body image features, this model could also explain how additional C-14 was created on the cloth.

While the Protonic model is a good start, there are a number of features of the Shroud it cannot explain. The perfect blood marks, for example, could not have been recorded by radiation, or by a simple combination of direct contact and radiation. And other subtle image qualities point to a more complex model than Rinaudo's.

Cloth-Collapse Theory of Image Formation

Dr. John Jackson, one of the founders of STURP who has studied the Shroud images for more than twenty-five years, proposed a method of image formation that accounted for more image features than any method had previously. After many years of studying the cloth and its images and participating in numerous experiments testing the many methods proposed to explain the Shroud's images, Jackson concluded that "we seem to have a situation where the set of observables is so restrictive that all hypotheses posed thus far must be excluded . . . often on the basis of multiple objections."[38]

Jackson was the first to incorporate into a proposal the findings of Dr. Gil Lavoie, who first explained why some of the blood marks on the head of the man in the Shroud had been displaced into the hair. Lavoie illustrated in figures 137–139 that the blood marks now seen in the hair all originated on the sides of the man's forehead and face. Lavoie placed a cloth over the Shroud's facial image, then traced the blood marks and cut them out of the cloth. He then draped the perforated cloth over a human face to clearly demonstrate that, when the Shroud was first draped over the man, these blood marks rested on the sides of his forehead and face. Only when the cloth was subsequently straightened or flattened did the location of these blood marks extend into the hair. While previous studies we've discussed showed that the blood marks and body images have been encoded somewhat differently, Dr. Lavoie and his associates demonstrated that the Shroud was in two different positions at the time these blood marks and body images were encoded.[39]

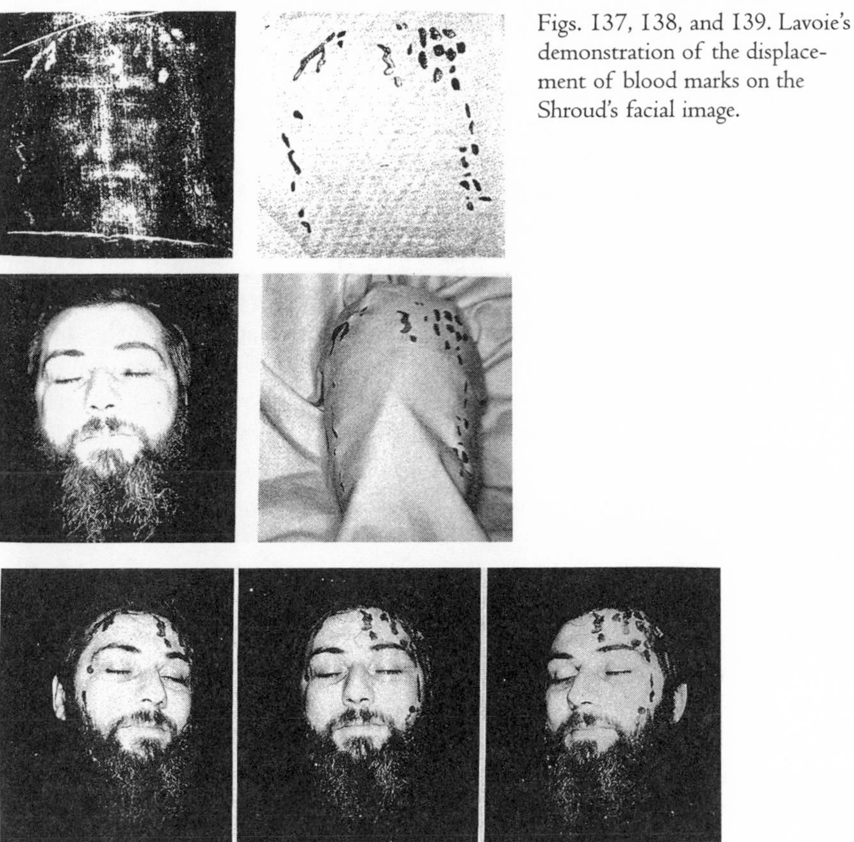

Figs. 137, 138, and 139. Lavoie's demonstration of the displacement of blood marks on the Shroud's facial image.

Jackson first considered the unidirectional nature of the image mechanism. He accounted for this vertical encoding by using nature's most constant vertical force—gravity.

Jackson was familiar with the numerous medical and scientific studies indicating that the Shroud wrapped a body from which radiation encoded the body image onto the cloth. He was also one of the first to notice that some internal skeletal features could be partially seen in the body images. In addition, his years of study led him to observe possible motion blurs under the chin, and near the neck, with two subtle blurs off the man's left eye and cheek, which are best seen on the photographic negative. He was also aware of the extensive scientific and medical characteristics of the Shroud's images that we have discussed in this book.

With these various conditions to consider, Jackson devised a simple theory that accounted for the Shroud's diverse characteristics with a method that is scientifically well posed and internally consistent. Under this method, physical objects as well as their physical environment (such as the cloth, air, gravity, radiation, chemical modification of cellulose) behave according to scientific laws. The only thing that reacts unconventionally is the body. Of course, something extraordinary must have occurred to the body to produce all the unprecedented and unique features of the Shroud's images.

Jackson's theory predicts that the Shroud's images would be encoded if the body became insubstantial and emitted ultraviolet light. As the cloth fell through the body region, each point on the cloth would receive a radiation dose in proportion to the time it was within the region. The parts of the cloth that were over the highest points of the supine body (for example, the tip of the nose) would receive the longest dose of radiation, while the parts of the cloth over the lowest points of the body would receive the least. Thus, the intensity of all points on the resultant body image on the two-dimensional cloth would be directly correlated to the distance that they originally were from the surface of the three-dimensional body. Furthermore, since the draped cloth fell by gravity, all points of the resultant body image would have aligned vertically with the corresponding body point below it. Even those parts of the body that were not initially touching the cloth, such as the sides of the nose, would be encoded in a three-dimensional and vertical direction onto the cloth.

This method could also encode the internal skeletal features observed on the frontal image near the surface of the body, while not encoding body features (such as the insides of the legs below the hands) more than four centimeters from the initially draped cloth. Since this image-encoding event occurred over a very short period of time, the cloth would only briefly receive the radiating stimulus and encode just those features nearest the body's surface and the cloth. After the brief image-encoding event was complete, the part of the cloth draped over the front of the man continued to fall to the ground by gravity.

In this theory, since only the frontal part of the cloth falls through the body, only the frontal image would be three-dimensional and vertically directional. The dorsal image is encoded by direct contact. Since radiation is emitted from the entire body, the amount of time that the frontal part of the image is exposed to the radiation would be equal to the dorsal part, thus leaving both images with approximately equal intensities, as is also found on the Shroud's body images.

Short-wave ultraviolet radiation is highly attenuated in air, being absorbed even over distances of a millimeter. Since the Shroud's linen fibers are 800 times denser than air, ultraviolet radiation would also be easily absorbed upon contact with the cloth's fibers, thus leaving a superficial image only two to three fibrils deep. Under the Cloth-Collapse Theory, since this attenuating zone in air is so small and the body surface so much larger, the radiation dose would not effectively accumulate on the cloth until it actually intersected the body surface. Therefore, only the body point directly below the corresponding point on the falling cloth would become encoded. This would occur at all locations on the Shroud's body image, thus making it highly resolved.[40]

Since the cloth is falling by gravity and the ultraviolet radiation is easily absorbed in air, the sides of the body would not be encoded, since they did not intersect or touch the cloth during the image-encoding process. However, the edges of the frontal and dorsal body that intersected the cloth while it collapsed would be encoded.

Jackson thinks that all dorsal body image features were encoded on the Shroud by direct contact. He also thinks the blood marks from the man's back, head, face, and the rest of his body were all encoded by direct contact. He states, "*Blood on the Shroud.* As the Shroud is initially draped over a body covered with blood, that blood is transferred naturally to the Shroud by direct contact."[41]

Recall also how Dr. Lavoie illustrated that the Shroud was in two different positions when the blood marks from the sides of the man's head were encoded onto the body image region of his hair. The Cloth-Collapse Theory has been the only method to date that can even claim to account for both the blood and body images at these mutually inconsistent locations. Under this method, as the Shroud first fell downward, it then flattened and moved laterally to some extent. As it moved laterally, the cloth took the blood marks, formed by direct contact, from the sides of the head and transferred them into the region occupied by the hair, thus encoding the hair and head images around the blood marks.

Rinaudo's Protonic Model of Image Formation and the Jackson's Cloth-Collapse Theory were the first to account for all of the basic features found on the body images of the Shroud. Yet they are not without their imperfections. Rinaudo's method does not explain how the blood marks became encoded and laterally displaced into the man's hair. It seems that only a cloth-collapse model can explain this lateral displacement.

Yet the radiation used in Jackson's model does not seem to be the most appropriate. Rinaudo's experiments using protons produced "acid oxidation and dehydration of the cellulose."[42] He and his associates were able to produce a straw-yellow color which developed over time and which resulted from conjugated carbonyl groups. All of these features are like the Shroud's body image. Rinaudo exhibited photomicrographs of these linen samples that looked remarkably like photomicrographs of the Shroud's body image.

Jackson did not publish any photomicrographs of his linen samples irradiated with ultraviolet light, nor does he claim that they became straw-yellow, were oxidized or dehydrated, or contained conjugated carbonyl groups like the Shroud samples or Rinaudo's samples. These same shortcomings, plus other problems relating to a superficial image, could also be found with low-energy, long-wave X rays,[43] which are next to ultraviolet on the electromagnetic spectrum. For these and several additional reasons, we will see that perhaps only protons or similar particles will produce all the chemical features found on the Shroud.

The next model that we will examine will account for all the Shroud's body image features and all the detailed characteristics of the man's various wounds, blood flows, and blood marks found on the cloth. It will also account for several other features, such as the coin images and flower images, as well as other important characteristics of the Shroud itself. Whereas the last two models had to make several assumptions to account for the Shroud's mutually exclusive features, this model makes only one assumption, yet this assumption is consistent with several historical sources.

Historically Consistent Method

This new theory of image formation is not only historically consistent, it makes certain predictions with which to judge the validity of this or related models. Moreover, these predictions can be evaluated in a future round of scientific examination of the Shroud. According to this method, the Shroud and its physical environment, such as gravity, the tomb, the hydrodynamics of air, chemical modification of cellulose, the structure of blood, et cetera, behave according to standard physics principles. The method itself, and what occurs to the body, will be presented in terms of currently accepted scientific principles—some of which are just being discovered. Yet, as pointed out by Dr. Jackson, who has tested and evaluated more methods of image formation of the Shroud than any other person, "The real test of any hypothesis is . . . its ability to explain observations, make predictions, and provide insight into how reality is constructed."[44]

Whether the Shroud is going on seven hundred years old or two thousand years old, everyone who has ever examined it has noted its remarkable condition. As STURP scientists Roger and Marion Gilbert remarked quite some time ago in an observation section of a scientific paper, "The cloth [Shroud] is in excellent condition, extremely soft and pliable with no apparent degradation of strength."[45] Needless to say, ancient cloth is rarely, if ever, described as "in excellent condition." The only friable parts are its topmost two to three fibers, which contain the superficial body image. If an image-forming method or event helped account for or contributed to these characteristics of the cloth itself, it should certainly be given due consideration.

For many years, numerous scientists and other experts have noted that although all conventional methods for causing the image on the Shroud failed for a variety of reasons, if an instantaneous burst of energy or radiation came from the body, it might explain the mutually inconsistent features of the Shroud's frontal and dorsal images. Yet these scientists were at a loss to explain how a body could do this or what type of radiation would work.

In 1994, Dr. Kitty Little a retired nuclear physicist from Britain's Atomic Energy Research Establishment in Harwell, stated what many other scientists previously observed regarding the Shroud, then added another important point: "Now it seemed almost certain that the image must have been caused by some sort of radiation. . . . However, there was one source of ionizing radiation that they did not try."[46] Dr. Little based her remark upon a number of experiments that she had conducted much earlier at the Atomic Energy Research Establishment in Harwell. Dr. Little further noted that, "Some scientists have suggested something in the way of a nuclear disintegration, acting almost instantaneously, as with the flash from a nuclear explosion."[47]

Professor Wesley McDonald, Elizabethtown College, summarized the viewpoints of many scientists who think that radiant energy coming from the body caused the Shroud's images: "Many scientists now describe this burst of energy as a pulsed laser beam caused by dematerialization of the body into energy in a millisecond."[48]

If we were to combine the contributions of these many scientists, along with Dr. Little's experiments and proposals, with the best parts of the models of Dr. Rinaudo and Dr. Jackson, we could get some astounding results: a method or event that explained *all* the chemical and physical features of the Shroud's body images and its blood marks, which no method has ever provided before. It could account for the cloth's excellent condition and durability, the friability of its image fibers, and its image distortions, as well as numerous other subtle features—like the possible coin, flower, and skeletal imaging—that no other method has discussed or attempted to account for. Moreover, it could even explain the controversial or erroneous C-14 dating of the Shroud that occurred a dozen years ago.

In 1950, long before she ever heard of the Shroud, Dr. Little irradiated several different cellulose fibers at the nuclear reactor in Harwell with radioactive illumination consisting of combinations of protons, alpha particles, neutrons, and gamma rays. As we shall see shortly, while the last two forms of radiation, neutrons and gamma rays (along with electrons) certainly affect the non-image parts of the cloth, the protons and alpha particles cause its coloration or body image. Using this illumination, Dr. Little was able to reproduce the straw-yellow color, which she learned subsequently is on the Shroud. The temperatures in the reactor's channels were as low as 40°C, so the radiation effects could be examined without the complication of heat degradation.[49]

Alpha particles, like protons, have short ranges in air and in solids; thus, both would evenly deposit their energy to produce the straw-yellow color—only on the topmost fibers of the cloth.[50] This particle radiation, like the neutrons and protons discussed earlier, would break many of the bonds of the molecular structure of the cellulose, but only in these topmost image fibers, thereby causing some of the single-bonded carbon atoms attached to hydrogen or oxygen to, thereafter, re-form with other carbon atoms into double-bonded, or conjugated, carbonyl groups. This occurs because carbon double-bonds re-form more rapidly and effectively than the other carbon combinations.[51] As demonstrated earlier with protons, these short-range, high-energy particles used by Little can also be irradiated on cloth without causing any initial visible effect, producing its oxidized, dehydrated, and conjugated carbonyl straw-yellow coloration over time or after the cloth has been artificially aged. Neutrons and gamma rays (along with electrons) will strengthen the cloth, making it more durable in a number of ways, while protons and alpha particles cause the body image fibers to become friable. The gamma rays will also initiate the cause of the coin and flower images. Further, if this radioactive illumination is combined with the Cloth-Collapse Theory, it would explain all the body image and other secondary image features encoded on the Shroud. It would also account for the blood marks, which have never been adequately or completely explained before.

Protons, electrons and neutrons are the main building blocks of matter. Alpha particles, consisting of two protons and two neutrons, behave like protons and are one of the simpler things that result from these building blocks. If the body in the Shroud dematerialized, a portion of these main particles and their products would be given off along with gamma radiation.[52] As the body dematerialized, it would lose its mass, allowing the cloth to fall through it and receive this particle radiation, resulting in all of the unique and unusual features found on the images and background of the Shroud.

As the cloth fell through the body region, the very penetrating neutrons, electrons, and gamma rays would pass completely through the cloth, without encoding body images. However, the easily absorbed protons and alpha particles would penetrate no further than the topmost two to three fibers of the cloth, resulting in a straw-yellow color that developed over time. The parts of the draped cloth that were closest to the body would have been the locations over the highest parts of the supine body. They would have fallen through the radiating body region longer than the other parts of the cloth. The next closest parts of the cloth would have received not quite so much radiation as the closest parts, and so on. Yet even parts of the cloth that were not originally touching the body would have fallen into the radiating body region and received some radiation. Thus, a perfectly encoded, three-dimensional frontal body image would have developed.

The directionality of the image would have been a straight-line, vertical direction from each point of the draped cloth to each part of the body immediately below it. Gravity would naturally encode this feature of the frontal body image.

Since the entire image encoding event is completed while the cloth is only partially through the body region, only internal skeletal features closest to the body's surface (e.g., the hands and the face) would become encoded. This explains why the man's internal organs are not encoded. Some of the skeletal images may be somewhat convuluted, such as the thumb, because the surface skin and tissue above it would be encoded ahead of it.

Under this method, the oxidized, dehydrated, straw-yellow fibers would lack any cementation or added pigments, powders, or foreign materials. They would be uniformly encoded and stable in water, and stable in heat up to temperatures that would produce scorches. This process would leave a negative image on the cloth with left/right and light/dark reversal. However, when the negative was developed into a positive, every detail of the body would be revealed because each detail was encoded vertically onto the cloth in three-dimensional correlation. The collapse of the cloth through the body, encoding each particular point directly below it, would provide the image's highly detailed resolution.

This Historically Consistent Method accounts for all the body image features on the Shroud and is the first method to explain the very different and extensive blood marks found throughout the body images, yet contained within the threads of the cloth itself. These unique blood marks have never been displayed so realistically on cloth, canvas, or any other surface; they appear on the cloth exactly as they would appear on a real human being with extensive wounds.

These wounds were inflicted by several different instruments (lancia, flagrum, nails, thorns or sharp pointed objects, cross-beam, fists, etcetera) and were incurred over a period of several hours while the body was in different positions. Yet all of these different kinds of blood flows, marks, and wounds—numbering 130 or more—are displayed *on and within* the cloth with perfect correspondence as to how they would look on a body that actually bled from each of these wounds.

Throughout the century, physicians and other experts have described them as anatomically precise, perfect, and pristine, with clean outlines or borders yet without any evidence of having been smeared. This is most apparent with the approximately one-hundred scourge marks: photographic enlargers and microscopes revealed upraised edges and indented centers with serum surrounding borders for each scourge mark. Ultraviolet fluorescent light also confirms that serum surrounds the borders of the scourge marks and wounds located throughout the body. This feature, along with many other findings and features of the blood and body, reveals that this is congealed or clotted. Dr. Barbet referred to the various blood marks as "mirror images" because they reflected on the cloth what they would have looked like while they were on the body. However, whereas an actual mirror would merely reflect the blood marks of the body, this cloth also *contains* the precise blood marks themselves. No scientists, physicians, artists, or others have ever been able to convey such anatomically precise and complete blood marks onto cloth by direct contact or by any means, or portray them on any kind of surface.

Interestingly, many of these blood marks would not even have been in contact with the cloth when it originally draped over the body, and others would have been in only partial contact. It is very unlikely that the blood flows and scourge marks on the small of the back would been in contact with the back of the cloth. Similarly, the scourge marks on the back of the upraised left leg could not have been in contact with the cloth below it.[53]

There are also scourge marks all over the man's lower back. Since the body is in rigor mortis with the right leg stretched out and pointing down and flat toward the ground, and since the left leg is upraised and also pointed similarly as the right foot, the lower part of the back will arch farther, preventing it and its scourge marks from contacting the underlying cloth. There are also scourge marks located above the shoulder blades and abrasions at the very top of the shoulders near the neck. Because the man's arms were placed over his groin, the top part of the shoulders and their accompanying scourges would not have been in contact with the underlying flat cloth. (A simple experiment of lying in these positions on a hard floor verifies the points of contact.)

In addition, some of the man's wounds may have been in only partial contact with the cloth. One example is the side wound, where the upper arm could have prevented the cloth from draping around the entire bloodstain. Similarly, the bloodstains on the back of the head cover a very wide area. Since the head is round, it would not be in direct contact with the flat cloth along the entire width of these wounds. Yet, all of the various blood marks on the Shroud have been encoded in the same precise, complete, and unique manner—whether they were originally in direct contact, partial contact, or no contact with the cloth.

Yet each of these blood marks obviously did come into contact with the cloth—intimate, direct contact. This is best seen in the approximately one hundred scourge marks found all over the body, where photographic enlargers, microscopic examination, and ultraviolet lighting reveal that each of the scourge marks on the cloth contain upraised edges with indented centers and borders surrounded by serum. Such contact seems even more intimate than what direct surface contact would leave. Only a cloth collapsing into a body could explain how these blood marks could have *acquired* such intimate contact, especially the blood marks that were in only partial contact or were not even initially in contact with the cloth. Moreover, only a passing-through mechanism could explain the marks' pristine condition with clean outlines and borders, yet without any evidence of having been broken or smeared. After acquiring such intimate contact between the entire wound and the cloth and after encoding it as a result of such intimate contact, had the cloth then been separated or removed from the wound by any model other than a passing-through model, its wounds could not have maintained their pristine appearance.

The Historically Consistent Method encodes all blood marks during the same process that it encodes the body images; however, the blood marks are encoded differently. As we know, the blood marks are found within the threads of the cloth itself. When the cloth falls through the body region, it comes into complete and intimate contact with all the blood marks that were below the cloth. This leaves a much more thorough and precise blood mark than mere direct contact could do with a stationary body and stationary cloth. Because the blood marks are attached to the body they do not fall through it; however, as the cloth passes through, the blood marks become encoded and embedded into the Shroud in the same configuration that they were in while they were on the body. (Alternatively, because the blood consists of the same molecules, atoms, DNA, et cetera, as the body, it, too, allows the cloth to pass through it, but once embedded into the cloth and no longer attached to the body, the blood does not disappear when the body does.) We cannot say for sure, but the Gospels do record Jesus speaking of his resurrection in the days before he died, and he speaks of his body and his shed blood as two distinct things.

The reason these blood marks are mirror images is that each blood mark in its complete configuration was transferred onto and into the cloth regardless of how congealed it was or whether it had started to scab. When the cloth was initially laid over the man, some of the blood that was in contact with the cloth could have been partially encoded onto it, especially if the blood was still semi-liquid. This would most likely be found with blood from the postmortem or most recent blood flows. Because they are so easily attenuated, the protons and alpha particles that encoded the body images would not have penetrated the blood marks on the skin or on the cloth. Therefore, body image would not be found underneath the blood marks on the Shroud.

This process would also encode the blood marks on the dorsal side. When the body dematerializes, a small vacuum would be created. This would draw or pull the dorsal cloth up a short distance into the body region, encoding all of the blood marks and the parts of the dorsal body image not initially touching, along with its skeletal features. All of these features would also be encoded if the body vanished or disappeared vertically in the same direction in which the cloth collapsed.[54] Since radiation is emitted from the entire body, the amount of time that the dorsal part of the cloth is exposed to the radiation would be equal or nearly equal to the frontal part, thus leaving both images with nearly equal intensities, as is also found on the Shroud's body images.

Extensive and unforgeable medical, scientific, archaeological, and historical evidence indicates that the Shroud is the burial garment of the historical Jesus Christ. In several different Gospel references where Jesus was first seen again early in the morning on Easter Sunday, he was not described as having any blood flows or blood marks on his body.[55] Mary Magdalene and two of the disciples even mistook him for the gardener or another man.[56] These initial, unbloodied appearances of Jesus on that Sunday could be a corroboration that the blood marks completely left the body and became embedded in the cloth, as stated in the Historically Consistent Model.

More than fifteen years ago, I wrote to four of the leading STURP scientists about the blood marks and body images becoming encoded under a passing-through mechanism (appendix G). Never did these or any other scientists assert that the anatomical exactness of the blood marks would not have been encoded or embedded under such a mechanism. In 1991, two years after Dr. Jackson first presented his cloth-collapse model, he subsequently stated in another article explaining his passing-through method, "It is also possible that the collapse hypothesis might explain Barbet's 'mirror image' effect of the blood stains."[57] Recently, Dr. Thaddeus Trenn, director of the Science and Religion Course Programme at the University of Toronto, has confirmed that the intact, unbroken, and sharply defined blood marks would have been transferred to the Shroud under such a model.[58]

This model, combined with the initial drape and fall of the cloth, also accounts for many other features of the Shroud's image, features which most methods have not attempted to explain. This can be seen clearly by the complete blood marks from the forehead and sides of the head becoming embedded into the cloth and then displaced into the hair. Since the cloth initially falls down then out, these blood marks, already in stationary contact with the draped cloth, embed completely into the cloth and are laterally displaced into the region where the hair is encoded as body image. No other method has ever accounted for this feature, as only a Cloth-Collapse Model can account for it.

As can be clearly seen on the photographic negative, positive, and three-dimensional images of the face, part of the beard is upturned. The most likely explanation for this and the gaps along the sides of the face is that a small chin band held the mouth closed. Notice also on the photographic negative that vertical lines run down from the chin, especially below the right side of the man's beard, which is not so upturned. As the part of the cloth lying over the chin fell and flattened, it would acquire such lines or motion blurs. Perhaps more lines would be left, before the radiation ceased, in the area that did not have to fall down and through as much beard before flattening. This would be the area immediately next to where the beard was turned up on the man's right side. Notice also the wide, rectangular area of body image below the man's chin and beard. This could have been caused by the cloth's coming into contact with and encoding the neck or throat area under this model, leaving this appearance after the cloth is straightened or flattened.

Notice also the odd-shaped feature encoded as body image next to the neck area and below the end of the length of hair on the man's left side. This could be a displaced hair image caused by the cloth moving under this method. There are also two faint body images in the blank space off the left side of the man's face, next to the eyebrow and cheek bone. They, too, could be from motion blur by the cloth in this region, but their faintness, and the lack of any such image on the right side, could be due to the chin band slowing or impeding the complete collapse or encoding of the overlaying cloth in this region. In addition, the small lateral distortion at the femoral quadriceps would most likely be encoded by a cloth collapsing.

Other subtle forms of distortion also exist on the Shroud image that are accounted for by this model. For example, this model also explains the length of the man's fingers. Under the Historically Consistent Method, the protons and alpha particles emanating from these surface bones became encoded as the cloth passed through this high portion of the supine body. Dr. Giles Carter was the first to observe that the man's fingers were bent; this position naturally remains from the crucifixion. After a two-dimensional cloth falls through and encodes curved fingers, when the cloth is then straightened or flattened, it results in a longer area of the cloth having been used to encode the fingers than if the fingers had been straight. A simple experiment with a cloth tape measure bears this out if you measure from the top of the wrist to the end of your bent fingers, and then measure again to the end of your straight fingers. The cloth tape will reveal that the first measurement is longer. Thus, the encoded fingers look somewhat longer when encoded under this method.

As the Shroud passed through the body region, the gamma rays, electrons and neutrons would also strengthen the cloth, which helps explain its excellent condition. As we have seen, the Shroud's linen consists of cellulose, which contains long-chain molecules. These molecular chains have repeating subunits that pass through crystalline, partly crystalline, and noncrystalline regions. (A crystalline region has a specific internal and symmetrically arranged structure.) Gamma rays, electrons, and neutrons (unlike protons and alpha particles) are long-range particles or radiation and easily pass through linen cloth. As they did, a very small fraction of them would have caused a limited number of molecular bonds to break and re-form in the noncrystalline regions, thus cross-linking these molecules and giving the cloth greater resistance to solubility, oxygenation, and chemical reactions. According to Dr.

Little, "Given a high crystallinity such as one would expect to find in good quality linen . . ." this type of cross-linking "would account for the lack of degradation and 'aging' that might be expected in a material two thousand years old, and that had been subjected to repeated handling and ill-treatment."[59] Dr. Little continued: "Such a reduced chemical activity would also account for the fact that although the Shroud was reported to be covered with mildew spores there were no mildew reactions, so that the fabric was unharmed."[60]

On the other hand, Dr. Little's experiments show that protons and alpha particles produced quite different results on cloth. These nonpenetrating particles distribute their energy entirely within the topmost two to three fibers of the cloth. This particle radiation, which produces the body image under the Historically Consistent Method, disrupted the ordered regions and shortened the molecular chains, greatly reducing the cloth's strength and causing the material to become friable—the same condition ascribed to the body image fibers by STURP and other scientists.[61] Summarizing, Dr. Little stated, "An instantaneous disintegration of the nuclei of the atoms in the body would account for the formation of the image, detail by detail, and the good state of preservation of the linen of the Shroud. It would seem to be the only mechanism whereby the straw-yellow colour could be produced—and the lemon-yellow colour of the serum deposits."[62]

The only other forms of radiation that possessed many of the necessary image-forming qualities under the Cloth-Collapse Model, X rays and ultraviolet light, could not produce these cloth-strengthening results, since they are nonpenetrating. In addition, X rays and ultraviolet light tend to distribute their energy exponentially. Protons and alpha particles, conversely, distribute their energy much more evenly and uniformly. The uniform body image fibers that are found throughout the frontal and dorsal body images on the Shroud may be another indication and confirmation that these types of particle radiation and this process caused its images.

Moreover, as the neutrons pass through the Shroud under the Historically Consistent Method, they also have another important effect on the cloth. As established in chapter 8, these neutrons will cause newly created C-14 isotopes to form. Experiments, calculations, and carbon dating performed on cloth samples by Rinaudo, Moroni and the Isotrace Radiocarbon Laboratory showed that if a first-century cloth sample was irradiated with a neutron flux and then artificially aged (or subjected to elevated temperatures for a relatively short time), the cloth would subsequently carbon date to medieval times when standard acid and base pretreatment cleaning processes were applied.

These experiments appear to duplicate exactly what happened to the Shroud. All of the previous evidence in this book consistently indicates that the Shroud is from the first century and that some form of radiation caused its images. Under the Historically Consistent Method, a flux of neutrons is one component of the radiation naturally given off by the dematerializing body. For many centuries thereafter, the cloth then aged naturally. Subsequently, it was heated in the fire of 1532. Both of these conditions would have provided ample opportunity for the additional C-14 to single- and/or double-bind to the unbound chemicals of the broken bonds created by the neutron flux and to embed itself into the cellulose. This would have allowed many of the additional C-14 atoms to have survived the standard pretreatment cleaning processes that the three Shroud samples also received, which were similar to the pretreatment cleaning processes used in the experiments above.

Neither Dr. Jackson's method nor any of the various artistic or natural methods already discussed can create additional C-14 in the Shroud; the only other method that can is Dr. Rinaudo's. Yet, again, there are several physical features that his method cannot account for, which can be explained only by a Cloth-Collapse Model. There are also other image features that his particular radiation cannot account for. The Historically Consistent Method is the only one that creates additional C-14 in the cloth and can encode all of the physical features of the Cloth-Collapse Model and all of the image features of the Shroud.

This Historically Consistent Model can also encode other features that no other method has accounted for or no forger is capable of encoding: the possible coin, teeth, and flower images observed on the Shroud. None was observed until the twentieth century with the advent of modern technology. Since the man's teeth would be at a high point on the supine body, they would be subtly encoded as body image as the cloth fell through the mouth. However, the coin and flower images would be more faintly encoded in a different manner. The coin and flower images are like the Shroud's body image in the sense that their images were left on the linen cloth and they appear thermally stable and stable in water to the same extent as the body image. Since these features were not observed until the 1980s, samples from these regions were not consciously removed by STURP in 1978, so their precise chemical structures are not known. However, they are not three-dimensional, nor as highly resolved or focused, and they lack the vertical directionality of the body image.

When some of the many neutrons that flew out of the body region hit the coin or flowers, they could have caused these objects to leave faint images on the cloth in several ways. For example, when a neutron hits the nucleus of copper, the primary component of ancient bronze coins, the nucleus can absorb the neutron and give off either a proton, alpha particle, deuterium, or a low-energy gamma ray. Each of these particles—protons, alpha particles, or deuterium—encodes superficial images on the cloth and, if they were given off the coin's surface, could encode the coin's features. Similarly, flowers contain trace amounts of heavier elements such as iron, calcium, and potassium. When any of the countless neutrons hit these three heavier elements, each could also absorb the neutrons and give off protons and alpha particles.[63] Any protons or alpha particles given off the flowers' surfaces would also encode a superficial image on the Shroud.

Gamma rays are also given off by all of these heavier elements and could also encode faint coin and flower images on the Shroud. These gamma rays would radiate at low energy, and when they hit the electrons in the heavier elements of the coin and flowers, their atoms could fluoresce long-wave X rays from the objects' surfaces, which, again, could also encode superficial images on the Shroud. They could also fluoresce as short-wave X rays, or even as visible light, and encode an image more than two or three fibrils deep at these particular locations, or fluoresce as ultraviolet and leave a superficial image. The most common element in ancient coins—copper—would be the best of all of the heavier elements above at absorbing the neutrons and giving off these gamma rays. However, the heavier elements found in flowers, iron and potassium, are also excellent at absorbing neutrons and giving off gamma rays. Calcium could also do this fairly well.[64]

If further research confirms the presence of the coin and flower images, they would certainly fit into this image-encoding theory. While both coin and flower images would be faint, the coin image would not necessarily be encoded much more intensely than the flowers. Many of the nonpenetrating X rays and ultraviolet rays would not escape from the coin, but would not have nearly as much difficulty with the flowers. Flowers have some tendency to leave images on cellulose. There are many examples of flowers having left images on paper. Unlike the coin, throughout this image-encoding process the flowers would have maintained or increased their original contact or closeness to the frontal and dorsal sides of the Shroud, where their images have been observed. The coin could have made only a faint image on the cloth from its original position over the eye. (There could easily have been another coin over the left eye, but it did not leave an impression that is clearly detectable at present.) It would have fallen through the body region faster than the larger, connected cloth after briefly and faintly encoding the letters, lituus, and motif of a Pontius Pilate lepton.

However, something else quite interesting is also revealed by studying the area of the man's eyes. Notice on the positive images of the face that the man's closed eyes appear normal, that is eyeballs or round curved objects appear under the eyelids. In fact, these round curved eyeballs appear at the same locations on the three-dimensional images of the face. (The coin itself would simply be too small and would lack sufficient relief to encode this type of three-dimensional image.) The overall area of the eyes is three-dimensional and vertically directional and possesses the photographic resolution of the body images. The facts indicate that the coin features were encoded in one manner and the eyeballs in another. Under the Historically Consistent Method, the coin's feature are faintly encoded at the beginning of the event. After it falls through the body, the cloth collapses through the same area encoding the eyelids and eyeballs (over the faint letters, lituus, and motif) in the same manner as it did the rest of the body image.

While Rinaudo's Protonic Method and Jackson's Cloth-Collapse Model were the only ones to date that accounted for all of the Shroud's basic body image features, they do not fully explain or account for all the features found on the Shroud. Rinaudo's method and research made immeasurable contributions to resolving the cause of the Shroud's image and its C-14 dating; however, his radiation does not strengthen the cloth, and his model can't encode the blood marks or displace them into the hair, encode the skeletal features, teeth, coin or flower images, or image distortions. Jackson's research and model are also invaluable because so many of the Shroud's features can only be explained by a cloth-collapsing model. Yet his radiation cannot create additional C-14, strengthen the cloth, encode coin or flower images or fully explain the body image color or chemistry. He also fails to realize or explain that such a model can encode and embed into the cloth all of the man's blood marks, encode all the dorsal body image features, and explain the elongated fingers and body image nuances at the sides of the face. Only the Historically Consistent Method can explain all image and non-image features found on the Shroud of Turin.

FUTURE RESEARCH AND TESTING

Future examination and testing obviously must be undertaken on the Shroud. The research recommended in this chapter is in addition to that mentioned in chapter 8. However, these recommendations cover just some of the important areas of research that must be conducted in another scientific examination of this important relic.

One of the initial ways the Shroud should be examined is to assess or measure its background and body images as much as feasible, regarding the oxidized, degraded state of its cellulose; the extent of its conjugated carbonyl groups; and its straw-yellow coloring. The spectral imaging techniques proposed by Dr. Grunfest could be applied to the entire Shroud to ascertain whether the molecular bonds of the cellulose had been broken and to what extent their chemicals had reattached or recombined. If the Shroud's body images were formed by the radiation indicated by the Historically Consistent Method, these new bonds should have developed on the Shroud's body images and/or background portions, only much more so on the body images. Perhaps the types of cross-linking in the cloth's non-crystalline background region that strengthen the linen might also be ascertainable by spectral imaging technology. These assessments or measurements should be compared to existing 2,000- and 650-year-old linen cloth samples.

Various other linen cloth samples should also be irradiated with various combinations of protons, alpha particles, neutrons and gamma rays, and then artificially aged. The forms of radiation from Rinaudo's and Jackson's methods should also be applied to cloth samples in varying amounts. The methods discussed in chapter 5 should also be tested. If they can produce superficial, straw-yellow fibers on cloth, the colored portions and backgrounds of these cloths should be examined and compared to the Shroud's on the state of their oxidized degraded cellulose, the extent of their conjugated carbonyl groups, their straw-yellow coloring, the extent to which their chemical bonds have broken and reattached, their cross-linking, uniformity of color of image fibers, et cetera.

If one of the combinations of protons, alpha particles, neutrons, and gamma rays irradiated onto these cloth samples can approximate the above characteristics found on the Shroud, then the amounts of neutron flux that could be found within this combination should be applied to other known first-century cloth samples.

These samples should then be artificially aged or heated and pretreated normally, then carbon dated. Then, if these cloths carbon date to medieval times, this would be one further and critical confirmation that the Shroud is from the first century and that the radiation, as explained in the Historically Consistent Method, encoded its image.

Both sides of the cloth should be examined for non-superficial images at locations where coin or flower images appear. The lack of such images on the back side or non-superficial images on the body image side would merely mean that short wavelength or visible light waves were not created from these objects. However, the presence of oxidized, degraded cellulose on the back side of such coin or flower images would indicate that the radiation from the Historically Consistent Method encoded these images on the Shroud. Similarly, the back of the cloth behind the frontal image should be examined for oxidized, degraded cellulose, for if the body gave off ultraviolet light instead of gamma rays, it would have superficially colored the back of the cloth as it passed through the body.

Ancient coins and flowers one to two days old should also be irradiated with various forms of radiation, especially the radiation in the Historically Consistent Method, with control cloth samples placed over them to see if they will absorb such radiation and fluoresce in other wavelengths and/or give off their own protons, alpha particles and/or deuterium from the neutrons within the radiation. The coin features over the right eye and the flower images observed on the Shroud should also be carefully examined by the spectral imaging technique to ascertain if they result from encoded fibers with similar chemical characteristics as the rest of the body images. If so, it could be another indication that the radiation of the Historically Consistent Method encoded these faint images.

Control cloths irradiated with ultraviolet light and x-rays would also indicate if these forms of radiation tend to deposit their energy superficially but exponentially, that is, more toward the end of their deposit. These and other types of radiation should be compared to the evenly distributed and uniform straw-yellow fibers produced by combinations of protons, neutrons, alpha particles, and gamma rays to see if they, too, match these features on the Shroud body image. Moreover, additional chemical studies should be undertaken on samples removed from the Shroud in order to understand its chemistry better, to further compare it to the control samples, and to better understand how to preserve the Shroud and its images.

The research and testing recommended at the end of chapter 8 showed how to ascertain whether the Shroud had been irradiated with neutrons and, if so, how to calculate the amount of additional C-14 created by this event, and to ascertain the cloth's true age. The testing suggested at the end of this chapter shows how to determine whether the Historically Consistent Method also caused the images on the Shroud.

This and even more extensive testing and examination should be conducted openly on the Shroud and its samples by an international group of scientists who possess an extensive knowledge of the Shroud and have studied all previous scientific articles and reports on the subject. These scientists must be willing to work as part of a team, and no one individual or segment of scientists should seek to monopolize the research and testing, but should seek to include all interdependent and complementary forms of nondestructive testing and examining of the cloth. Full disclosure of all methods, data, measurements, findings, and conclusions should also be published in peer-reviewed scientific journals for the worldwide public.

EVIDENCE FOR THE RESURRECTION

For all the reasons discussed in the last three chapters, future research and testing is clearly required in connection with and on the Shroud; however, there is presently a great deal of evidence for the resurrection that can be derived from studying the cloth and its images.

From the extensive medical, chemical, and physical evidence we know the Shroud contained the body of a dead man. From centuries of observation and experiments, we know that dead bodies do not leave such unprecedented body images or blood marks. We also know for many reasons that these features could not have been forged in any way. The most sophisticated technology available today is not capable of duplicating the Shroud's qualities nor the process that created them. As Shroud scholar and author Frank Tribbe has stated, "If the markings on this Shroud resulted from some other natural event or process, then that event or process must have been totally unique, for it never occurred previously and has not been repeated to the knowledge of our civilization."[65]

The evidence in this and previous chapters clearly indicates that radiation caused the body images on the Shroud. This radiation came from the length and width of a real human corpse, including the internal parts of his body. Radiation does not naturally come from a dead body, and if we were to start a fire under a corpse or make it radiate in some way, we would not only create additional problems with the body, blood, and cloth, we still couldn't come close to to making this kind of unique image on a cloth. Moreover, the radiation was vertically directional and encoded through space. Radiation coming from a corpse in such an unprecedented and unique manner is evidence of and consistent with the resurrection.

Only a cloth collapsing through a wounded body giving off uniform radiant energy can explain the Shroud's more than twenty body image features, along with the more than one hundred blood marks. We saw in this chapter how this method not only can encode the mutually inconsistent primary body image features, but also the distracting and misregistered blood marks and body image features caused by the cloth's collapsing motion. Furthermore, this method not only explains how each of the complete and coagulated blood marks that formed naturally on a human got embedded into the cloth, but also how they separated from the body, leaving the original smooth surfaces between the wounds and the skin unbroken and intact on the cloth. Obviously the body has left the cloth. Obviously, each of the numerous wounds once had intimate contact with the cloth. However the cloth could not have been removed from the body by any human means without breakng or smearing many, if not all, these blood marks. Since there are no decomposition stains of any kind on the cloth, this body had to have left it in a unique manner within two to three days.

The completely embedded blood marks in Jesus' burial shroud are also consistent with the historical descriptions of Jesus' appearance following his resurrection, since none of the witnesses who saw him on Easter Sunday mention any blood marks on his body, let alone partial or broken blood marks. These facts, along with the image-encoding event and the body exiting the cloth within two to three days of dealth, are all consistent with and indicative of the resurrection.

Furthermore, only a cloth collapsing through a body giving off particle radiation consisting of a neutron flux can explain the Shroud's faint coin and flower images, its cloth strengthening features, and the creation of additional C-14. If a neutron flux irradiated the Shroud, this *alone* would have been an unprecedented event that could only be explained by the resurrection. Even today, to produce a neutron flux requires the facilities of a nuclear laboratory, yet the neutron flux cannot be made to come from a body let alone the length, width, and depth of a dead body. Similarly, if a neutron flux is shown to have occurred in the limestone walls of Jesus' tomb (which has been closed for centuries), it would clearly have been an unprecedented event that is consistent with and indicative of the resurrection having caused it.

In addition, if particle radiation consisting of the basic building blocks of matter irradiated the Shroud, this, too, would be an unprecedented event that is consistent with the resurrection, for like a neutron flux, this radiation can only be produced today at a nuclear laboratory. For these particles to have radiated from the length, width, and depth of a dead body would constitute additional evidence for the resurrection.

Unlike other failed methods, the Historically Consistent Method makes only one assumption—that the body disappeared or dematerialized. With this dematerialization, particle radiation is given off naturally and the cloth falls through the radiant body, thereby encoding or causing every one of the Shroud's numerous image and nonimage features summarized above. This is one of the real strengths to this method, for all of the historical sources that describe Jesus' burial also declare that his body disappeared without any human assistance. This disappearance was a startling occurrence to the people of this historical time.

At Jesus' resurrection, his body clearly disappeared from his Shroud and from his burial tomb. Moreover, after the resurrection event, the same historical Jesus was described by these same historical sources as having the ability to pass through walls, appear in other forms, and make his body vanish.[66] Of all the remarkable or miraculous things attributed to Jesus, he was never described as passing through walls or making his body change or vanish until after his resurrection.

As we traced the evolution and development of the Historically Consistent Method and showed how it naturally explains and accounts for every single one of the Shroud's numerous, incompatible and unprecedented features, we only referred to this method generally and accurately as a historically consistent method, without ever giving it a specific name. That is because, although scientists are just beginning to understand this method, it already had a name—the resurrection of the historical Jesus Christ.

As a further point of corroboration, the evidence shows the body images, blood marks, and the image-encoding event, none of which can be forged or occur naturally, all occurred to the historical Jesus Christ. It is important for the reader to understand that most of the above evidence already exists. Future testing could refute this, or add additional evidence to confirm it.

CHAPTER ELEVEN

UNLIMITED, WORLDWIDE OPPORTUNITY

CORRELATION WITH HISTORICAL ACCOUNTS

The vast majority of the evidence derived from the Shroud matches in every minute detail, and in ways that cannot be forged or occur naturally, all of the physical events, including location, time, place, and participants, that occurred to the historical Jesus Christ during his crucifixion, death, burial, and resurrection. Since these events are described in several books of the New Testament, these historical sources should be examined as objectively as possible and compared to other historical sources in order to measure their authenticity and reliability.[1]

In the following table, we compare New Testament manuscripts to those of the classical Greek and Roman eras, which run from approximately 800 B.C. to A.D. 476, the period before, during, and after the writings and events of the New Testament. In order to copy or publish such writings, they had to be written by hand. In fact, these writings or manuscripts are the primary sources for our knowledge of the history of these periods. This table compares the Gospels and other books of the New Testament manuscripts with the major works of every major writer of the Greek and Roman eras in an attempt to ascertain the accuracy and reliability of these manuscripts today.

As the accompanying table will show, we may feel more confident of the textual authenticity of the New Testament than we can for any of the works of the classical age of history. Indeed, as John Warwick Montgomery, Dean of Jurisprudence at the Simon Greenleaf School of Law and former Director of Studies for the International Institute of Human Rights, has stated, "To be skeptical of the resultant text of the New Testament books is to allow all of classical antiquity to slip into obscurity, for no documents of the ancient period are as well attested bibliographically as the New Testament."[2]

The table below is a listing of the major writers and their major works from the classical age of history. These great works are listed along with the dates of their original composition, the number of manuscripts or copies in existence today, the dates of our earliest copies, and the time span between the earliest existing copy and the original composition of the work.

Textual Attestation of Ancient Authors

Author	Work composed	Manuscripts/Papyri	Earliest	Time lapse
Homer, poet (Greek) lived: ca B.C. 700				
Iliad	ca. B.C. 800?	188 manuscripts	ca. A.D. 950	1750 years
		372 papyri	ca. B.C. 50	750 years
Odyssey	ca. B.C. 700?	80 manuscripts	ca. A.D. 950	1650 years
		30 papyri	ca. B.C. 50	650 years
Hesiod, poet (Greek) lived: ca. B.C. 700				
Theogony	ca. B.C. 700?	79 manuscripts	ca. A.D. 950	1650 years
		32 papyri	ca A.D. 50	750 years
Works & Days	ca. B.C. 700?	260 manuscripts	ca A.D. 950	1650 years
		22 papyri	ca A.D. 50	750 years
Aeschylus, tragic poet (Greek) lived: B.C. 525–456				
7 Tragedies	B.C. 484–456	30 manuscripts	ca. A.D. 1050.	1500 years
Herodotus, historian (Greek) lived: B.C. 480–425				
Histories	ca. B.C. 450–430	9 manuscripts	ca. A.D. 950	1380 years
		18 papyri	ca. A.D. 50	480 years
Sophocles, tragic poet (Greek) lived: B.C. 496–406				
7 Tragedies	ca. B.C. 441–406	200+ manuscripts	ca. A.D. 950	1350 years
Thucydides, historian (Greek) lived: ca. B.C. 460–400				
Peloponnesian War	ca. B.C. 425-400	70+ manuscripts	ca. A.D. 950	1350 years
Euripides, tragic poet (Greek) lived: ca. B.C. 485–406				
21 Tragedies	ca. B.C. 438–406	43 manuscripts	ca. A.D. 1150	1550 years
Aristophanes, comic poet (Greek) lived: ca. B.C. 450–385				
11 Comedies	ca. B.C. 427–388	6 major manuscripts	ca. A.D. 1000	1400 years
		numerous papyri	ca. A.D. 250	750 years
Lysias, orator (Greek) lived: ca. B.C. 459–380				
Orations	B.C. 407-380	10+ manuscripts	ca. A.D. 1150	1500 years
Xenophon, historian (Greek) lived: ca. B.C. 430–354				
Hellenica	to B.C. 362	12 major manuscripts	ca. A.D. 1350.	1700 years
		3 papyri	ca. A.D. 50	500 years
Plato, philosopher (Greek) lived: ca. B.C. 429–347				
Dialogues	B.C. 393–350	7 manuscripts	ca. A.D. 900	1250 years
Demosthenes, orator (Greek) lived: B.C. 384–322				
Orations	B.C. 363–340	200+ manuscripts	ca. A.D. 900	1250 years
Aristotle, philosopher (Greek) lived: B.C. 384–322				
Metaphysics	B.C. 348–320	3 major manuscripts	ca. A.D. 950	1280 years

Author	Work composed	Manuscripts/Papyri	Earliest	Time lapse
Euclid, mathematician (Greek)				
Elements	B.C. 300	6 major manuscripts	ca A.D. 888.	1175 years
Epicurus, philosopher (Greek) lived: B.C. 342–271				
Letters	B.C. 300	6 manuscripts	ca A.D. 1150.	1450 years
Polybius, historian (Greek) lived: B.C. 203–120				
Universal History	to B.C. 145	22+ manuscripts	ca A.D. 950.	1100 years
Cicero, orator (Latin) lived: B.C. 106–43				
Orations in Catilinam	B.C. 63	13+ manuscripts	ca A.D. 850	900 years
De Oratore	B.C. 55	6+ manuscripts	ca A.D. 850	900 years
De Republica	B.C. 51	1 manuscript	ca A.D. 350	400 years
De Officiis	B.C. 44	7+ manuscripts	ca A.D. 800	850 years
Catullus, poet (Latin) lived: ca. B.C. 84–54				
Poems	B.C. 60–54	4 manuscripts	ca A.D . 1300	1350 years
Lucretius, poet (Latin) lived: ca. B.C. 94–55				
De Rerum Natura	ca. B.C. 60	10 manuscripts	ca A.D. 800	60 years
Sallust, historian (Latin) lived: B.C. 86–34				
Bellum Catilinae	B.C. 43	20 manuscripts	ca. A.D. 850	900 years
Vergil, poet (Latin) lived: B.C. 70–19				
Aeneid	B.C. 42–19	19 major manuscripts	ca. A.D. 150	170 years
Horace, poet (Latin) lived: B.C. 65–8				
Poems	B.C. 35–13	250 manuscripts	ca. A.D. 850	875 years
Livy, historian (Latin) lived: B.C. 59–A.D. 17				
Histories	B.C. 29–A.D. 11	15 manuscripts	ca. A.D. 350	350 years
Propertius, poet (Latin) lived: ca. B.C. 50–16				
Elegies	B.C. 33–A.D. 16	18 manuscripts	ca. A.D. 1200	1215 years
Tibullus, poet (Latin) lived: ca. B.C. 48–19				
Elegies	ca. B.C. 28	9 major manuscripts	ca A.D. 1050	1080 years
Ovid, poet (Latin) lived: B.C. 46–A.D. 17				
Metamorphoses	A.D. 8	17+ manuscripts	ca A.D. 860	850 years
Lucan, poet (Latin) lived: A.D. 39–65				
Pharsalia	A.D. 62	17+ manuscripts	ca A.D. 850	800 years
Seneca the Younger, philosopher, poet, administrator (Latin) lived: ca. B.C. 5–A.D. 65				
Moral Letters	A.D. 64–65	14 manuscripts	9th century	800 years
Pliny the Elder, natural historian (Latin) lived: A.D. 23–79				
Natural History	A.D. 77	200 manuscripts	5th century	400 years

Author	Work composed	Manuscripts/Papyri	Earliest	Time lapse
Martial, composer of epigrams (Latin) lived: ca. A.D. 40–104 *Epigrams*	A.D. 80–102	14 manuscripts	9th century	800 years
Josephus, historian (Greek) lived: ca. A.D.. 37–100 *Jewish Antiquities*	A.D. 93–94	8 manuscripts	ca. A.D. 1050	950 years
Pliny the Younger, civil servant and imperial administrator (Latin) lived: A.D. 61–113 *Letters*	A.D. 96–109	14 major manuscripts	6th century	450 years
Tacitus, historian (Latin) lived: ca. A.D. 55–117 *Annals*	ca. A.D. 115–117	36 manuscripts	ca. A.D. 850	730 years
Plutarch, historian-biographer (Greek) lived: ca. A.D. 46–125 *Parallel Lives*	A.D. 105–115	11+ manuscripts	ca. A.D. 1100	1000 years
Suetonius, historian-biographer (Latin) lived: ca. A.D. 69–140 *Lives of the 12 Caesars*	ca. A.D. 121	16+ manuscripts	ca. A.D. 850 .	730 years
Florus, historian (Roman) lived: ca. A.D. 120 *Epitome of Rome*	ca. A.D. 120	27+ manuscripts	ca. A.D. 850 .	730 years
Juvenal, satiric poet (Latin) lived: ca. A.D. 60–130 *Satires*	A.D. 110–130	17 manuscripts	A.D. 875–900	750 years
Claudius Ptolemy, mathematician, astronomer, geographer (Latin) *Mathematike Syntaxis*	A.D. 121–151	6 manuscripts	9th century	700 years
Appian, historian (Greek) lived: ca. A.D. 160 *Roman History*	ca. A.D. 160	8+ manuscripts	ca. A.D. 1100	840 years
Galen, philosphical and medical writer (Latin) lived: A.D. 129–199 *De usu partium*	A.D. 129–199	9 manuscripts	ca. A.D. 1050	750 years
NEW TESTAMENT (Totals as of 1972)	ca. A.D. 50–140	85 papyri 268 majuscule mss. 2792 minuscule mss.	ca. A.D. 110 ca. A.D. 250–350 A.D. 835	20 years 100+ years 800 years

Without some qualification the figures given above could be misleading. Note that for some authors over two hundred manuscripts are indicated, while twenty or fewer are listed for others. This reflects figures available to the researcher; the figures were drawn fundamentally from the most recent scholarly editions of the Greek or Latin texts in question and the information presented in each instance by the editor, who indicates the manuscripts he or she has employed to construct the text there printed. Of concern to such editors is not the total number of manuscripts available, but the major older manuscripts offering significant variant readings of the text. In the case of most authors, numerous manuscripts survive from centuries following the earliest extant manuscripts, but these can easily be grouped into families and shown to have been copied from extant earlier manuscripts, so that they do not enter into account in the establishment of the most authoritative text of an author.

These manuscripts were kept on either parchment or papyri. Parchment is a much higher grade of paper, made from carefully processed sheepskin or other animal skin. Papyri, which is where the modern term *paper* comes from, is probably the oldest form of paper we have. It was made from reed grown in marshes. In time, it generally rotted everywhere except in arid climates such as Egypt. The papyri are always fragmentary, as is the case also with the earliest surviving papyrus fragment of the New Testament, a fragment containing only the Gospel of John 18:31-33, 37–38; as the Gospel of John is supposed to have been composed about A.D. 90, this papyrus fragment is only about twenty years later than the text of which it is a copy.

The terms *majuscule* and *minuscule* refer to styles of handwriting. Majuscule used larger, capital or uncial letters throughout. Minuscule was a script of smaller letters, and eventually superseded the former style.

In addition to manuscripts and papyri of the Greek and Latin texts of ancient authors and of the New Testament, there are manuscript anthologies containing short citations of ancient authors, lectionaries containing short biblical passages, and early translations of the New Testament into Latin, Coptic, Syriac, and other languages used by Christians in the later ancient world. These have not been counted in the table listing.

Professor Carl W. Conrad, Washington University, Department of Classics, researched and prepared the preceding table.

As can be clearly seen for all of the Greek and Roman works of antiquity, the earliest copies that we have were written many centuries after the original work. In every instance, the earliest extant copies were written hundreds and, at times, more than a millennium after the original work was first composed. And of these centuries-old copies, the number of available manuscripts are very few in comparison to the Gospels and the rest of the New Testament. F. F. Bruce, a New Testament scholar at the University of Edinburgh, is certainly correct when he states, "There is no body of ancient literature in the world which enjoys such a wealth of good textual attestation as does the Gospels and the New Testament."[3] We can certainly feel more confident that the New Testament accurately reflects the contents of the original documents than we can of any other ancient works of literature.

In numerous references in several books comprising the New Testament, Jesus appears in the flesh to his disciples and others following his crucifixion and death. He appeared to the women returning from the tomb; to the apostles; to the Emmaus disciples; to James; to Paul; and to more than five hundred other people.[4] These appearances of the resurrected Jesus completely changed the disciples. Fearful and despondent following the crucifixion and death of their leader, they became emboldened enough to preach and testify throughout the land of Jesus' resurrection and of his works and deeds. They preached and witnessed these accounts at great risk to themselves. They were arrested, persecuted, and killed because of their testimony, yet none ever recanted. From being afraid to be identified with Jesus, they were transformed into some of the most dedicated and well-known witnesses in history. Such transformation on the part of the disciples could not have taken place as a result of a lie or fabrication. Just like the bibliographical information above corroborates the contents of the New Testament, the conduct of the disciples confirms or corroborates the first-hand accounts of seeing and touching the resurrected Jesus Christ.

The earliest known written accounts of Jesus' resurrection begin appearing within about twenty years of the event, or within the lifetimes of the numerous witnesses to this event.[5] If such accounts were incorrect or inaccurate, one would expect them to have been refuted by the participants and witnesses, including hostile witnesses. Yet there was no such refutation of any of these historical sources, which further corroborates these accounts. The establishment of Christian churches throughout the land also occurred during the disciples' lifetimes. At the very center of the preaching of the early Christian church was also the resurrection of Jesus Christ. This was not a later legend added by future generations, but was central to the original teachings of the original witnesses.

These historical sources were the most widely circulated documents of their day. They were so widespread that even without these existing ancient manuscripts, practically the entire New Testament could be reconstructed just from their quotations that still survive today in commentaries and sermons from the first three centuries alone. In addition, unlike other ancient writings, they were also translated into several foreign languages beginning at a time contemporaneous to the originals.

Archaeology has also authenticated the accuracy of these historical sources by verifying countless detailed historical statements contained in them. The list would be too exhaustive to cover, but archaeology has confirmed the existence of markets, temples, aqueducts, public officials, dates of service, et cetera, exactly as they are mentioned in the New Testament. Archaeology has often corrected numerous assertions against the New Testament's accuracy by subsequently confirming the preciseness of their contents and the inaccuracy of the conflicting sources. Archaeology has consistently revealed biblical accuracy to the smallest detail on countless items, customs, individuals, titles, physical locations, et cetera, of the type that eyewitness contemporary accounts would consistently contain.[6]

These historical sources also report that, after the witnesses saw the resurrected Jesus, his body could transform itself to pass through solid objects and vanish. They also describe the transfiguration of the historical Jesus Christ on a mountain where his face shone like the sun and his clothes became dazzling white. This type of radiation did not harm Jesus' body or immediately affect his clothes.[7] Interestingly, these are the same types of features that scientists have independently concluded, after twenty-five years of investigation, could be among the principal causes of the formation of all the unique features found on the images on the Shroud, which the original eyewitnesses and authors of the Gospels could never have anticipated.

Because of the development of modern professions such as archaeology, science, and medicine, we can sometimes know more in the twentieth century about subjects and events in the past than those who lived during those actual times. Because of such developments, we may very well know more today about the crucifixion and resurrection of the historical Jesus Christ than those who lived in the first century. And this knowledge should only continue to grow as these and other professions continue to advance and develop, and wide-ranging testing is conducted on the Shroud. Because we have both the recently developed evidence from the Shroud and the historical evidence from the first century, we actually have a strong and very interesting case for the crucifixion and resurrection of the historical Jesus Christ. This results in, perhaps, the most dynamic, startling, and scientifically advanced evidence in the world today combined with the most well -attested and well documented evidence in all of ancient history.

SUMMATION

The events of the crucifixion and resurrection of the historical Jesus Christ and their evidence in the body and blood images, along with many other aspects of the Shroud, are unparalleled in history. As we have seen in a variety of ways, the numerous features found on the Shroud could not have been seen or forged and could not have occurred naturally. The combination of evidence found from the Shroud cannot be matched with any other known or hypothetical series of events. The Shroud of Turin is easily the most unique relic in all of history. The crucifixion and resurrection of the historical Jesus Christ are the most unique events purported to have occurred in history. The intricate relationship between the Shroud and the resurrection should be the subject of intense, earnest, and objective study by people throughout the world and should serve as a basis for further scientific, medical, and other examination by experts from around the world.

This medical, scientific, archaeological, and historical evidence not only provides us with a great deal of information about the body images and the blood marks, their characteristics, and how they were encoded, but the evidence also relates directly to the age and origin of the cloth, the events depicted thereon, who performed these acts, where the acts were performed, and the identity of the victim. This evidence relates not only to the authenticity of the Shroud but also necessarily to the crucifixion and resurrection of the historical Jesus Christ. For the first time in history, the worldwide public has an unprecedented opportunity to look at the question of the occurrence of the crucifixion and resurrection of the historical Jesus Christ in an impartial, comprehensive manner.

From the anatomically flawless representations of a body whose only distortions are consistent with a cloth draped over it; the elimination by scientific means of all other artificial body models such as statues, bas reliefs, engraved lines, et cetera, as a cause of or being represented on the image; the bruises and swelling on the body; the contraction of the thumbs; the identification of serum surrounding many of the wounds; the arterial, venous, and capillary bleeding; the intimately and microscopically precise representation of each scourge mark as they would occur on a human body; the anatomically flawless wounds of coagulated blood located on the cloth precisely as they would appear on a human body, which are found from head to feet, on both the front and back; and the identifications of human hemoglobin, human albumin, human whole blood serum, and human immunoglobulins: all clearly show the image is that of a human male.

The swelling found on both cheeks; the triangular wound on the right cheek; and the bruised, swollen, deviated, and possibly dislocated nose show the victim was beaten about the face. The more than thirty wounds on the top and middle of the back of the head and on the top, middle, and sides of the forehead indicate that his head was pierced with many sharp objects. Such wounds are what one would expect if the victim had been fitted with a crown of thorns. Israeli experts have also identified numerous images and pollens from thorns on the Shroud.

There are 100 to 120 wounds, in dumbbell-shaped patterns running parallel and diagonal across the body in groups of two or three. Photographic enlargement, microscopic examination, and ultraviolet lighting show that the skin was torn open and that clot formation and retraction occurred. These wounds are signs of a savage scourging, probably by two separate individuals.

Two broad excoriated areas present across the victim's shoulder blades are consistent with a large, rough burden being placed upon the shoulders. Many crucifixion victims were forced to carry their own crossbeams to the execution site. Scratches, lesions, and abrasions found on the front of the man's knees, along with dirt on the knee regions and nose, suggest that the victim endured several falls.

Pathologists and physicians are convinced the victim was crucified. This is apparent from the nail wounds in the wrists; the blood flows down the arms at the angles formed by a body moving repeatedly in a seesaw motion; the nail wounds in the feet; the abnormally expanded ribcage; the enlarged pectoral muscles; the upraised left leg; the beating and scourging (which often preceded crucifixions); and the vertical position of the victim at the time of death. Even the blood chemistry and color are consistent with a crucifixion and its preceding tortures.

All medical authorities who have studied the Shroud agree the victim is dead. This is most apparent from the findings of rigor mortis present on numerous places throughout the body, such as the left leg, thighs, buttocks, torso, thumb, head, expanded ribcage, and the pectoral muscles. His death is further indicated by the postmortem blood flow on the foot and the postmortem side wound. Neither the unusual mixture of fluids that escaped from the side wound, nor the manner of flow, would have been present if the victim had not already been dead. This postmortem flow from the side wound consists of blood and a watery fluid. Unlike most crucifixion victims, it was not necessary to break this victim's legs to kill him.

Numerous signs indicate that radiation coming from this dead body produced its images. This can be seen in the many similarities between the Shroud's light scorches and its body images; that particle radiation and artificial aging produces the advanced yellowing and degraded cellulose with a conjugated carbonyl molecular structure like that on the Shroud; that this advanced yellowing and chemically altered degraded cellulose is found throughout the length and width of the frontal and dorsal body images; that these body images are perfectly coordinated and synchronized with about 130 human wounds of various types scattered throughout both sides of the body; that nonmagnified, correctly encoded internal skeletal features are also observed on the frontal and dorsal body images; that the three-dimensional distances from a body to a cloth draped over it which have been encoded in a vertically directional manner have only been explained by scientists by bursts of collimated radiation from the body to the cloth; that the body image's high resolution, three-dimensionality, and vertical directionality have only been simulated by scientists using a mechanism in which light or radiation was attenuated and then traveled in a vertical straight-line direction from a model face while it was focused in a camera; that radiation from a body would act uniformly over skin, hair, bones, teeth, coins, or flowers; that radiation can produce all of the Shroud's body image features; that only a method involving radiation emanating uniformly throughout the length, width, and depth of a human body has explained all of the Shroud's numerous features.

Something miraculous occurred to this individual. We have seen that the images on the Shroud literally defy the laws of chemistry and physics as we understand them. Despite centuries of efforts from people across the earth, all proposed mechanisms for creating the image on the Shroud have failed to account for the unprecedented characteristics of the body images and blood marks until the recent development of the Historically Consistent Method that was developed by combining research from scientists throughout the world on all aspects of the body images and blood marks on the Shroud. This theory states that if a body instantaneously dematerialized or disappeared, particle radiation would be given off naturally and all the unique features found on the Shroud's body images and blood marks would occur. As detailed in chapter 10, this method accounts for all of the Shroud's body image features produced by radiation in the last paragraph, as well as a negative body image with left/right and light/dark reversal that is superficial, thermally stable, and stable in water; lacks two-dimensional directionality and saturation; whose fibers are uniformly encoded, do not show any cementation, and have no added pigments, powders, or other added materials.

This method also explains less obvious features on the Shroud, such as the blood marks being displaced into the hair; the blank spot at the groin; the skeletal features and the teeth images; the length of the fingers; the lateral distortions at the hips; the sharp body edges, yet a lack of images along the length of the man's sides and the top of his head. This historically consistent method also explains how the image gaps along the sides of the face, along with the motion blurs in one of the gaps and below the beard, were developed in combination with the chin band. It also explains how faint coin and flower images could have been encoded on the Shroud.

This method further encodes the Shroud's images through space, including details of the body surfaces and blood marks that were not originally in contact with the cloth; while embedding into the linen all the complete wounds and blood flows from the body of various forms and sizes, and in various stages of congealment. All the wounds on the body, obviously, once had intimate contact with the cloth; yet, to have removed the cloth from the body by any human means would have broken or smeared many, if not all, of the blood marks. This historically consistent method also explains how the blood marks separated from the body, leaving the original smooth surfaces between the wounds and the skin intact and unbroken on the cloth. This is consistent with the historical descriptions of Jesus' appearances following his resurrection since none of the witnesses who saw him on Easter morning mention any blood marks on his body, let alone partial or broken blood marks. No decomposition stains have been found anywhere on the Shroud; this means that the body not only left or disappeared from the cloth in an extraordinary manner, it did so within two to three days of having been placed within it.

The particle radiation within this method is produced naturally when the body disappears, which is completely consistent with the historical accounts of Jesus' burial and resurrection. This radiation produces an image over time whose fibers become friable, yet strengthens the rest of the cloth. The particle radiation, consisting of the basic building blocks of matter, contains a neutron flux, which creates additional C-14 throughout the cloth. Neither a neutron flux nor particle radiation could be produced until the twentieth century by a nuclear laboratory. Yet, such a laboratory cannot make either form of radiation radiate uniformly throughout the length, width, and depth of a body. For either to have irradiated the Shroud from a corpse would have been an unprecedented event that could only be explained by the resurrection.

From the transmission photos taken of the entire cloth, the STURP scientists thought the Shroud was much older than the patches that were sewn onto the cloth in 1534. The scientists thought the Shroud appeared much older than the 1350s, when it first surfaced in Europe. After studying X-radiographs of the Shroud and the crudeness of its weave, and comparing it to the more modern backing cloth and patches sewn onto it, John Tyrer, head of textile investigations at the Manchester Testing House for more than twenty-five years, also concluded the Shroud was much older than the Middle Ages. From the comparison of the features found on the Shroud image with the features found on many religious icons, it appears that the Shroud existed prior to the thirteenth century and as early as the sixth century.

By comparing the Shroud's image features with those found on two Justinian II coins, it appears the Shroud existed prior to A.D. 695. The evidence that the crucifixion was likely performed by Roman military guards indicates that the crucifixion occurred before A.D. 315, since Emperor Constantine banned such crucifixions throughout the Roman Empire at that time. The evidence that the Roman executioners allowed an individual burial of a Jewish crucifixion victim in Palestine is an indication that these events took place before A.D. 66. The possible appearance of a Pontius Pilate lepton minted between A.D. 29 and 32 over the right eye of the man in the Shroud relates quite specifically to the first century A.D. The image-causing event that accounts for all of the unprecedented body image features and blood marks is consistent with the first century historical accounts of the resurrection of Jesus Christ.

The instruments used on this crucifixion victim were those commonly used by the Roman military guard, which performed such executions for the government. The scourge marks on the man match in detail the size and shape of marks from the Roman *flagrum*. The side wound matches the head of a Roman *lancea*, which is the very instrument described in the Gospels as having been used by the Roman guard on Jesus. Mocking and tormenting of crucifixion victims by the Roman military guard was common; this victim appears to have endured this treatment in the form of a crown of thorns placed over his head.

Further indications of a Roman execution are that an individual burial was allowed for this victim, something the Romans did permit, and similarities between the man in the Shroud and Yehohanan, another Roman crucifixion victim in Jerusalem. The evidence that the Shroud victim's crucifixion took place in Jerusalem in the first century is another indication that it was performed by the Romans, since they alone, as military rulers in the area at this time, had the power of execution. The victim was not a Roman citizen, however, for the form of scourging with a *flagrum* that he received was forbidden to be performed on Roman citizens.

There are many indications that the man was Jewish. His physiognomy is Jewish. He had a beard, and his hair was parted in the middle and fell to his shoulders, traits found in Jewish men of antiquity. Perhaps the most strikingly Jewish feature is the possible appearance of a pigtail, also found among Jewish men of antiquity. The man in the Shroud appears to have been buried with a chin band around his jaw, a proper Jewish burial custom. In addition, his burial posture matches that of skeletons found at the first -century Jewish community at Qumran. The use of a single linen shroud is also consistent with ancient Jewish burial practices, as is the custom of not washing the body of a victim of violent death in which blood that flowed during life and after death is present.

The events that the Shroud victim endured appeared to have taken place in the Middle East. The crown of thorns that was placed on the man's head appears to have been a full crown as was worn in the East, unlike the wreathlet type found in the Western world. One of these thorny plants, which grows only in Israel and parts of the Middle East, left its pollen, and possibly its image, on the Shroud. The numerous indications that the victim was a Jew denote a Middle Eastern location since that is where most Jews resided in the first century. Limestone found in burial tombs in Jerusalem matches the limestone found on the Shroud. The cloth appears to have been made in Jerusalem, since the vast majority of pollens found on the Shroud grow there. Using coins at Jewish burials was a common practice in Second Temple Jerusalem, as was the occasional custom of placing them over the victim's eyes. The flowers in bloom around the victim collectively grow only in Jerusalem and bloom only in the spring.

That Jesus Christ is the man in the Shroud can be ascertained from a number of signifying features. The victim appears to have been a Jew who was executed by the Romans, in Judea or Jerusalem, during spring in the first century—the exact circumstances of Jesus' crucifixion. Moreover, this victim, like Jesus, was scourged with a Roman *flagrum,* and beaten about the face. A crown of thorns was placed over his head. Shoulder abrasions, as if he had carried a heavy, rough object, are apparent. He suffered cuts from falls. He was nailed, not tied, to a cross. His legs were not broken as was usual with crucifixion victims. After he was dead, he was stabbed in the side. The instrument used was the very weapon that the Gospel of John states had been used on Jesus. Blood and water flowed from the wound. He received an individual burial, in a linen shroud. The burial, however, was provisional and was performed by individuals who had an intimate knowledge of Jewish burial customs. Microscopic traces of limestone on his burial shroud match limestone samples from the same rock shelf in which Jesus was buried. He remained in the cloth no longer than two or three days; then a miraculous event occurred that caused his body to leave the burial cloth by supernatural means.

Several historical books in the New Testament contain accounts of the crucifixion and resurrection of the historical Jesus Christ, which detail the aforementioned occurrences. When compared in an objective, historical, and bibliographic manner to all other works of ancient history, these historical books are the best attested sources in all of antiquity. Not only are there far more ancient manuscripts of these books still in existence, but they were written far closer to the time of the original than any other sources of ancient history. They were also written and translated in numerous languages, and quoted in many other ancient documents, more so than any other sources in history.

These accounts appeared during the lifetimes of the witnesses to these events. If such accounts were incorrect or inaccurate, one would expect them to have been refuted by the participants in and witnesses of these events. Yet history shows that these witnesses became emboldened attestants of the resurrected Jesus upon seeing him after his crucifixion and death. These witnesses never recanted their testimony in spite of persecution and torture; many of them suffered horrible deaths as a result of their testimony. The first written accounts not only specifically list and identify individuals and groups as witnesses, but they also state that more than five hundred others saw the resurrected Jesus, and that most of them were still alive at the time these accounts were first written down.

That these witnesses were testifying to the crucifixion and resurrection of the historical Jesus Christ is also confirmed by the beginning of Christian churches at this time whose tenets centered on the crucifixion and resurrection of the same individual. In addition, the day of Jesus' resurrection, Sunday, now replaced the centuries-long Sabbath day that began at sundown on Friday evening and ended on sundown Saturday evening.

Archaeology has also substantiated the accuracy of these historical sources by confirming countless detailed historical statements contained in them. Archaeology has never controverted these historical sources and has often corrected assertions against the New Testament's accuracy. Archaeology has consistently revealed biblical accuracy to the smallest detail on countless items, customs, individuals, titles, physical locations, et cetera, of the type that eyewitness contemporary accounts would consistently contain.

These historical accounts in the New Testament also report that after the witnesses saw the resurrected Jesus, his body could transform itself to pass through solid objects and vanish. These same accounts also state that radiation emanated from Jesus at his transfiguration while in the presence of God, but that it did not harm him or affect his clothing in any immediate way. Astonishingly, these are the same types of features that scientists have independently concluded, after twenty-five years of scientific investigation, to be principal causes of the formation of all the unique features found on the body images and blood marks on the Shroud of Turin. These are just some of the numerous details contained in these historical accounts that have been confirmed independently by scientific discoveries made almost 2,000 years later on the Shroud of Turin and which could never have been anticipated by the witnesses therein or the authors of these historical sources.

In light of these historical sources being the best attested sources of ancient history; their confirmation by the subsequent conduct of the eyewitnesses; their confirmation by other independent contemporaneous acts; by their lack of refutation while first appearing during the lifetimes of hundreds of the witnesses and participants; by their contents prophesying and predicting these events ranging from a short time to many centuries prior to the crucifixion and resurrection; by the direct and indirect archaeological confirmations of their contents and the events described; and by the subsequent scientific, medical, and historical corroboration thousands of years later in a manner completely unimagined by the original witnesses and authors, we can state that there are no sources like this in all of history which are so remarkable and accurate. Like the Shroud, these sources cannot be ignored nor put aside.

THE SHROUD TODAY

It is easy to wonder if the Shroud of Turin was not preserved and intended for these times. So many of its secrets have come to light only in the past twenty-two years. Were it not for the space program and the meteoric rise of computers in the latter part of this century, we might still be ignorant of this treasure trove of evidence. And, as with the scientific and medical evidence, most of the historical and archaeological evidence, too, remained undiscovered until the last two decades, with much of it coming to light only in the last few years. For this reason and others, the general public also remains largely unaware of the extensive archaeological and historical evidence. This evidence was revealed through the disciplines of archaeology, botany, textile analysis, numismatology, palynology, limestone analysis, art history, ethnology, exegesis, bibliography, and studies of Middle Eastern and Jewish burial customs as well as Roman weapons and crucifixion techniques.

Some of the most important evidence has only come to light in the past decade. The explanation of why the Shroud's fold marks correlate with historical descriptions of the Mandylion was first suggested in 1993. The discovery that the Shroud contains images (not visible to the naked eye) of more than twenty flowers in bloom that were placed around the deceased, are collectively native only to Jerusalem, and bloom during the Easter season, was just confirmed in 1997. The argument that there were no body images on the Shroud in its early years in Jerusalem was not developed and published until 2000. The publication of the study herein that the scriptural accounts of John actually refer to Jesus as being "prepared for burial" was not published until 2000. Nor was the observation made until this year that those who prepared the man in the Shroud for burial (based on their not washing the body, using a chin band, and placing coins over the victim's eyes) apparently possessed intricate knowledge of Jewish burial practices and customs that were to be applied in exceptional circumstances. Moreover, the substantive scientific challenge and discrediting of the 1988 radiocarbon date of the Shroud did not begin to develop until the last two years. Furthermore, the inimitable image-causing method, which accounts for all of the unprecedented features of the body images and blood marks and which is consistent with the historical accounts of the resurrection of Jesus Christ in the first century, was not publicized until 2000.

All of these archaeological and historical findings, as well as the recent scientific ones, are completely consistent with the countless scientific and medical findings described in the first part of this book. As a result of these numerous, corroborative examinations, the Shroud of Turin has now become the most intensely studied relic in the history of the world. Well over a quarter of a million scientific man hours alone have been devoted to it. This does not include time spent by the numerous pathologists, physicians, and various people from the fields of archaeology and history. Many, such as Paul Vignon, who originated the iconographic theory, spent their lifetimes investigating the Shroud. Pierre Barbet, who began the modern medical examination of the wounds on the man in the Shroud, continued his study for a lifetime, as has Dr. Robert Bucklin, the leading Shroud medical examiner today. Numerous STURP scientists, who thought they had signed up for only a five-day examination of the cloth in 1978, have spent the rest of their lives investigating it. Ian Wilson's work, containing the Shroud/Mandylion/Image of Edessa Theory, and other historical contributions, took ten years to research and write. Max Frei's investigative work with pollens covered a ten-year period. Father Filas's research with the Pontius Pilate leptons depicted over the Shroud victim's eyes consumed several years. This book alone has taken almost twenty years to compile and write. Yet, the above represent but a few examples of years of investigation from individuals and organizations around the world.

The equipment used to study the Shroud has been the best that science and technology can offer. In fact, only with the application of such sophisticated technology were most of the secrets of the Shroud revealed. Not only has the latest and finest in scientific technology been applied to the study of the Shroud, the scientists and researchers who conducted these Shroud studies are of the highest caliber in the world, coming from such prestigious institutions as the Jet Propulsion Laboratory, Los Alamos Scientific Laboratory, Sandia Laboratory, NASA, U.S. Air Force, Zurich Criminal Police, and renowned universities such as Harvard, Yale, Oxford, Cambridge, and the Sorbonne. Never before has a greater collection of scientists and researchers ever studied such a relic, over a greater number of years, covering a wider variety of professional fields, and from all parts of the world!

In light of all the above, it is time for people throughout the world to become aware of this extensive and previously unknown evidence concerning the Shroud of Turin. This evidence is far-ranging, consistent, and compelling, and it allows us to now evaluate the Shroud on the basis of extensive and objective evidence, something which previous generations and the present worldwide public has never had the opportunity to do before. This evidence has been gathered from the general fields of science, medicine, archaeology, and history; it relates not only to the authenticity of the Shroud, but necessarily to the crucifixion and resurrection of Jesus Christ. Whether or not the literal crucifixion and resurrection of the historical Jesus actually occurred could have more implications and relevance to us than any events in history.

EVIDENCE OF HISTORICAL EVENTS

After a comprehensive analysis of the evidence pertaining to the Shroud performed more than fifteen years ago and based on far less evidence than is known today, archaeologist William Meacham stated:

> Applying standards of proof no more stringent than those employed in other archaeological/historical identifications, one is led, I submit, to an almost inescapable conclusion about the Shroud of Turin: it is the very piece of linen described in the biblical accounts as being used to enfold the body of Christ. The pattern of data revealed by the Shroud is unquestionably unique, it concurs in every detail with the record of Christ's death and burial, and it is unfakeable.[8]

Certainly, if such extensive evidence existed for the identification of an ancient city site, or for the conquests of a Pharaoh or the events surrounding the death of an emperor, or for the occurrence of any other historical events, no one would object. Indeed, in order to evaluate the authenticity of the Shroud and the events depicted thereon, the same criteria that are used for any other historical identifications should be used.

Some might say that a case for the crucifixion and resurrection of the historical Jesus Christ should be held to a higher standard of proof than other historical identifications. In an ideal world, perhaps; but in reality, just because one matter is more important than another doesn't mean we'll have more evidence with which to decide the more important question. Whenever we make a decision in life, we always wish to have as much evidence as possible. However, we always have to do the best we can with the evidence we have, and never give up trying to acquire more evidence. Fortunately, we now have a great deal of evidence with which to answer these historical questions.

The primary source of our knowledge of most events in history is actually the written, contemporaneous accounts of these events. Archaeology usually provides only the context to such events by confirming such things as the customs or culture surrounding the events. It is rare that archaeology provides evidence of the actual event itself. To ascertain information about historical events—who were the participants, what were their roles, what was the outcome, when did the events specifically occur, where did they occur, what were their causes, what were the consequences, et cetera—invariably, the sources containing the witnesses' or authors' contemporaneous accounts of these and related events are the best sources for this information. In fact, until the advent of photography in the mid–nineteenth century, the above were our only sources of documented or recorded history. The bulk of our knowledge of historical events prior to this century is based primarily, if not solely, on the written contemporaneous accounts of these events.

In the case of the Shroud of Turin, however, we have an archaeological object that contains direct evidence of the crucifixion and resurrection of the historical Jesus Christ. It also contains evidence that specifically confirms the exact circumstances such as time, place, location, participants, et cetera, of the written historical accounts of these events.

All of this evidence is of a very sophisticated nature, containing extensive scientific and medical evidence published independently in numerous scientific journals and other periodicals. If we were to study any important event in history, whether it be Caesar's crossing of the Rubicon, the posting by Martin Luther of the 95 theses on the door at Wittenberg, or practically any other historical event prior to this century, we would not find such extensive scientific evidence to prove their occurrences.

Interestingly, in the historical accounts of Jesus' crucifixion and resurrection, his burial garment is mentioned only incidentally as the object in which his crucified corpse was wrapped and buried, and from which it disappeared a couple of days later. However, all of the numerous eyewitness accounts of the resurrected Jesus found in the historical accounts are based upon the subsequent sight, touch, and feel of the individual named Jesus. In fact, some of these witnesses were so incredulous of what they saw that they felt compelled to touch this individual and his wounds. Two thousand years later, these historical accounts have now been specifically corroborated, in minute detail, and in ways that cannot be forged or occur naturally, by medical, scientific, archaeological, and historical evidence derived from a comprehensive examination of this individual's burial cloth. The physical evidence was developed in a strikingly unanticipated manner, separate from the testimony of the witnesses.

No event in history has more extensive evidence, from a wider variety of fields, with more strikingly independent and comprehensive corroboration for its occurrence, than does the crucifixion and resurrection of the historical Jesus Christ. While the written, contemporaneous accounts were our primary sources of documented or recorded history until the advent of photography, it must be further noted that the crucifixion and resurrection of Jesus Christ not only have the best attested, contemporaneous, written accounts in all of ancient history, but also "photographic" evidence far greater in quality than anything that the most advanced scientific technology of today can produce.

ABSOLUTE AND RELATIVE EVIDENCE

All of this evidence provides us with some extremely unique and significant opportunities. This generation can now objectively examine some of life's most fundamental questions based not only upon the best documented sources of ancient history, but also upon newly discovered, independent, and comprehensive medical and scientific evidence. There are certain fundamental issues and decisions that all of us have to face in life. The most obvious issue we all have to face is our certain death, the consequences of which we have all wondered about. As a corollary to death's certainty, among the questions we also face is whether the crucifixion and resurrection of the historical Jesus Christ occurred. There are a number of other fundamental issues that the evidence in this book forces humanity to consider from a new and objective viewpoint. These issues relate not only to historical matters, but also to a host of philosophical, religious, cultural, social, and political questions. Attempts to resolve these issues have had enormous affect on both our past and our present, and promise to continue to significantly affect future generations throughout the world. Attempted resolutions actually form the bases of the laws, morals, values, and philosophies underlying various societies and cultures. Religious thought and belief have served as the foundations of Western and Eastern cultures throughout their histories.

Decisions concerning religious beliefs should be made earnestly, and with complete objectivity. We are reminded of Yves Delage, the agnostic, who responded to criticism of his 1902 presentation to the French Academy of Science with this statement: "I have been faithful to the true spirit of Science in treating this question, intent only on the truth, not concerned in the least whether the truth would affect the interests of any religious party." After reviewing all the evidence, Gary Habermas, who has coauthored two previous books on the Shroud, states, "Intellectual honesty causes me to say that if the same historical and scientific evidence existed for some other religious figure, I would be sufficiently challenged to investigate it. For instance, if the Shroud seemed to be that of Mohammad instead of Jesus, with all the attending evidence, I would be bothered, but I would have to face the facts."[9] While these two men come from very different points in the religious or philosophical spectrum, they each reflect the correct attitude for examining the evidence on this subject.

If the crucifixion and resurrection of the historical Jesus Christ occurred, these events could have more implications and relevance to each of us than any other events in history. As we have seen, there is a tremendous amount of medical, scientific, archaeological, and historical evidence to support and corroborate the occurrence of these events.

For the first time in history, we have a unique opportunity to answer some of the most fundamental questions in life, while having an unprecedented and extensive amount of evidence with which to answer them. Although there is not complete, absolute proof, there is still an enormous amount of evidence with which to answer these questions. However, we must also realize that rarely in life are conclusions absolute. If the only decisions we let ourselves make were those that are absolutely 100 percent certain, our lives would be tedious, dull, and simple. We would never take a job, seek an education, buy a house, get married, et cetera, for we would not have absolute certainty of the outcome of our decision. In short, we could not live this way. Almost all of our decisions in life are based upon relative evidence, or evidence that logically relates to a question and leads to a conclusion, but not an absolute conclusion. Almost all evidence is relative.

Surprisingly, some scientists, including STURP scientists, have failed to distinguish between absolute and relative evidence. A few scientists have expressed the illogical view that unless one has absolute proof of a proposition in connection with the Shroud and the historical Jesus Christ, one cannot make such an assertion or conclusion; for them, such a connection would be "beyond the scope of science." The essence of this thinking has been most clearly expressed by one scientist who does not connect the man in the Shroud and Jesus because "there is no scientific test for Jesus since we don't have his dental records."[10] Yet, even though we do not have the dental records of King Tut, Cleopatra, or any figure in history prior to the invention of X-ray technology, that has not been necessary to make historical identifications of them or their garments, or to draw conclusions about them. To place an absolute standard on historical events is ludicrous since we don't do it for anything else in life. It's impossible to achieve this standard for historical events, and we do not even attempt to do so. For example, one could go to the site of the Battle of Hastings, excavate the skeletal remains of soldiers, and scientifically examine them, yet you still couldn't tell scientifically from the bones which side they fought on, who the individuals were, who won the battle, what were its causes, its consequences, et cetera. It would all be "beyond the scope of science." Yet, there's still a lot of valuable relative evidence from other sources as to what happened, when, where, why, to whom, et cetera.

The fact that we do not have absolute proof of the crucifixion and resurrection of the historical Jesus Christ or that we cannot duplicate the Shroud image or the resurrection is not the fault of the evidence. If anything, the fault lies with the limitations of science and humankind. There are many things that science cannot duplicate but that obviously occurred. The examples are endless and range from things as large as the revolution and rotation of planets, the mechanics of comets, black holes, the solar system, and the universe to things that are so tiny they are invisible, such as molecules, atoms, quarks, and neutrinos. It took centuries upon centuries for science to even *discover* the existence of these things, and we are far from having 100 percent proof of how they operate or of ever duplicating them. There are still many things that we have yet to even discover.

As with the numerous examples above, we cannot duplicate the Shroud image characteristics or its image-forming event. All the extensive and sophisticated scientific equipment and examination could do was discover the many varied, unique features of the Shroud. Even though we did not cause and cannot duplicate these features—just as we cannot cause or duplicate comets, black holes, quarks, neutrinos, et cetera, they obviously occurred.

It is not within the realm of possibilities to re-create or duplicate the original Battle of Hastings. All of the participants, their passions, motives, horses, weapons, clothing, and even the terrain, have long since disappeared. The overriding causes and consequences of this historic event can never be repeated as they existed at the time of the original event. All that we could do would be to reenact aspects of the battle scene, knowing that we were not duplicating one exact circumstance of the original or of the circumstances surrounding it.

We presently have extensive medical, scientific, archaeological, and historical evidence to support the crucifixion and resurrection. Furthermore, this evidence pertains to Jesus Christ, who claimed he was the Son of God and who predicted these very events would occur. These events were also first mentioned in the Old Testament many centuries prior to their occurrences. All of these facts are consistent and corroborative and supportive of the crucifixion and resurrection, and with God's causing them (see appendix I).

The fact that we do not have absolute scientific proof of a proposition should never stop us from deciding a question based on all the available evidence. Rarely ever do we have an important decision to make where we have 100 percent absolute scientific evidence. (Actually, we do not even have 100 percent absolute evidence that this victim is a male, for the hands are folded over the groin, yet we have a lot of relative evidence to conclude that the victim is a man.)

Science itself is not absolute. Think to what degree scientific thought has changed in the last century, from matters as small as quarks, neutrinos, and atoms to matters as large as black holes and the universe. It has also changed on countless matters in between, as has the field of medicine; until the discovery of the germ theory in the nineteenth century, a patient often had a better chance of survival without the assistance of a physician. If science or medicine were absolute, their thoughts and conclusions would never change. Science, medicine, and anything else conducted by humans is relative. Frequently, scientific and medical conclusions change. What science and medicine usually do is accept a conclusion based on the best available evidence until a more likely conclusion based on new or better evidence takes its place.

That is precisely the criterion that should be exercised whenever a determination of any kind is being made. This is all the more so when the evidence and the determinations are not the domain of any one field and involve a number of different fields and specialties within several fields. The scientific and medical evidence is welcomed and valued in this determination because such evidence is based on empirical and scientific principles, and neither this nor any other evidence needs to be absolute before it can be considered valuable. If we are going to think in absolute terms concerning the decisions to be made from the evidence in this book, the proper place for the absolute value is actually in the equation itself. The one absolutely certain fact we all must face is that every single one of us is going to die. Therefore, the real question becomes, Do we want to take advantage of this unique historical opportunity to consider all the new and comprehensive evidence pertaining to these fundamental issues and decisions, which necessarily result from and pertain to the absolute certainty of our death?

Most decisions are actually made upon a preponderance of the evidence. This means that a decision is based on the position with the best evidence after examining all the available and relevant evidence. That is the correct criterion, and it is the best anyone can do whenever he or she makes a determination. In this case, we can certainly base our decision on a comprehensive and corroborating amount of medical, scientific, archaeological, and historical evidence.

UNLIMITED, WORLDWIDE OPPORTUNITY

The failure of one segment of science at the beginning of this century, and the failure on the part of the Shroud's owners thereafter, to seek a scientific examination of the cloth had the effect of postponing such an examination until the latter part of the twentieth century. Perhaps this was a blessing in disguise, for the development of space-age and computerized technology at the highest scientific levels had recently advanced by that time to the point where they could be applied to the Shroud. This application, in turn, would reveal so many of the Shroud's secrets that had lain dormant for centuries. The inaccuracies of Walter McCrone and the controversial radiocarbon dating of the Shroud in the years soon following the first comprehensive scientific examination also only served, ironically, to better bring the entire subject of the Shroud and the dramatic scientific discoveries made from it to the attention of the largest worldwide audience that it could ever have—all occurring at the end of the millennium.

There exists today an unprecedented, unlimited opportunity for worldwide awareness, which previous generations in history have never possessed. The Shroud of Turin will have been exhibited to the world twice within two years, in 1998 and 2000. These exhibits will gain the Shroud far more attention from both the media and public than it has ever received. Moreover, as a result of this worldwide attention, people throughout the world will have access to the extensive scientific, medical, archaeological, and historical evidence that has been acquired from the first wide-ranging scientific examination of the cloth in 1978 and from two decades of continued research by experts from a variety of fields, including some of the most startling evidence acquired just this year.

Since all of this extensive evidence lays before the worldwide public for the first time, and it relates directly to fundamental decisions that all of us face, everyone should be able to access and analyze this evidence. The Shroud itself, and its evidence, will undoubtedly be examined further by a variety of professionals. New evidence will develop and evolve continually over time. The evidence as it presently exists, and the evidence and information yet to be acquired will continue to affect present and future generations worldwide. At present, some governments censure information and material with religious implications. They, like the rest of the world, are just becoming aware of this unique opportunity, the extensiveness and importance of all this new information, and how it could greatly affect the welfare of their citizens. No government, society, culture, group, or individual should ever prevent or interfere in any way with an individual's or the public's right and opportunity to read and analyze this evidence—for much is at stake concerning their future.

This extensive evidence was developed by some of the finest scientists, physicians, mathematicians, archaeologists, and other professionals, using the most sophisticated equipment and techniques, from some of the most prestigious institutions throughout the world. Most of this evidence could not even be seen, let alone forged, and has been gradually disclosed only with the advancement of scientific, photographic, and computer technology. For the first time, we have a comprehensive set of evidence with which to answer some of the most fundamental questions in life.

This unlimited opportunity—beginning at the dawn of the new millennium—extends far beyond the fact the Shroud is being exhibited twice within two years, or that the resulting attention will allow the world to learn of all the evidence derived from its study and examination, and to resolve humankind's most fundamental questions and differences. Many philosophical, religious, scientific, and historical issues have been publicly debated, and even fought over, throughout history. However, not one of these issues contains such far-reaching evidence, and can affect one's life in as significant a manner, as the historical events and their substantiating evidence that are the subjects of this book. The complete unlimitedness of this worldwide opportunity can be understood fully only when each individual realizes the impact that this astounding evidence can have on his or her own life, not only today but in the future. It is much, much more than an opportunity of a lifetime—it's the opportunity of an eternity.

APPENDIX A

THE VINLAND MAP

The ink on the Vinland map consists of two layers, one of which is a yellow-brown line formed of ingredients that were absorbed into the parchment, and another consisting of the spontaneously formed surface remains of nonabsorbable black particles. In medieval manuscripts a similar characteristic is common and well-known and requires no special explanation.[1] Dr. McCrone examined twenty-nine ink particles that his staff removed from the map and found that all of the yellow-brown particles largely consisted of a mineral commonly known as anatase. In fact, he found they contained "up to 50 percent" anatase.[2] Anatase is a crystalline form of titanium dioxide that is rare in nature and was not commercially produced until 1920. Dr. McCrone further found that the black particles in the ink did not contain anatase. Dr. McCrone stressed that the anatase on the map was precipitated rather than a ground material and stated "the presence of anatase as a precipitated pigment was impossible before about 1920."[3] He proclaimed that the possibility of the Vinland Map being authentic was analogous to Admiral Nelson's flagship at Trafalgar being a hovercraft. Dr. McCrone asserted that a forger had used a yellow-brown ink to give the map an antique appearance and then applied a black line down the middle of the yellow lines.

This alone is not a proper basis for concluding that the ink was modern. Anatase produced from titanium in the medieval formula for ink would be identical to that of modern anatase since the processes by which they are formed are chemically identical. Both would be in the precipitated form.[4] However, a proper basis for distinguishing between modern anatase and that from medieval times would be to measure the amount of contaminants in the ink.

Dr. McCrone's own analysis reveals the presence of a variety of inorganic elements on the map. In some cases these elements are present at levels comparable to or exceeding those of titanium, together with organic carbon and various other elements in the ink, for a total of sixteen different elements.[5] Dr. George Painter, a retired scholar at the British Museum, states "such massive impurities are inconceivable in the chemically pure modern titanium dioxide pigment preparations. They are, on the contrary, decisive for a medieval ink, of an unmodern, unpurified and grossly contaminated composition retained from the natural constituents of its mineral sources."[6] For some unknown reason, Dr. McCrone completely ignored the significance of these sixteen contaminants in his analysis of whether the ink was medieval or modern in origin.

Scientists from the University of California at Davis were subsequently able to examine the entire Vinland Map using various nondestructive techniques. The scientists took numerous photographs of the document in reflected light, transmitted light, and long and short-wave ultraviolet light, as well as microphotographs at over one hundred specific locations on the map. In addition, they used a piece of equipment called a cyclotron to fire a beam of accelerated protons through the map, thereby generating X rays from which all elements present in the ink and parchment could be identified and quantified, a technique that has been rigorously tested and successfully applied to hundreds of documents since the early 1980s.

The scientists at Cal-Davis found that the map contained only trace amounts of titanium, well over one thousand times less than the amount claimed by Dr. McCrone. Moreover, these trace amounts are completely consistent with those that occur in nature.

Comparative analysis of documents from the same period (c. 1440), including the Gutenberg Bible, show that titanium occurred naturally in greater amounts on those documents than that detected on the Vinland Map. The faded yellow lines that Dr. McCrone contended were made of up to 50 percent anatase and applied by a forger to give an antique appearance, were found to have almost no anatase. The amounts of anatase were found to be so minute that they could not possibly be explained as a consciously added ingredient.

Another example of Dr. McCrone's erroneous findings on the Vinland Map can be seen in the discovery by scientists at Cal-Davis that trace amounts of anatase are present in the black ink at various locations on the map. Dr. McCrone had also stated that anatase was not present in the black ink at various locations on the map and that a forger had used black ink to line over the yellow-brown ink. Dr. McCrone's findings turned out to be based on a single anatase-free microparticle sample.[7] Such an erroneous and isolated finding necessarily played a crucial role in Dr. McCrone's double-ink theory, for he concluded that two different inks with two different applications were used. Not only was anatase found in various locations in both the black and yellow-brown inks, but the two layers of ink were found to coincide within "about 0.1 mm at over 100 locations examined microscopically."[8] This is "a conformity certainly beyond the powers of human hand or eye, and caused naturally by different absorption on the ink constituents, without human intervention."[9] In short, no human being could have possibly drawn one line over the other like that so perfectly.

The utilization of the cyclotron by the Cal-Davis researchers also identified five additional trace elements, bringing to twenty-one the total number of contaminants found in the ink. All twenty-one of these contaminants are natural and common in unpurified medieval mineral ores, but unthinkable for modern refined anatase.[10] Dr. George Painter, the last surviving author of the original Yale University publication, points out that on the basis of the overall scientific evidence, the chemical constitution of the ink can no longer be used as an argument against the map's authenticity.

APPENDIX B

JESUS' BODY WAS NOT WASHED BUT WAS PREPARED FOR BURIAL

Some scriptural scholars assume that Jesus' body was washed after his crucifixion because some translations of John 19:40 state that Jesus was buried in accordance with Jewish burial customs, which typically call for washing. Yet such a translation of John 19:40 is imprecise. To be as precise as possible in translating this verse, we should use the original Greek. What follows is original research obtained by the author. All the ancient Greek references discussed in this appendix were located and translated by Professor Carl Conrad of the Department of Classics at Washington University, St. Louis, Missouri.

Several similar words were used in the Greek language to refer to burial, including the verb ἐνταφιάζω, which means "to bury" or "prepare for burial," the Greek noun ἐνταφιασμός, which means "burial" or "preparation for burial," and the Greek noun ἐνταφιατής, which means "one who prepares for burial" or "one who buries" or "burier." The precise word utilized depended upon the context. In John 19:40, the Greek verb ἐνταφιάζειν is used in the infinitive form. Since a hurried burial of Jesus is mentioned throughout the Gospels, as is the fact that the burial process had not been completed, the context would clearly call for the translation of "prepare for burial."[1] This translation is supported by the lack of any reference in the Gospels to Jesus' dead body being washed, which would have violated Jewish burial custom because he died a violent death and his body contained blood that flowed during life and after death. In a similar manner, Matthew 26:12 also uses the infinitive form of the verb ἐνταφιάσαι, and indeed this interpretation is the only one that makes sense. In Matthew 26:12, Jesus defends the action of the woman who poured ointment on him just before the Last Supper when he says: "In pouring this ointment on my body she had done it *to prepare* me for *burial*" (italics added).

Plutarch, who lived during the time the New Testament was written, used the verb ἐνταφιάσαι in a figurative sense in a passage from an essay entitled "On the Eating of Meat." He described the preparation of a fish for cooking as follows: "mixing up olive oil, wine, honey, fish-sauce and vinegar with Syrian and Arabian condiments, just as if we were really *preparing* a corpse thus for *burial*. . . ." Similarly, in his "Fragments in the Psalms" (44.9,10), Origen (ca. A.D. 100–200) wrote that: ". . . Jesus *was prepared for burial* by Joseph of Arimathea with myrrh and aloes."

Gregory of Nyssa (fourth century A.D.) used the verb in a citation of Mark 14:8 in his commentary on the Song of Songs (6.189), when he noted that ". . . myrrh is useful for the *preparation* of bodies for *burial*" (ἐνταφιασμόν). A few lines later he wrote: "for the one who accepted death on our behalf and that myrrh rubbed onto the very flesh of the Lord for *burial-preparation* (ἐνταφιάζω) . . . ." Gregory also stated in 6.290: ". . . just as the sublime gospel says that by means of these things [spices] the *burial-preparation* [ἐνταφιασμός] took place for the one who tasted death on our behalf."

Elsewhere, in his "Sermon Against the Wealthy" (9.30), Basilus Theologus cautioned: "So be careful; *prepare* yourself for *burial*" (ἐνταφίασον). Similarly in his "Sermons on Psalms" (29.405), he wrote: ". . . the evangelist John has taught us, saying that he was *prepared for burial* [ἐντεταφιάσθαι] by Joseph of Arimathea with myrrh and aloes." Another fourth-century theologian, Gregory Nazianzenus, explained: "And if you are Nicodemus, the god-fearer who comes by night, *prepare* him for *burial* with balms . . . " (36.656).

In several other sources where "prepare for burial" is not the precise contextual translation, the term is used to mean "to embalm," or "embalmers." Historically, embalming has always been a treatment with balsams, spices, et cetera (and today with chemicals and drugs) that was performed to preserve a body from decay. In other words, embalming refers to preparation of a body for final showing and burial. This translation can be found in the *Greek Anthology* (11.125), which was written sometime between 300 B.C. and A.D. 300. There the verb and the agent-noun both appear in a clever little poem about a physician, Krateas, and an *embalmer*, Damon, who formed a partnership arrangement. Damon would filch broad linen strips from shrouds and send them to his friend Krateas to use for bandaging; Krateas in turn would send Damon whole bodies to be treated for *embalming*. Another example is found in the documentary papyri, *Papyrus of Cologne* (2.112), which reads ". . . take the body of the elder to the *embalmers* (τοῖς ἐνταφιασταῖς). . . ." This passage clearly refers to a process of preparing and preserving the body.

In the book of Genesis in the Greek Old Testament, the Septuagint, the verb is used in two forms and the noun ἐνταφιαστής is used in two forms: "And Joseph ordered the servants, the *embalmers, to embalm* his father, and the *embalmers embalmed* Israel" (Genesis 50:2). In Mark 14:8, the noun form ἐνταφιασμόν is used: "She did what she was able to do: She has anointed my body in advance for *burial*." Even here, where the exact translation of "prepare for burial" is not found, the passage clearly alludes to the idea of an action taken in advance of burial. This is the same description as used for the woman's actions referred to earlier in Matthew 26:12, which obviously translates as "prepare for burial."

APPENDIX C

SCRIPTURAL DESCRIPTION OF JESUS' BURIAL CLOTHS AND THE SHROUD

Some critics have asserted that the Shroud cannot be authentic because of alleged inconsistencies with the biblical description of Jesus' burial cloths and a shroud. These alleged discrepancies concern the interpretation of some versions of descriptions of the burial cloths in John 19:40 and 20:5–7. However, there are no inconsistencies between the Shroud and the synoptic Gospels (or any other part of the New Testament or Old Testament). A closer examination reveals there also is no disparity in the gospel of John.

When describing the burial cloths of Jesus, the first three Gospels clearly refer to Jesus being laid in a sindon or shroud at burial (Matthew 27:59; Mark 15:46; Luke 23:53). The alleged inconsistency between the Shroud and John's descriptions of burial cloths arose because some translators (e.g., New English Bible, New International Version) incorrectly translated the word *othonia,* found in John 19:40 and 20:5–7, to mean "narrow bands" or "strips of linen."

In fact, *othonia* can refer to cloths of all sizes and shapes. In an excellent examination of this question, Rev. Andre Feuillet, an exegetical expert and docent at the Catholic Institute of Paris, points out the various meanings of *othonia* found in the Greek dictionaries and lexicons and explains how the word was used by ancient Greek writers.[1] In these sources, the term could mean a boat sail, a light tunic, or cloths in general, as well as narrow bands. The ancient Greek writer Dioskorides not only used *othonia* to mean a sheet but also coupled it with the verb *eneilein* (envelop in a sheet, *eneilesas othonio*), which is the verb that Mark chose to describe the burial of Jesus in a shroud (*eneilesente sindoni*).

Use of the term *othonia* found in the Greek Old Testament, or Septuagint, excludes the "narrow band" definition. In Hosea 2:9, *othonia* refers to linen fabric, as distinct from the wool fabric worn by the symbolic wife of Yahweh. Because of her infidelities, Yahweh threatens to take away the wool and the linen with which she covers her nakedness. Again, in the Book of Judges 14:12–13, thirty linen garments or undergarments are referred to with the word *othonia.*

To restrict the definition of *othonia* to "narrow bands" would put John in direct contradiction with the synoptic Gospels, which specifically state that Jesus was buried in a linen cloth or shroud (Matthew 27:59–60; Mark 15:46; Luke 23:53). Of particular interest is the use of the word *othonia* in Luke 24:12, where he refers to what he had previously described as a sindon or shroud in 23:53.[2] The only use that Luke can possibly intend for this term is the sindon that he mentioned earlier. In a broader context, Luke may also be referring to the *keiriai,* or the *soudarion,* when using the term *othonia* to mean sindon in 24:12; if so, his meaning would encompass linen cloths in general, including a sindon. Such an interpretation is the only logical one, and the same interpretation can be applied to the use of *othonia* by John, as well.

As pointed out by Rev. Feuillet and the Benedictine scholar Maurus Green,[3] the translation of *othonia* in John 19:40 as "narrow bands" is recent. Previously, it had always been translated in all languages as "linen cloths" or "linen." It was not until 1879 that *othonia* was first translated incorrectly by two French exegetes to mean "narrow bands" or "bandages."

The translation of *othonia* as narrow bands would easily lead one to think that Jesus' body had been mummified. Unlike the Egyptians, who disemboweled and pickled corpses before wrapping them in narrow linen bands, the Jews did not mummify corpses. Such a burial would contradict John's own statement in 19:40 that Jesus was prepared for burial in accordance with Jewish burial customs. Although the hands and feet of Lazarus are bound by bands when he is described as coming forth from the tomb in John 11:44, Lazarus obviously is able to shuffle out on his own, something he would not be able to do if his whole body were wrapped like a mummy. Moreover, these bands are not referred to as *othonia* by John, but are called *keiriai,* and the verb used is *deo,* which always means "to bind or tie," never "to swathe."[4] Similarly, the young man of Nain, in Luke 7:15, would not have sat up at the word of Jesus had he been completely swathed, nor could have the corpse of Tabitha mentioned in Acts 9:41. For all these various reasons, there is no inconsistency with the use of a shroud and the burial of Jesus as described in the Gospels.

APPENDIX D

VERONICA'S VEIL

According to archaeologist William Meacham, the recorded tradition of the miraculous imprinting of Jesus' image onto cloth developed first in the Byzantine Empire. He cites the famous historian Gibbon, who wrote that these "images made without hands" were prevalent in various cities of the Eastern empire, and one of these Byzantine icons was well known in Rome by the sixteenth century. In the West, a similar tradition arose, called (The) "Veronica," a woman who supposedly wiped the face of Jesus with her veil and found a facial imprint left on her cloth. Meacham believes these traditions are grounded in the Shroud image and evolved into images of the living Christ to correspond with the conventions of iconography originating in early Byzantine history.[1]

The story of the Veronica tradition has a long and somewhat confusing history. The legend first appeared in a work titled *Acts of Pilate,* which dates from the second or fourth century. There, Veronica (sometimes called Berenice) is identified as the Hemorrhissa of the Gospels (Matthew 9:20–22; Mark 5:25–34; Luke 8:43–48), who is called upon to testify at Jesus' trial that he healed her of an issue of blood she had suffered from for a dozen years.[2] In the *Apocritus* of Macarius of Magnesia, a fourth-century work, and the writings of other authors, Veronica is described as having erected in her yard a bronze statue of Jesus in the act of healing her.[3] Around the beginning of the seventh century, the legend changed somewhat, as seen in the *Cura Sanitatis Tiberii.* In this version, the Roman emperor Tiberius sent a messenger, Volusian, to Jerusalem to beg Jesus to cure the emperor's leprosy. While in Jerusalem, Volusian learned of Jesus' execution, which led to Pilate's recall and punishment. Volusian also heard of Veronica, who possessed a portrait of Jesus. He forced her to come to Rome, where Tiberius was cured by the image.[4] In a later work of probably the seventh or eighth century, titled *The Death of Pilate,* Veronica, knowing the Savior would soon leave her, decided to have a portrait of the Lord painted. On the way to the painter's, she met Jesus, who learned of her errand, touched the canvas to his face, and miraculously imprinted his image.[5]

The next stage of the legend is linked with the famous Abgar story (discussed in chapter 7), which was more widely known than the Veronica during the Middle Ages. The Veronica story is probably the Roman version of the exchange of letters between the sick King Abgar of Edessa and Jesus, with the Roman story asserting that Jesus sent the king his portrait miraculously painted on cloth as well as a promise of healing and an apostle to convert his people. The Veronica image and the Image of Edessa both depicted Jesus alive before the Passion.

In about 1150, Veronica is described as having a *sudarium,* or head cloth, with

Jesus' image impressed upon it "when his sweat became as drops of blood falling to the ground," an obvious reference to what preceded the Passion in the agony at the Garden of Gethsemane (Luke 22:44). This change in the story reflected the growing interest in the Passion encouraged by the Franciscans, who invented the Stations of the Cross and the new artistic emphasis on the dead Christ.[6] By the late Middle Ages, the legend had been modified again so that the cloth bearing the image of Jesus was obtained by Veronica when she emerged from the crowd and wiped his face as Jesus was led toward his crucifixion by the soldiers along the Via Dolorosa. This last modification of the story was the invention of Parisian "Miracle" playwrights who wanted to add drama to their stagings of the Passion. This version of the legend became so popular that it can still be seen as one of the fifteen stations of the Cross in Catholic churches all over the world.

Indeed, just as the legend evolved over time, contemporaneous descriptions of the veil also seemed to change. For fifteenth- and sixteenth-century artists such as Albert Dürer, William of Cologne, and Quentin Massys, the Veronica portrayed Jesus with a disembodied head that was crowned with thorns and suffering—a depiction that was consistent with the sagas of the Via Dolorosa and shared some obvious similarities with the Shroud image.[7] Earlier, some thirteenth-century authors described the Veronica as representing Jesus from the chest upward, and not with a disembodied head.[8] Illuminators of some thirteenth-century manuscripts, such as Matthew Paris's *Chronica Majorna* and the Westminster Psalter, showed the Veronica image with a neck and also a body.[9].

The story as we have it today did not become popular until the fifteenth century. From that point onward, two types of Veronicas existed: the nonsuffering Byzantine image and the suffering image with sweat and blood. But what is known about the Roman relic called the Veronica? This cloth apparently came to Rome about A.D. 700 and was at St. Peter's at some point. Did it have an image? Monsignor Joseph Wilpert examined the cloth in the early twentieth century and found stains but no clear image.[10] What became of the Byzantine icon—another cloth associated with the Veronica story—that was famous in Rome in the sixteenth century? It disappeared when the troops of Charles V sacked Rome and the Vatican in 1527 and never appeared again.[11]

Because of the numerous similarities between the legend of Veronica's veil and the earlier, more popular Abgar story, as well as the marked resemblance between descriptions and illustrations of the veil and the Shroud/Mandylion, several scholars have asserted that the Shroud had a strong influence on the development of the story of Veronica and her veil.[12]

APPENDIX E

BACTERIAL AND FUNGAL COATING

Dr. Garza-Valdes, who proposed that bacteria and/or fungi caused the image on the Shroud, has also proposed that they altered the radiocarbon date of the cloth from the first century to medieval times.[1] Many of the same problems that were pointed out in the previous discussion of bacteria and fungi in chapter 5 are also applicable here. Not one of the many STURP scientists who have examined the entire Shroud and its fibrils, including threads from the radiocarbon sample site, have confirmed Dr. Garza's findings. Neither did Dr. McCrone, who has also examined a variety of fibrils from throughout the Shroud. Furthermore, STURP chemist Alan Adler and Walter McCrone both went to Texas at Dr. Garza's invitation to see Dr. Garza's own observations and techniques; however, they did not confirm his findings. In addition, they disagreed with and questioned his interpretations, methods, and techniques.[2] Moreover, the very latest examination by Fourier Transform Infrared (FTIR) Microspectrophotometry of thirty-four samples taken from various locations throughout the Shroud did not reveal the existence of this bioplastic coating with bacteria and/or fungi.[3]

As stated earlier, STURP found particles of bacteria, fungi, pollen, dust, wax, starch, and many other substances in minute amounts on the Shroud; it just did not find them in the large amounts claimed by Dr. Garza. Dr. Adler has stated that if the Shroud were coated with the amount of bacteria and fungi claimed by Dr. Garza, it would be brittle and would shatter if you dropped it.[4] However, the Shroud is quite flexible and its fibers can be cut with scissors.

Recently, Dr. John Jackson and Bryan Walsh, statistician and executive director, Richmond Shroud of Turin Center, independently calculated that the areal mass density of the samples removed from the Shroud in 1988 were approximately the same as the rest of the cloth.[5] Walsh also calculated that the areal mass density of the Shroud was approximately the same as other similarly woven linen. If the Shroud samples from the radiocarbon site contained as much bioplastic coating as claimed by Garza, 60 to 70 percent of the samples' carbon would have to come from this coating.[6] However, that would cause the areal mass density of the sample to be much greater than the rest of the cloth, but that is not the case with the Shroud. Alternatively, Dr. Garza cannot claim that the radiocarbon site weighs approximately the same as the rest of the Shroud because the bioplastic coating is everywhere on the cloth, since the cloth itself has the same approximate areal mass density as other similarly woven linen.

Dr. Garza has been claiming the existence of a bioplastic coating of bacteria and/or fungi on the Shroud of Turin since 1993; however, he has never published his scientific work in a peer-reviewed scientific journal, demonstrating his methods, controls, techniques, and specific findings. Furthermore, scientists who have examined the same and additional samples from the Shroud not only failed to confirm his finding, but disagreed with his methods and interpretations. Such publication would certainly allow his claims to be better analyzed. In a recent paper by Dr. Harry Gove, Dr. Garza, and two others, it is claimed that examinations by Dr. Garza with only an optical microscope have shown that coatings also cover the surfaces of the thread fibrils of other ancient textile samples. The authors speculate that these coatings were deposited by bacteria and fungi over time;[7] however, they concede, "Such coatings have not been previously observed nor confirmed by other investigators."[8]

Recently, some controversy has also arisen over the radiocarbon samples and blood samples that were removed from the Shroud and given to Dr. Garza by Giovanni Riggi, the technician who removed the samples from the Shroud when it was carbon dated in 1988. At that time, Mr. Riggi cut a much larger piece than was needed for radiocarbon dating purposes. He cut this piece approximately in half and made three samples from the top half for the radiocarbon dating laboratories (with a small piece from the bottom half completing the third sample) and retained the bottom half for himself. The retention of this sample, which was more than twice the size of the sample given to each of the carbon dating laboratories, was not part of the protocol and was not authorized by the archbishop's office, according to Cardinal Saldarini, who was the Archbishop of Turin from 1990 to 1999. Mr. Riggi also took some blood samples from the back of the head at this same time. The taking of these blood samples was not part of the protocol nor was permission officially given from the owner, the Vatican, or the official custodian, the office of the Archbishop of Turin.[9] Mr. Riggi claims that taking the blood samples was "in line with the 1978 proposal and that it was done in the presence of the official caretakers."[10] Archbishop Ballestrero, the Archbishop of Turin at that time, was also present, at least while the radiocarbon sample was removed, but has not commented on this matter, nor has the new Archbishop Poletto.

Some of these samples were given to Dr. Garza; two of the blood samples from the back of the head were also given to Dr. Victor Tryon. Both of these men were informed that these were genuine samples removed from the Shroud of Turin. Cardinal Saldarini stated that since the taking of these samples was not authorized, and he did not know, himself, that they came from the Shroud, the Vatican would not recognize research results from these samples.

While there certainly may be a question of whether official permission was granted to Mr. Riggi to remove these particular blood samples, or to retain part of the radiocarbon sample, no one has really disputed or stated that these samples did not come from the Shroud. Dr. Garza's samples were even examined by Dr. Adler and Dr. McCrone, both of whom are very familiar with Shroud samples, and while they disagree with many aspects of Dr. Garza's research, they did not dispute that his samples came from the Shroud.

Cardinal Saldarini's disagreement over the authorization for the removal or retention of the samples and recognition of their subsequent study actually related to research that has made very little impact or contribution to date on the important questions concerning the Shroud. As stated earlier, the bioplastic coating of bacteria and fungi could not have caused the image on the Shroud, nor is there any evidence for their existence to the extent claimed by Garza, or that they had any effect on the radiocarbon dating of the Shroud.

Dr. Tryon actually has quite a ways to go before he can conclusively establish, on the basis of the DNA and chromosome studies alone, the overall points that the victim is a human male. However, both of these overall points are already established, and Dr. Tryon's evidence will add very little to their validity. It is known that many people throughout the centuries have handled the Shroud—as caretakers, or for exhibitions, or repairs, or examinations. Well over fifty different artists have been documented to have sanctified their paintings by laying them on the Shroud. Anyone who has handled the Shroud could have left his or her DNA on it. In light of this, it is easy to see why current scientific consensus requires at least twelve different blood samples from various areas on the cloth to confirm that the DNA belongs to the man who was wrapped in the Shroud.[11] However, even if Dr. Tryon had these additional samples and all of them also contained X and Y chromosomes, it would be only one more part of an entire body of medical, chemical, physical, mathematical, archaeological, historical, and visual evidence to indicate the cloth contained a man. As stated in the always colorful and expressive language of Alan Adler, "nobody ever thought the guy in the Shroud was a shaved orangutan anyway."[12]

The most sensational aspect of research concerning blood samples from the Shroud has to do with the subject of cloning. Countless people have asked whether another Jesus Christ could be cloned from the blood contained on the Shroud. Assuming that the blood on the Shroud is that of Jesus, the answer to the question is still certainly not. This question has arisen, in part, because some genes have been "cloned" during the above research in Texas.[13] Blood is actually a poor source of DNA. Red blood cells do not contain a nucleus. DNA can be acquired only from white blood cells; however, there are comparatively few white cells in blood. Moreover, DNA starts deteriorating soon after death. Blood that is six hundred to two thousand years old would contain very little of the DNA segments of the original molecule. When the famous sheep Dolly was cloned, *living* cells from another animal were used.

The human genome consists of 80,000 to 100,000 genes formed by 3 billion base pairs. What Dr. Tryon's team did was to replicate three short segments of three genes which provided them with only about 700 to 750 base pairs.[14] Even if there are a few more segments of genes that can be replicated, they are obviously way short of obtaining a complete set of genes and/or their base pairs. Moreover, they would all have to be aligned in the correct sequence.

APPENDIX F

FIRE SIMULATION MODEL

Comparison between the experimental conditions used in Kouznetsov's simulation and those described in report of Moroni and associates.

Parameters	Kouznetsov's and others[a]	Bettinelli's and others
Temperature (°C)	200	200
Pressure (atm.)	0.003[b]	atmospheric
Duration (minutes)	90	90
CO_2 (%)	0.03	0.03
CO (μg/m³)	60	60
H_2O demin. (μg/m³)	20	20
Ag (μg l^{-1})	0.8–1.45	1.5
Weight of sample (mg)	3.5–5.5[c]	43–75
Volume of cell (cm³)	100	325

[a]D. A. Kouznetsov, A. A. Ivanov, P. R. Veletsky. "Effect of fires and biofractionation of carbon isotopes on results of radiocarbon dating of old textiles: the Shroud of Turin," *Journal of Archaeological Science*, 23, (1996): 109–121.
[b]J. P. Jackson; Review for ACS Publishing Committee (September 1995).
[c]Letter of Dr. Kouznetsov to Dr. Novelli, dated August 6, 1996.

Chart and footnotes taken verbatim from M. Moroni, F. Barbesino, M. Bettinelli. "Verification of an Hypothesis of Radiocarbon Rejuvenation," *Third International Congress on the Shroud of Turin*, Turin, Italy, June 5–7, 1998.

Beginning in 1993, Russian scientist Dmitri Kouznetsov asserted, among other things, that when he and his associates exposed a first-century cloth sample to his fire simulation model, that the C-14 and C-13 content in the cloth not only greatly increased but each reached its respective peaks after approximately one and a half to two hours' exposure.[1] Scientists from the NSF Arizona AMS carbon dating facility in Tucson rejected or questioned this, as well as many other assertions made by Kouznetsov. In a similar experiment, they found no such increase in the cloth's isotopic content when they measured after fifteen and a half hours' exposure.[2] Other scientists have studied and rejected Kouznetsov's assertions on other grounds.[3]

Kouznetsov and American scientist John Jackson have responded that the Russian data could indicate that the isotopic increases at one and a half to two hours could be temporary nonequilibrium situations, and the Arizona data could represent equilibrium, late-term conditions.[4] Jackson is constructing a theoretical model of how such an isotopic increase could occur through exchanges resulting from the fire.[5] Yet, it must be noted that this model requires a preequilibrium step with the interaction of atmospheric carbon dioxide with the air and the body of the cloth. Ensuing chemistry must then also occur that leads to the kinetic isotope effect taking place thereafter, along with some rate-limiting step occurring immediately after that.[6] Obviously, such an occurrence would be extremely unusual and needs to be demonstrated and confirmed by experiment.

It must also be noted that Moroni et al. tested Kouznetsov's fire model experimentally and did not find the large increases in isotopic content that Kouznetsov claimed after one and a half hours. Moroni's experiments showed that Kouznetsov's fire model yielded only a 160 year younger result; this sample was not even pretreated following the experiment and prior to its dating. Neither did Moroni and his associates find such increases in their own fire model, which was also quite similar to Kouznetsov's, both of which can be seen in the above chart. Moroni's fire model results yielded a 300 year younger result with a pretreatment method that removed 37 percent of the original sample. Moroni even presoaked the shrouds used in both fire models in a very diluted solution of aloe and myrrh before placing the Lyma mummy sample within the forty-eight-fold cloths. This would only introduce further sources of carbon for the sample to receive, yet the results obtained were quite modest.

APPENDIX G

LETTER TO STURP SCIENTISTS

MARK ANTONACCI
ATTORNEY AT LAW
3550 McKELVEY RD., SUITE 204D
BRIDGETON, MISSOURI 63044
314-739-8885

January 19, 1984

Al Adler
% Western Connecticut
State University
Chemistry Department
181 White Street
Danbury, Connectucut 06810

Dear Mr. Adler:

I'm an attorney in St. Louis who's planning an address to fellow attorneys on the subject of the Shroud of Turin in early April. I've been studying the subject in my spare time whenever possible.

I've been studying the image forming process, and would like to ask you what you think of an idea I had. If the body in the Shroud passed through the cloth either during (or after or before) the image making process this would account for the unbroked blood stains and clots, and for the anatomical exactness of the blood flows and scourge marks.

If the body passed through the cloth during the image making process, it would also account for the three dimensional characteristic of the image, as well as explain the problem of density shading inherent with each theory thus far.

I believe the passing through of the body during the image making process could apply to both the scorch theory and the Pellicori hypothesis. While the cloth was laying over the body, those parts of the cloth in contact with the body were receiving the stimulus for a longer period of time and reflect this in the density of the image. While the body passed through, these parts of the body that were not in contact then make contact for a very brief moment, and thus a less dense image.

I realize that according to this hypothesis there could be a problem accounting for the superficial nature of the image. One possible way to account for this would be a low volume or degree of stimulus emanating from the body, combined with an extremely quick passing through of the body. The only other guess I would have would be if the stimulus from the body, or the stimulus' penetration, ceases upon contact with the cloth. Is there any scientific principle that would account for such ceasing?

I believe the theory of how the body passing through the cloth could account for the exactness of the blood flows and marks is self-explanatory. There would still, perhaps, require an accounting for why the blood did not spread out in an inexact manner where there was contact with the cloth for the period of time between first covering the body and when the body passed through. I'm not sure if an accounting for these locations at that time period is necessary or not, because I really don't have a good idea of the stiffness or coarseness of the Shroud, or of the nature of the blood marks or flows that were actually in contact with the cloth when it was laid over the body, or for how long they set before contact, or for how long they were in contact.

I realize a testing of such a "passing through" hypothesis is impossible, but could you please tell me if you think that might hypothetically answer some of the problems with the image forming process, and the blood line characteristics. I'm thinking about including this in my address, but wanted to get your thoughts first. On January 10th & 11th, I wrote two letters with the same information as this letter to Messrs. Jackson, Jumper, and Pellicori, and am awaiting a response from them.

Thank you very much for your time and consideration.

Sincerely,

Mark Antonacci

APPENDIX H

THE SUDARIUM OF OVIEDO

A very interesting cloth has been kept in Oviedo, Spain, since the eighth century. It is much smaller than the Shroud (approximately 84 x 54 cm, or 2'9¾" x 1'8⅞") and is reputed to be the sudarium, or the cloth that had been over Jesus' head when he was buried, and found rolled or folded up in a place by itself on Easter morning (John 20:7). Several investigators think that this cloth was folded or placed over Jesus' head following his death, after he was taken down from the cross. It was then placed by itself in the tomb when Jesus was buried in a shroud.

The Oviedo Cloth does not contain an image, but does have bloodstains and pollens. Some investigators think that some of the bloodstains on the Oviedo Cloth show a real similarity to or congruence with the bloodstains on the back of the head of the man in the Shroud.[1] The blood on the Oviedo Cloth is also reported to be human blood of the group AB.[2] Like the Shroud, pollen grains from *Gundelia tournefortii*, an insect-pollinated species of a thorny plant that grows only in Israel and parts of the Middle East, have also been identified on the Oviedo Cloth.[3] Moreover, no pollen was found on this cloth connecting it to Constantinople, France, Italy, or Europe.[4]

This plain woven cloth has been reported to have been carbon dated by two different laboratories to A.D. 679 and A.D. 710.[5] The co-director of one of these laboratories is reported to have written that its samples "were received supposedly as CO_2 gas in ampoules that leaked air and so were unusable."[6] If this cloth was the sudarium and has similarities to the Shroud, it is possible that it was exposed to a neutron flux in the tomb at the same time that the Shroud was. The Shroud, being more centrally located in the tomb and closer to the source of radiation, could have received more neutron flux and additional C-14 than the sudarium. Like the Shroud, the Oviedo Cloth should be examined for Ca-40 + Ca-41 and for Cl-35 + Cl-36. If the presence of either the first two elements and/or the last two elements can be found on this cloth, the amount of neutron flux and the amount of additional C-14 created in the cloth could be determined.

The Oviedo Cloth should also be examined for limestone to see if it matches the limestone found on the Shroud and the same rock shelf in which Jesus was buried. This limestone should also be examined for Ca-40 +Ca-41 in the same manner and for the same purpose that the Oviedo Cloth, the Shroud, and its limestone are examined and tested. Similarly, a small amount of the plentiful blood from the Oviedo Cloth should be tested to see if any of its Fe-56 was converted to Cr-53, and if so, how much neutron flux was involved. Like the blood on the Shroud, a very small amount should also be carbon dated to confirm the above results. In addition, its blood should be examined and tested in the same ways in which the blood from the Shroud has been examined, in order to see how identical the blood from these two cloths are.

Should the above tests be performed and a neutron flux be found to have occurred to this cloth, and its true age also be determined to be the first century, the authenticity of the sudarium and further evidence for the resurrection of the historical Jesus could be established. However, should the Oviedo Cloth be confirmed as being from around the seventh or eighth century, it would not affect the extensive evidence found throughout the rest of this book.

APPENDIX I

PROPHECIES OF NEW TESTAMENT EVENTS

Another form of corroboration that the crucifixion and resurrection of the historical Jesus Christ occurred and that God caused them may be found from the numerous Old Testament prophecies and Jesus' own predictions of these events in the New Testament.

For Jesus' predictions, see Matthew 16:21; 17:9, 22–23; 20:18, 19; 26:32; 27:63; Mark 8:31; 9:9, 31; 10:33–34; 14:28, 58; Luke 9:22; 18:33; John 2:9–21; 16:16–22. For Old Testament prophecies related to Jesus' crucifixion and resurrection (source for all verses is the New American Standard Bible.):

Even my close friend, in whom I trusted;
Who ate my bread,
Has lifted up his heel against me. (Ps. 41:9)

And while He was still speaking, behold, Judas, one of the twelve, came up, accompanied by a great multitude with swords and clubs, from the chief priests and elders of the people. Now he who was betraying Him gave them a sign, saying "Whomever I shall Kiss, He is the one, seize Him." (Matt 26:47, 48. See also 27:3,5,7. See also Mark 14:43-47; Luke 22:47, 48; John 18:2,3)

And I said to them, "If it is good in your sight, give me my wages; but if not, never mind!" So they weighed out thirty shekels of silver, as my wages. Then the Lord said to me, "Throw it to the potter, that magnificent price at which I was valued by them." So I took the thirty shekels of silver and threw them to the potter in the house of the Lord. (Zech. 11:12, 13.)

Then when Judas, who had betrayed Him, saw He had been condemned, he felt remorse and returned the thirty pieces of silver to the chief priests and elders. . . . And he threw the pieces of silver into the sanctuary and departed. . . . And they counseled together and with the money bought the Potter's Field as a burial place for strangers. (Matt 27:3, 5, 7. See also Matt. 26:15, Mark 14:10,11.)

Malicious witnesses rise up;
They ask me of things that I do not know. (Ps. 35:11)

Now the chief priests and the whole Council kept trying to obtain false testimony against Jesus, in order that they might put Him to death; they did not find any, even though many false witnesses came forward. . . (Matt 26:59, 60. See also Mark 14:56-59.)

He was oppressed and He was afflicted,
Yet he did not open His mouth. (Is. 53:7)

And while He was being accused by the chief priests and elders, He made no answer. (Matt 27:12. See also Mark 14:60–61.)

I gave My back to those who strike Me,
And My cheeks to those who pluck out the beard;
I did not cover My face from humiliation and spitting. (Is. 50:6)

Then they spat in His face and beat Him with their fists; and others slapped Him. (Matt 26:67. See also Mark 14:65; 15:19; Luke 22:63; John 18:22.)
. . . and after having Jesus scourged, he delivered Him over to be crucified. (Mark 15:15. See Matt. 27:36; John 19:1.)

. . . They pierced my hands and my feet.
I can count all my bones.
They look, they stare at me. (Ps. 22:16–17)

And when they came to the place called The Skull, there they crucified Him. . . . And the people stood by, looking on . . . (Luke 23:33, 35)

They divide my garments among them.
And for my clothing, they cast lots. (Ps. 22:18)

And they crucified Him, and divided up his garments among themselves, casting lots for them, to decide what each should take. (Mark 15:24. See also Matt 27:35; Luke 23:34; John 19:23–24)

My loved ones and my friends stand aloof from my plague;
And my kinsmen stand afar off. (Ps. 38:11)

And all His acquaintances and the women who accompanied Him from Galilee, were standing at a distance, seeing these things. (Luke 23:49. See also Matt. 27:55–56; Mark 15:40)

All who see me sneer at me;
They separate with the lip, they wag the head, saying,
"Commit yourself to the Lord; let him deliver him;
Let Him rescue him, because He delights in him." (Ps. 22:7–8. See also Ps. 109:25.)

And those passing by were hurling abuse at Him, wagging their heads, and saying, ". . . if You are the Son of God, come down from the cross." In the same way the chief priests also, along with the scribes and elders, were mocking Him and saying, " . . . He trusts in God; Let Him deliver Him now, if He takes pleasure in Him; for He said, 'I am the Son of God'" (Matt 27:39–43)

They also gave me gall for my food,
And for my thirst they gave me vinegar to drink. (Ps. 69:21)

They game him wine to drink mingled with gall; and after tasting it, He was unwilling to drink. . . . And immediately one of them ran, and taking a sponge, he filled it with sour wine [vinegar], and put it on a reed, and gave Him a drink. (Matt 27:34, 48. See also Mark 15:23,36; Luke 23:36; John 19:28–30.)

. . . Because He poured out Himself to death, And was numbered with the transgressors. (Is. 53:12)	At that time two robbers were crucified with Him, one on the right and one on the left. (Matt. 27:38. See also Mark 15:27; Luke 23:47–48.)
"And it will come about in that day," declares the Lord God, "That I shall make the sun go down at noon And make the earth dark in broad daylight." (Amos 8:9)	And when the sixth hour [noon] had come, darkness fell over the whole land until the ninth hour. (Mark 15:33. See also Matt 27:45; Luke 23:44–45.)
My God, my God, why hast Thou forsaken me? (Ps. 22:1)	And at the ninth hour Jesus cried out with a loud voice, "Eloi, Eloi lama sabachthani?" which is translated, "My God, my God, why hast thou forsaken Me." (Mark 15:34. See also Matt 27:46)
Into Thy hand I commit my spirit. (Ps. 31:5)	And Jesus, crying out with a loud voice, said, "Father, into Thy hands I commit My spirit." And having said this, He breathed His last. (Luke 23:46. See also Matt 27:50.)
He keeps all his bones; Not one of them is broken. (Ps. 34:20)	but coming to Jesus, when they saw that He was already dead, they did not break His legs. (John 19:33)
. . . [T]hey will look on Me whom they have pierced. . . (Zech. 12:10)	but one of the soldiers pierced His side with a spear, (John 19:34)
I am poured out like water, And all my bones are out of joint; My heart is like wax; It is melted within me. (Ps. 22:14)	. . . and immediately there came out blood and water. (John 19:34)
"And I will pour out on the house of David and on the inhabitants of Jerusalem, the Spirit of grace and of supplication, so that they will look on Me whom they have pierced; and they will mourn for Him, as one mourns for an only son, and they will weep bitterly over Him, like the bitter weeping over a first-born. (Zech. 12: 10)	Now when the centurion saw what had happened, he began praising God, saying, "Certainly this man was innocent." And all the multitudes who came together for this spectacle, when they observed what had happened, began to return, beating their breasts. (Luke 23:47–48)

But he was pierced through for our transgressions, He was crushed for our iniquities; The chastening for our well-being fell upon Him, And by His scourging we are healed. (Is. 53:5)	And He himself bore our sins in His body on the cross, that we might die to sin and live to righteousness; for by His wounds you were healed. (1Peter 2:24)
. . . Yet He Himself bore the sin of many, And interceded for the transgressors. (Is. 53:12)	For this is My blood of the covenant, which is poured out for many for forgiveness of sins. (Matt 26:28. See also Mark 10:45; John 3:17)
His grave was assigned with wicked men, Yet He was with a rich man in His death . . . (Is. 53:9)	. . . there came a rich man from Arimathea, named Joseph, who. . . went to Pilate and asked for the body of Jesus . . . And Joseph took the body and wrapped it in a clean linen cloth, and laid it in his own new tomb, which he had hewn out in the rock . . . (Matt 27:57-60.)
For Thou wilt not abandon my soul to Sheol; Neither wilt Thou allow Thy Holy One to undergo decay. (Ps. 16:10)	. . . He was neither abandoned to Hades, nor did His flesh suffer decay. (Acts 2:31.)
But the Lord was pleased To crush Him, putting Him to grief; If He would render Himself as a guilt offering, He will see His offspring, He will prolong His days, And the good pleasure of the Lord will prosper in His hand. (Is. 53:10)	. . . Him who raised Jesus our Lord from the dead. He who was delivered up because of our transgressions, and was raised because of our justification. (Romans 4:24–25.)
Arise, shine; for your light has come, and the glory of the Lord has risen upon you. For behold, darkness will cover the earth, And deep darkness the peoples; But the Lord will rise upon you, And His glory will appear upon you. And nations will come to your light, And kings to the brightness of your rising. (Is. 60:1-3)	And He is the radiance of His glory and the exact representation of His nature . . . (Heb 1:3) These are in accordance with the working of the strength of His might which He brought about in Christ, when He raised Him from the dead, and seated Him at His right hand in heavenly places, far above all rule and authority and power and dominion, and every name that is named, not only in this age, but also in the one to come. (Eph 1:19-21.)

APPENDIX J

COMPREHENSIVE LIST OF THE SHROUD'S UNIQUE CHARACTERISTICS

Let's review the unusual characteristics that would have to have been accounted for by a medieval forger in any credible explanation of the how the body images, blood marks, and other features were created on the Shroud of Turin. Any forger responsible for the image would have to have been able to:

- Encode the image on only the most superficial fibrils of the cloth's threads;
- Transfer an image so low in contrast that it fades into the background when an observer stands within six feet of it;
- Create an image that is pressure-independent so that both frontal and dorsal body images are encoded with the same intensity, even though the dorsal side of the cloth would have had the full weight of a body lying on top of it;
- Use an image-forming mechanism that operates uniformly regardless of what lies beneath it, i.e., over diverse substances such as skin, hair and, possibly, coins, flowers, teeth, and bones;
- Encode the thousands of body image fibrils with the same intensity;
- Create an image that is not composed of any particles or foreign materials of any kinds, with the individual joints of its individual fibrils remaining distinct and visible;
- Create an image that is not soluble in water, remains stable when subjected to high temperatures, and does not demonstrate signs of matting, capillarity, saturation, or diffusion into the image-forming fibrils;
- Encode an image that lacks any evidence of two-dimensional directionality;
- Compose a yellowed body image out of chemically degraded cellulose with conjugated carbonyls that has resulted from processes associated with dehydration and oxidation;
- Encode the front and back full-length images on cloth of a real human being in rigor mortis;
- Incorporate specific effects of a draped cloth that fell through a body region—such as blood marks displaced into the hair, motion blurs at the side of the face and in the neck/throat region and below the hair, along with elongated fingers;
- Encode a superficial, resolved, and three-dimensional image of the closed eye over the different and invisible features of a coin;
- Transfer the blood marks before encoding the body image, yet still place them in the appropriate locations and ensure that the blood marks are not altered when the body image is later transferred onto the cloth;
- Create actual blood marks with actual serum around the edges of the various wounds;

- Reproduce blood marks incurred at different times with different instruments that correspond with both arterial and venous bleeding;
- Encode blood marks on the cloth in exactly the form and shape that develop from wounds on human skin;
- Embed into the cloth the various blood marks leaving the original smooth surfaces between the skin and the blood intact;
- Remove the cloth from the body within two to three days without breaking or smearing the numerous blood marks;
- Employ a mechanism that transfers distance information through space in vertical, straight-line paths;
- Produce an image that is a vague negative when observed by the naked eye, but with highly focused and finely resolved details that become visible only when photographed, at which point the negative turns into a positive image with light/dark and left/right reversed;
- Encode accurately proportioned, three-dimensional information on a two-dimensional surface that directly corresponds to the distances between a body and cloth;
- Include realistic details of scourge marks so minute that they are invisible to the naked eye and can be seen only with cameras, photographic enlargers, microscopes, and ultraviolet lighting;
- Encode a line representing the narrow lesion of the side wound that corresponds to the shape of the lancea used by Roman executioners in such a manner that the line would not be visible with the eye and could not be seen until the development of computer imaging technology 600 years later;
- Distribute an array of pollens onto the Shroud beneath the linen's threads and fibers that reflected its manufacture and history in Jerusalem and Turkey. To do this successfully, the forger would have to not only be a pollen expert, but also anticipate development of the theory that emerged 600 years later which asserts the Shroud, Mandylion, and Image of Edessa are the same cloth;
- Encode the subtle appearance of Judean plants in the off-image area of the Shroud that would not be seen for more than six centuries;
- Place microscopic samples of dirt and limestone at the foot of the man in the Shroud that match the limestone found in Jerusalem, but which would not be visible for centuries;
- Encode actual whole blood and watery fluid at the side wound and the small of the back in a uniquely realistic manner and also encode this and all other clotted bloodstains on the Shroud so that they remain red and do not darken over time like all other actual blood;
- Encode the appearance of a Pontius Pilate lepton over the right eye of the man so that only when photography, photographic enlargers and three-dimensional reliefs are invented 600 years later, the motif, letters and outline of the coin can be ascertained. The forger would not only have to anticipate this technology, but also the development of the field of archaeology and the discovery in the late

twentieth century that coins were used in burials in Jerusalem and the surrounding area between the first century B.C. and the first century A.D.

- Encode the wound on the cloth at the man's left side so that when the image was photographed 500 years later, the wound would be located in the precisely correct location on the man's right side so that blood and water would escape from the victim if he received a postmortem wound at this location.

To encode these features, our forger would not only have to have understood advanced scientific principles, but also have possessed a knowledge of anatomy and medicine that was centuries ahead of his time. Obviously, it would have been impossible for him to have possessed such knowledge and understanding, but even if he had, somehow, he still couldn't have seen any of these numerous features to know if he was getting them right. The technology needed to visualize them would not be developed for another five to six hundred years.

- How could a medieval artist have displayed a knowledge of physiology that would not be known until centuries later?
- How could an artist paint without showing any evidence of directionality?
- How could an artist encode three-dimensional information (on a two-dimensional surface) that directly corresponds to the distance between a body and a cloth?
- How could a medieval artist include details that are undetectable with the human eye and become visible only under ultraviolet light, or only through a microscope, or only on three-dimensional reconstructions, or only with the most advanced, twentieth-century computer scanning devices?

NOTES

CHAPTER ONE

1. While a few major newspapers of the time carried articles about the Shroud and news of Pia's startling photographs, no book or newspaper published the photographs during 1898 or 1899, as photos in news columns had not yet begun to appear in newspapers. Nor were the photographs displayed in any special photo sections of the papers. Only two magazines carried his photos, one of which contained a very poor representation of the full-length figure, and the other only representing the face. Coverage in the American press was extremely sparse, as column space was at a premium due to the ongoing Spanish-American War and the invasion of Cuba. J. Walsh, *The Shroud* (New York: Random House, 1963), pp. 177–178.
2. E. A. Wuenschel, *Self-Portrait of Christ* (Esopus, N.Y.: Holy Shroud Guild, 1957), p. 18; Walsh, pp. 92, 96–97; I. Wilson, *The Shroud of Turin,* rev. ed. (Garden City, N.Y.: Doubleday Image Books, 1979), pp. 32–33.
3. Wuenschel, pp. 18–25; Walsh, pp. 98–101; I. Wilson, p. 33.
4. Wuenschel, pp. 18–20; Walsh, pp. 98, 101–102.
5. Wuenschel, pp. 18, 25; Walsh, pp. 92–93, 102; F. Tribbe, *Portrait of Jesus?* (New York: Stein and Day/Publishers, 1983), p. 178. See also Wilson, p. 33.
6. Wuenschel, p. 25; Tribbe, p. 178; Wilson, p. 33.
7. Wuenschel, pp. 25–26; Walsh, pp. 106–107; Tribbe, p. 178.
8. Wuenschel, p. 26; Walsh, p. 107; Tribbe, p. 178.
9. Wuenschel, p. 26; Walsh, 107; Tribbe, p. 178.
10. These sources were acquired and shared by Joe Marino, a former Benedictine monk, who maintained his collection for more than twenty years in St. Louis, MO.
11. E. J. Jumper and R. W. Mottern, "Scientific Investigation of the Shroud of Turin," *Applied Optics* 19.12 (1980): 1909–12.
12. J. H. Heller, *Report on the Shroud of Turin* (New York: Houghton Mifflin, 1983), pp. 109, 115–116; L. Gonella, "Scientific Investigation of the Shroud of Turin: Problems, Results and Methodological Lessons," in *Turin Shroud—Image of Christ?* (Hong Kong: Cosmos Printing Press Ltd., 1987), pp. 29–40; and B. M. Schwortz, "Mapping of Research Test-Point Areas on the Shroud of Turin," in *Proceedings IEEE International Conference* (Seattle: IEEE, 1982), pp. 538–547.
13. Heller, *Report.*
14. Ibid.
15. Ibid.

CHAPTER TWO

1. F.C. Tribbe, *Portrait of Jesus?* (New York: Stein and Day/Publishers, 1983), chapter 6; W. Meacham, "The Authentication of the Turin Shroud: An Issue in Archaeological Epistemology," *Current Anthropology* 24.3, June 1983, p. 285; E. A. Wuenschel, *Self-Portrait of Christ* (Esopus, N.Y.: Holy Shroud Guild, 1957), p. 34; I. Wilson, *The Shroud of Turin,* rev. ed. (Garden City, N.Y.: Doubleday Image Books, 1979), chapter 3; *La Santa Sindone,* Nelle Richerche Moderne, Acts of the First National Congress of Studies, Turin 1939; along with the Acts of the First International Congress (1950); in the anastatic reedition of 1980 published by Marietti, Milan; the *Shroud Spectrum International,* Nashville, Indiana, Dorothy Crispino; Father Peter Rinaldi, S.D.B., founder, and Rev. Adam J. Otterbein, C.S.S.R., president, Holy Shroud Guild; International Centre of Sindolonology–Turin, Regional Sicilian Delegation; see also

footnote references throughout this and other chapters. These comprehensive studies (by individuals who have examined the cloth itself; have examined the body image and bloodstained fibrils; have studied the highly detailed photographs in different lighting, from different distances, and in different dimensions; have studied each other's works, and have published themselves) contrast with the efforts of Dr. Michael Baden, a forensic pathologist from Queens County, New York, who merely looked at pictures of the Shroud in a magazine and who is the only medical person to have voiced doubts about the human quality of the man in the Shroud. R.W. Rehin, Jr., "The Shroud of Turin," *Medical World News* (December 22, 1980): pp. 40–50.

2. P. Barbet, *A Doctor at Calvary* (Garden City, New York: Doubleday Image Books, 1963); D. Willis, cited in Wilson, *The Shroud of Turin;* G. Judica-Cordiglia, cited in G. Caselli, "Ascertainments of Modern Medicine on the Imprints of the Holy Sindon," in *La Santa Sindone;* R. W. Hynek, *True Likeness* (London and New York: Sheed and Ward, 1951); F. T. Zugibe, *The Cross and the Shroud,* rev. ed. (New York: Paragon House, 1988); and G. R. Lavoie, "The Medical Aspects of the Shroud of Turin as Seen by a Practicing Physician," in *The Mystery of the Shroud of Turin an Interdisciplinary Symposium,* video, Elizabethtown, Penn.: (Elizabethtown College, February 15, 1986).
3. Barbet, *Calvary,* pp. 21, 91; R. Bucklin, "The Shroud of Turin: Viewpoint of a Forensic Pathologist," *Shroud Spectrum International* 1.5 (December 1982): 3–10; R. Bucklin, in *The Silent Witness;* directed by David W. Rolfe, produced by Screenpro Films (London), 1978; G. Judica-Cordiglia, cited in Caselli, "Ascertainments"; Hynek, *True Likeness;* and P.L.B. Bollone, "A Pathologist Observes the Shroud of Turin," in *La Sindone E La Scienze,* Acts of the International Congress of Sindonology, Second Edition, Turin, Centro Internazionale di Sindonologia, Edizoni Paoline, 1978.
4. Bucklin, "Viewpoint of a Forensic Pathologist"; and J. H. Heller, *Report on the Shroud of Turin* (New York: Houghton Mifflin, 1983); See also S. F. Pellicori and M. Evans, "The Shroud of Turin Through the Microscope," *Archaeology* (January/February 1981): 32–43.
5. S. Rodante, "The Coronation of Thorns in the Light of the Shroud," *Shroud Spectrum International* 1 (December 1981): 4–24, Translated and reprinted from *Sindon* 24 (Oct. 1976).
6. Bucklin, "Viewpoint of a Forensic Pathologist"; R. Bucklin and J. Gambescia, in Shroud of Turin Research Project *Update* 3.1 (Fall 1981); Rodante, "The Coronation of Thorns in the Light of the Shroud"; Barbet, *Calvary;* P. Barbet, "Proof of the Authenticity of the Shroud in the Bloodstains" (1950), *Shroud Spectrum International* 1 (December 1981), excerpt reprinted from *Sindon* 14–15, (December 1970): p. 31; Caselli, "Ascertainments"; Bollone, "A Pathologist Observes"; J. M. Cameron, "The Pathologist and the Shroud" in *Face to Face with the Turin Shroud,* ed. P. Jennings (London: Mayhaus-McCrimmon, 1978), 57–59; Zugibe, *The Cross and the Shroud;* and D. Willis, cited in Wilson, *Shroud of Turin.*
7. Bucklin, "Viewpoint of a Forensic Pathologist"; R. Bucklin, "An Autopsy on the Man of the Shroud," Acts du IIIeme symposium scientifique International—Nice, 1997, Cetre International d' Etudes sur le Linceul de Turin, pp. 99–101; Barbet, *Calvary;* Bollone, "A Pathologist Observes"; Wilson, *The Shroud of Turin;* and E. W. Massey, "An Interpretation of the Hand and Arm Markings of the Shroud of Turin," *The Hand* 12.1 (1980).
8. W. Meacham, "The Authentication of the Turin Shroud: An Issue in Archaeological Epistemology," [100 scourge marks]; Barbet, *Calvary* [100–120]; G. R. Habermas and K. E. Stevenson, *Verdict on the Shroud* (Ann Arbor, Mich.: Servant Books, 1981) [90–120]; Wilson, *Shroud of Turin* [90–120]; Zugibe, *The Cross and the Shroud* [100–120]; Tribbe, *Portrait of Jesus?* [120]; P.L.B. Bollone, "A Pathologist Observes" [over 100]; and M. Straiton, "Evidence That the Body was Placed in the Holy Shroud after Death Had Occurred," in *La Sindone Scienza E Fede,* Acts of the II National Congress of Sindonology in Bologna, 1981, CLUEB, 1983.
9. Barbet, *Calvary;* Bucklin, "Viewpoint of a Forensic Pathologist"; Don Lynn (image enhancement specialist and STURP scientist), personal communication, 1985; and Vern Miller, STURP photographer, personal communication, December 1997.

10. Meacham, "Authentication"; I. Wilson, *Shroud of Turin* and *The Mysterious Shroud* (Garden City, NY: Doubleday, 1986); T. Humber, *The Sacred Shroud* (New York: Pocket Books, 1974); and R. K. Wilcox, *Shroud* (New York: Bantam Books, 1979).
11. Miller and S. F. Pellicori, "Ultraviolet Fluorescence Photography of the Shroud of Turin," *Journal of Biological Photography* 49.3 (July 1981); K. F. Weaver, "The Mystery of the Shroud," *National Geographic* 157.6 (June 1980); Heller, *Report*; and L. A. Schwalbe and R. N. Rogers, "Physics and Chemistry of the Shroud of Turin," *Analytica Chimica Acta* 135 (1982).
12. Bucklin, "Viewpoint of a Forensic Pathologist," p. 8, and "The Shroud of Turin: A Pathologist's Viewpoint," in *Legal Medicine Annual* (Philadelphia: W. B. Saunders, 1982), p. 38.
13. Meacham, "Authentication," p. 290; Wilson, *Shroud of Turin* and *Mysterious Shroud;* Barbet, *Calvary;* Humber, *The Sacred Shroud;* and Wilcox, *Shroud.*
14. Bucklin, "Viewpoint of a Forensic Pathologist," and "An Autopsy," p. 100; Barbet, *Calvary;* Willis, cited in Wilson, *Shroud of Turin;* Caselli, "Ascertainments"; Bollone, "A Pathologist Observes"; Hynek, *True Likeness;* and Cameron, "The Pathologist."
15. Barbet, *Calvary,* p. 98; Willis, cited in Wilson, *Shroud of Turin;* Cameron, cited in Wilson, *The Mysterious Shroud;* and Wuenschel, *Self-Portrait.*
16. G. Judica-Cordiglia, cited in Wilson, *Shroud of Turin,* p. 39, and in Barbet, *Calvary,* p. 97; Lavoie, "Medical Aspects"; and Barbet, *Calvary.*
17. Miller and Pellicori, "Ultraviolet Fluorescent Photography"; and Heller, *Report.*
18. Heller, *Report,* pp. 112, 152; S. F. Pellicori "Ultraviolet Fluorescent Photography"; Evans, "Through the Microscope"; J. Jackson, cited in C. Murphy, "Shreds of Evidence," *Harper's* 263.1578 (November 1981): 42–65. See also Wilson, *The Blood and the Shroud* (New York: The Free Press, 1998), p. 95.
19. Cameron, "The Pathologist," p. 58; personal communication with St. Louis medical examiner Dr. Mary Case, April 4, 1984; R. Bucklin, "Postmortem Changes and the Shroud of Turin," *Shroud Spectrum International* 14 (March 1985): 3–6, and "The Legal and Medical Aspects of the Trial and Death of Christ," *Medicine, Science and the Law* 10.1 (1970); Barbet, *Calvary,* p. 159; and Lavoie, "Medical Aspects".
20. Barbet, *Calvary,* p. 135; Caselli, "Ascertainments"; Dr. L. Gedda as stated in Caselli, "Ascertainments"; and Cameron, "The Pathologist."
21. Barbet, *Calvary,* p. 135.
22. Miller and S. F. Pellicori, "Ultraviolet Fluorescent Photography"; A. Battaglini, "Considerations on the Feet of the Man in the Shroud," *Shroud Spectrum International* 9 (December 1983): p. 6; Vignon, as stated in an excerpt from *Shroud Spectrum International,* No. 9, p. 6; and Barbet, *Calvary.*
23. Bucklin, "Viewpoint of a Forensic Pathologist"; Barbet, *Calvary;* Battaglini, "Considerations"; Judica-Cordiglia, excerpt in *Shroud Spectrum International* 9, (December 1983): 6; Caselli; "Ascertainments"; Bollone, "A Pathologist Observes"; Lavoie, "Medical Aspects"; and Vignon, "The Problem of the Holy Shroud," *Science America* 93.163 (1937): 162–64.
24. Bucklin, "Viewpoint of a Forensic Pathologist," p. 8.
25. Barbet, *Calvary,* pp. 125, 183; Caselli; "Ascertainments"; Judica Cordiglia, excerpted in *Shroud Spectrum International* 9 (December 1983): 6; and Vignon, "Problem."
26. Barbet, *Calvary,* pp. 121, 125–26; Caselli, "Ascertainments"; Willis, "Did He Die on the Cross?" *Ampleforth Journal* 74 (1969); Hynek, *True Likeness;* Straiton, "Evidence That the Body"; and Vignon, "Problem."
27. John 20:25, 27.
28. According to A. O'Rahilly, *The Crucified* (Dublin, Ireland: Kingdom Books, 1985), p. 137: "In Greek the word *cheir* usually means not hand but arm. So Homer, *Iliad* xi.252; Hesiod, *Theog.* 150; Euripides, *Iph.* 1404; etc. In Xenophon, *Anab.* i.5,8, it means wrist (bracelets on their wrists). Hippocrates uses *akre cheir* for forearm. Arm is probably the meaning in Heb 12:12."; Tribbe, *Portrait of Jesus?;* F. T. Zugibe, *The Cross and the Shroud* also cites Zorell's *Lexicon Herbraicum et Armaicum* and Lidell-Scott's *Greek-English Lexicon,* 7th ed. (New York: Harper Brothers, 1883).

29. V. Tzaferis, "Crucifixion—The Archaeological Evidence," *Biblical Archaeology Review* 11 (January/February 1985): 44–53, p. 52.
30. Tzaferis, "Crucifixion"; and N. Haas, "Anthropological Observations on the Skeletal Remains from Giv'at Ha-mivtar," *Israel Exploration Journal* 20.1/2 (1970): 8–59.
31. Psalms 34:20 and Numbers 9:12. See also John 19:36.
32. Barbet, *Calvary*. Alternatively, Dr. Zugibe suggests the nail could have passed through the wrist "through a space created by four other carpal bones. . . where "the trunk of the median nerve would most likely be damaged by this path." F. Zugibe, *The Cross and the Shroud*, p. 63, and "Pierre Barbet Revisited," http://www.shroud.com/zugibe.htm, p.7.
33. P. Vignon, *The Shroud of Christ*, translated from the French, Westminster, 1902, (New Hyde Park, NY: University Books), 1970; Vignon, "The Problem of the Holy Shroud," 1937; Barbet, *Les Cinq Plaies du Christ* (the Five Wounds of Christ), 1935; Barbet, *A Doctor at Calvary*, first published in 1950; Caselli, "Ascertainments," and Masera, "The Work of Medical Jurisprudence on the Marks of the Holy Shroud," in *La Santa Sindone* (Scotti di Di Pietro, S.D.B.), Nelle Richerche Moderne, Atti dei Convegni Di Studio, Torino 1939, Roma e Torino 1950; Willis, "Did He Die on the Cross?" 1969; Willis, "False Prophet and the Holy Shroud," *The Tablet*, June 13, 1970; Bucklin, "The Legal and Medical Aspects;" 1970; Rodante, "The Coronation of Thorns in the Light of the Shroud," 1976; Bollone, "A Pathologist Observes the Shroud of Turin," 1978.
34. Heller, *Report*, pp. 215–16; and Heller and Adler, "A Chemical Investigation of the Shroud of Turin," *Can. Soc. Forens. Sci. J.*, Vol. 14, No. 3 (1981), p. 92.
35. Rodante, "Coronation of Thorns," and Caselli, "Ascertainments."
36. Rodante, "Coronation of Thorns." Dr. Rodante's work was greatly influenced by that of Dr. Caselli, who earlier had studied all of the wounds thoroughly.
37. S. Rodante, "Coronation of Thorns," p. 8.
38. Ibid.
39. Ibid. F. La Cava, *La passione e la morle de N.S. Ges'u Cristo illustrate dalla scienze medica* (Naples: D'Auria, 1953); and Zugibe, *The Cross and the Shroud*, pp. 24–27.
40. Dr. P. Scotti, as cited by Dr. Alan Adler in "The Turin Shroud Lecture," Department of Chemistry, Queen Mary College, London, July 20, 1984; Adler, "Chemical Investigation on the Shroud of Turin" in *The Mystery of the Shroud of Turin Interdisciplinary Symposium* video, Elizabethtown, Penn.: Elizabethtown College, February 15, 1986.
41. Adler, "Lecture" and "Chemical Investigations"; Heller, *Report*, p. 185; Heller, and Adler, "A Chemical Investigation"; Adler, as stated in T. Case, *The Shroud of Turin and the C-14 Dating Fiasco*, (Cincinnati, Oh.: White Horse Press, 1996); Jumper, as stated in Adler, "Chemical Investigations"; Gonella, "Scientific Investigation of the Shroud of Turin: Problems, Results and Methodological Lessons," in *Turin Shroud—Image of Christ?* (Hong Kong: Cosmos Printing Press Ltd., 1987), pp. 29–40; and Miller and Pellicori, "Ultraviolet Fluorescence."
42. Heller and Adler, "A Chemical Investigation"; Adler, "Lecture" and "Chemical Investigations"; and Heller, *Report*, pp. 185–86.
43. Miller and Pellicori, "Ultraviolet Fluorescent Photography," p. 85.
44. Adler, "Lecture" and "Chemical Investigations;" and personal communication with Alan Adler, July 18, 1985.
45. Miller and Pellicori, "Ultraviolet Fluorescent Photography," pp. 76-83; and K. F. Weaver, "The Mystery of the Shroud." A. D. Adler, "The Origin and Nature of Blood on the Turin Shroud," [Excerpts from lecture of the Dept. of Anatomy, Univ. of Hong Kong, March 3, 1986] in *Turin Shroud—Image of Christ?*, pp. 57–59.
46. Heller and Adler; "A Chemical Investigation"; Adler, "Lecture" and "Chemical Investigations"; Heller, *Report;* Jumper, Adler, Jackson, Pellicori, Heller, and Druzik, "A Comprehensive Examination of the Various Stains and Images on the Shroud of Turin," in *ACS Advances in Chemistry No. 205 Archaeological Chemistry III*, American Chemical Society (1984).
47. A. D. Adler, "Updating Recent Studies on the Shroud of Turin," *Archaeological Chemistry: Organic, Inorganic, and Biochemical Analyses*, Mary Virginia Orna, ed. American Chemical Society (1996), pp. 223–228.
48. A. D. Adler, "The Nature of the Body Images on the Shroud of Turin," *Shroud of*

Turin International Research Conference, Richmond, Va., June 18–20, 1999, p. 2, http://www.shroud.com/; see also Adler, "The Nature and Origin," p. 57.

49. See for example Vignon, *The Shroud of Christ,* 1902; Barbet, *Les Cinq Plaies du Christ* (The Five Wounds of Christ), 1935; Caselli, "Ascertainments of Modern Medicine on the Imprints of the Holy Sindon," 1939; Barbet, *Calvary,* 1950; Bulst, *The Shroud of Turin* (Milwaukee: Bruce, 1957); Wuenschel, *Self-Portrait of Christ,* 1957; D. Willis, "Did He Die on the Cross?" (1969); and "False Prophet and the Holy Shroud," 1970; Bucklin, "The Legal and Medical Aspects," 1970; Rodante, "Coronation of Thorns," 1976.
50. Barbet, *Calvary,* p. 16.
51. Ibid. p. 29; Yves Delage proposed testing the cloth in 1902, as did scientists at the Sorbonne through Baron Manno, the former president of the 1898 exhibition. Scientists and physicians through Vignon also requested such tests. The 1939 Congress of the Italian Commission and the 1950 International Congress at Rome further passed resolutions to this effect. Wuenschel, *Self Portrait;* Caselli, "Ascertainments"; and Walsh, *The Shroud* (New York: Random House, 1963).
52. S. F. Pellicori, "Spectral Properties of the Shroud of Turin," *Applied Optics* 19.12 (1980): 1913–20.
53. Heller, *Report;* and Adler, "Lecture" and "Chemical Investigations."
54. P. L. B. Bollone and A. Gaglio, "Technical Immune—Enzymatic Application of the Shroud Drawings," presented at Third National Meeting of Studies on the Shroud, October 13–14, 1984; P. L. B. Bollone, M. Jorio, and A. L. Massaro, "Defining the Blood Group Identified on the Shroud," in *La Sindone Scienza E Fede,* p. 178; P. L. B. Bollone, M. Jorio, and A. L. Massaro, "Identification of the Traces of Human Blood on the Shroud," *Shroud Spectrum International* 6 (March 1983): 3–6; and P. L. B. Bollone and A. Gaglio, "Demonstration of Blood, Aloes and Myrrh on the Holy Shroud with Immunofluorescence Techniques," *Shroud Spectrum International* 13 (December 1984). See also Adler, "Lecture" and "Chemical Investigations."
55. *The Mysterious Man of the Shroud,* directed by Terry Landeau, CBS documentary, aired April 1, 1997; L. A. Garza-Valdes, *The DNA of God?* (London: Hodder & Stoughton, 1998), pp. 41, 42.
56. M. Borkan, "Ecce Homo? Science and the Authenticity of the Turin Shroud," *Vertices,* Winter 1995, Duke University Undergraduate Publications.
57. A. D. Adler, "The Origin and Nature" and "Chemical Investigations"; Jumper et al., "Comprehensive Examination."
58. Heller, *Report,* p. 210.
59. Wilcox, *Shroud,* p. 64. The one physician who comes closest to duplicating some of the blood characteristics is Gil Lavoie, a staunch advocate of the authenticity of the Shroud, but his samples do not duplicate all the blood marks or the characteristics of the wounds on the Shroud, nor does he claim they do. His experiments indicate that the blood marks transfer onto cloth best if they are first in the vertical position, allowing some of the serum to run down and away from the wound. See G. R. Lavoie, B. B. Lavoie, V. J. Donovan, and J. S. Ballas, "Blood on the Shroud of Turin: Part 2," *Shroud Spectrum International* 8 (September 1983): 2–10; and Lavoie, *Unlocking the Secrets of the Shroud* (Allen, Tex.: Thomas More, 1998).
60. Barbet, *Calvary;* Caselli, "Ascertainments"; Straiton, "Evidence That the Body"; Hynek, *True Likeness;* Lavoie, "Medical Aspects"; and R. Bucklin, "An Autopsy," p. 99. Experiments with volunteers hanging in a crucifixion position have also produced similar results. See experiments of artist/scientist Isabel Piczek in Wilson, *The Blood and the Shroud,* p. 23, and those of Dr. Frederick Zugibe in *The Cross and the Shroud,* p. 108.
61. Barbet, *Calvary;* W. D. Edwards, W. J. Gabel, F. E. Hosmer. "On the Physical Death of Jesus Christ," *JAMA* 255.11, (March 21, 1986); Hynek, *True Likeness;* Wuenschel, *Self-Portrait;* and Tzaferis, "Crucifixion."
62. Bucklin, "Viewpoint of a Forensic Pathologist," p. 8, and "An Autopsy," p. 99; Barbet, *Calvary,* p. 144; Caselli; "Ascertainments"; Willis, "Did He Die?"; Hynek, *True Likeness;* and Bollone, "A Pathologist Observes."
63. A. F. Sava, "The Wound in the Side of Christ," *Catholic Biblical Quarterly* 19 (1957): 343–347; A.F. Sava, "The Blood and Water from the Side of Christ," *American*

Ecclesiastical Review 138 (1958): 341–45; Bucklin, "Viewpoint of a Forensic Pathologist"; Barbet, *Calvary;* Judica-Cordiglia, cited in Bucklin, supra; and L. L. Gomez, "Legal Medical Study of the Wound on the Side of Christ," in *La Santa Sindone,* Nelle Richerche Moderne, Acts of the First National Congress of Studies, Turin 1939.

64. Bucklin, "Viewpoint of a Forensic Pathologist," pp. 8, 9.
65. Bucklin, "Postmortem Changes," "Legal and Medical Aspects," "Viewpoint of a Forensic Pathologist," "A Pathologist's Viewpoint," and "An Autopsy"; Caselli, "Ascertainments," Cameron, "The Pathologist," Hynek, *The True Likeness* and "The Real Cause of Crucifixion Death and Cadaveric Rigidity," in *La Santa Sindone;* Masera "The Work of Medical Jurisprudence on the Marks of the Holy Shroud," in *La Santa Sindone;* Straiton, "Evidence That the Body," Zugibe, *The Cross and the Shroud;* Bollone, "A Pathologist Observes"; Barbet, *Calvary;* Bruckner, "Some Observations on the Medical Aspects of the Shroud of Turin," *Linacre Quarterly* 54 (May 1987): 58–62; Lavoie, "The Medical Aspects"; Dr. William Drake, chief of Pathology, Missouri Baptist Hospital, personal communications in 1985 and 1986.
66. Bucklin, "Postmortem Changes"; personal communications with pathologist Dr. William Drake, 1985–86; personal communication with medical examiner Dr. Mary Case, April 4, 1984; personal communication with pathologist Dr. Dan McKeel in April 1984; Zugibe, *The Cross and the Shroud,* p. 131; and Brucker, "Some Observations."
67. Caselli, "Ascertainments"; Bucklin, "Postmortem Changes"; Brucker, "Some Observations"; Zugibe, *The Cross and the Shroud;* and personal communication with medical examiner Dr. Mary Case, April 4, 1984.
68. Bucklin, "Postmortem Changes," pp. 3–6, "Legal and Medical Aspects," and "An Autopsy," pp. 99–100; Barbet, *Calvary,* p. 122; personal communications with pathologist Dr. William Drake, 1985–86; Bollone, "The Pathologist Observes"; Straiton, "Evidence That the Body"; Hynek, *True Likeness;* Brucker, "Some Observations"; Zugibe, *The Cross and the Shroud;* Lavoie, "Medical Aspects"; and Caselli, "Ascertainments."
69. Caselli, "Ascertainments"; personal communications with pathologist Dr. William Drake, 1985–86; Hynek, *True Likeness;* Barbet, *Calvary,* p. 87; Brucker, "Some Observations"; and Lavoie, "Medical Aspects."
70. Bucklin, "Postmortem Changes"; Hynek, *True Likeness;* Straiton, "Evidence That the Body"; and Zugibe, *The Cross and the Shroud.*
71. Barbet, *Calvary;* Caselli, "Ascertainments"; Straiton, "Evidence That the Body"; Hynek, *The True Likeness;* and Lavoie, "The Medical Aspects."
72. Caselli, "Ascertainments"; personal communications with pathologist Dr. William Drake, 1985–1986; Cameron, "The Pathologist"; Hynek, *True Likeness;* Straiton, "Evidence That the Body"; and Barbet, *Calvary.*
73. Barbet, *Calvary,* p. 119; Heller, *Report;* R. Bucklin, "A Pathologist Looks at the Shroud of Turin," in *La Sindone E La Scienza,* p. 117, and "Legal and Medical Aspects"; Bollone, "A Pathologist Observes"; personal communications with pathologist Dr. William Drake, 1985–1986; M. Bocca, E. Messina, and S. Salvi. "Comments on the Wounds of Anatomic-Function of a Nailed Hand, with Reference to the Sindon of Turin," *La Sindone E La Scienze,* 1983; and Lavoie, "Medical Aspects."
74. Barbet, *Calvary* [110 lbs]; Wilson, *Shroud of Turin* [100 lbs]; Edwards, Gabel, and Hosmer, "On the Physical Death," [75–125 lbs]; Bucklin, "Legal and Medical Aspects" [80 lbs]; and Humber, *The Sacred Shroud* [80 lbs].
75. Barbet, *Calvary,* p. 119; Casselli, "Ascertainments"; Bucklin, "Legal and Medical Aspects"; Edwards, Gabel, and Hosmer, "On the Physical Death"; P. J. Smith, "Appendix II" in Barbet, *Calvary,* p. 119.
76. Bucklin and Gambescia, in *Update;* Bucklin, "Legal and Medical Aspects," p. 24; Caselli, "Ascertainments"; Smith, "Appendix II" in Barbet, *Calvary,* p. 119; C. T. Davis, "The Crucifixion of Jesus: The Passion of Christ from a Medical Point of View," *Arizona Medical* (1965): 183–87; Bloomquist, "A Doctor Looks at Crucifixion," *Christian Herald* (March 1964): 46–48.
77. La Cava, "La Passione"; Rodante, "Coronation of Thorns." See also Zugibe, *The Cross and the Shroud,* pp. 24–27.

CHAPTER THREE

1. E. J. Jumper, A. D. Adler, J. P. Jackson, S. F. Pellicori, J. H. Heller, and J. R. Druzik, "A Comprehensive Examination of the Various Stains and Images on the Shroud of Turin," *ACS Advances in Chemistry No. 205 Archaeological Chemistry III,* J. B. Lambert, ed. American Chemical Society (1984), pp. 447–476; J. H. Heller, *Report on the Shroud of Turin* (New York: Houghton Mifflin, 1983); L. A. Schwalbe and R. N. Rogers, "Physics and Chemistry of the Shroud of Turin," *Analytica Chimica Acta* 135 (1982): 3–49; J. H. Heller and A. D. Adler, "A Chemical Investigation of the Shroud of Turin," *Can. Soc. Forens. Sci. J.* 14.3 (1981): 81–103; L. Gonella, "Scientific Investigation of the Shroud of Turin: Problems, Results and Methodological Lessons," *Turin Shroud—The Image of Christ?* (Hong Kong: Cosmos Printing Press Ltd., 1987); S. F. Pellicori and M. Evans, "The Shroud of Turin Through the Microscope," *Archaeology* (January/February 1981): 32–43; W. C. McCrone and C. Skirius, "Light Microscopical Study of the Turin 'Shroud,' I," *Microscope* 28 (1980): 105–113; and W. C. McCrone, "Light Microscopical Study of the Turin 'Shroud,' II," Microscope 28 (1980): 115–128.
2. Heller, *Report;* G. R. Habermas and K. E. Stevenson, *Verdict on the Shroud* (Ann Arbor, Mich.: Servant Books, 1981); Gonella, "Scientific Investigation"; Schwalbe and Rogers, "Physics and Chemistry"; and personal communications with Don Lynn and Jean Lorre (NASA and STURP scientists), June 2, 1984 and March 16, 1999.
3. Personal communications with Don Lynn, August 21, 1989, and Jean Lorre, March 16, 1999.
4. Heller, *Report,* pp. 38–39; J. P. Jackson, E. J. Jumper, B. Mottern, and K. E. Stevenson, "The Three-Dimensional Image on Jesus' Burial Cloth," *Proceedings of the 1977 United States Conference of Research on the Shroud of Turin* (Albuquerque, N.M.: Holy Shroud Guild, Mar. 1977), 74–94; J. P. Jackson, E. J. Jumper, and W. R. Ercoline, "Correlation of Image Intensity on the Turin Shroud with the 3-D Structure of a Human Body Shape," *Applied Optics* 23.14 (July 1984): 2244–2270; and J. P. Jackson, E. J. Jumper, and W. R. Ercoline, "Three Dimensional Characteristics of the Shroud Image," *IEEE 1982 Proceedings of the International Conference on Cybernetics and Society* (October 1982): 559–575.
5. P. Vignon, *The Shroud of Christ,* translated from the French, Westminster 1902, (New Hyde Park, NY: University Books, 1970).
6. Jackson et al., "Three-Dimensional Image," p. 75, and "Correlation."
7. Jackson et al., "Three-Dimensional Image," p. 77.
8. C. Avis, D. Lynn, J. Lorre, S. Lavoie, J. Clark, E. Armstrong, and J. Addington, "Image Processing of the Shroud of Turin," *IEEE 1982 Proceedings of the International Conference on Cybernetics and Society* (October 1982): 554–558.
9. G. Tamburelli, "Some Results in the Processing of the Holy Shroud of Turin," *IEEE Transactions on Pattern Analysis and Machine Intelligence* PAMI-3.6 (November 1981): 670–676; G. Tamburelli, "Reading the Shroud, Called the Fifth Gospel, with the Aid of the Computer," *Shroud Spectrum International* 2 (March 1982): 3–11; G. Tamburelli, "An Image Resurrection of the Man of the Shroud," *Shroud Spectrum International* 15 (June 1985): 3–6.
10. E. J. Jumper, "Considerations of Molecular Diffusion and Radiation as an Image Formation Process on the Shroud," *Proceedings of the 1977 United States Conference of Research on the Shroud of Turin* (Albuquerque, NM: Holy Shroud Guild, March 1977), pp. 182–189; J. P. Jackson, "A Problem of Resolution Posed by the Existence of a Three-Dimensional Image on the Shroud of Turin," in *Proceedings* 223–233; Jackson et al., "Three-Dimensional Image," p. 91.
11. Another reason the face displays more three-dimensional features is that, in computer language, there is more "noise," or interference, on other parts of the body, ("noise" caused by the water and fire stains, patches, and blood flows on the Shroud). Only a few bloodstains appear on the face, in the forehead area and just below the right cheek. The marks made by water, fire, and blood don't contain any distance information and were most likely applied by direct contact. Consequently, because of various stains and much less "relief" in nonfacial areas, less three-dimensional information is found on the body of the man on the Shroud.
12. J. P. Jackson, "The Vertical Alignment of

the Frontal Image," *Shroud Spectrum International* 32/33 (1989): 3–26; see also Jackson et al., "Three-Dimensional Image," p. 83; and W. R. Ercoline, R. C. Downs, Jr., and J. P. Jackson, "Examination of the Turin Shroud for Image Distortions," *IEEE 1982 Proceedings of the International Conference on Cybernetics and Society* (October 1982): 576–579.

13. Jackson et al., "Three-Dimensional Image," p. 83; Jackson, "Vertical Alignment."
14. Jackson, "Vertical Alignment"; Jackson et al., "Three-Dimensional Image." p. 83.
15. As quoted in W. McDonald, "Science and the Shroud, *The World and I* (Oct. 1986): pp. 420–428, 426.
16. Ercoline et al., "Examination for Image Distortions."
17. Ibid.
18. Jumper et al., "Comprehensive Examination"; J. P. Jackson, "Blood and Possible Images of Blood on the Shroud," *Shroud Spectrum International,* 24 (September 1987): 3–11, p. 6.
19. Jumper et al., "Comprehensive Examination," p. 460.
20. Ibid., p. 456; Heller and Adler, "A Chemical Investigation," p. 98; Schwalbe and Rogers, "Physics and Chemistry," p. 14; Gonella; "Scientific Investigation"; Jumper, "An Overview of the Testing Performed by the Shroud of Turin Research Project with a Summary of Results," *IEEE 1982 Proceedings of the International Conference on Cybernetics and Society* (October 1982): 535–537; E. J. Jumper, "Science, the Shroud and the Public," in *Mystery of the Shroud of Turin: An Interdisciplinary Symposium,* video, Elizabethtown, PA: Elizabethtown College, February 15, 1986; A. D. Adler, "The Turin Shroud Lecture," Department of Chemistry, Queen Mary College, London, July 20, 1984; and A. D. Adler, "Chemical Investigations on the Shroud of Turin," *Mystery.*
21. Jumper et al., "Comprehensive Examination"; Adler, "Lecture," "Chemical Investigations," and personal communications on August 5, and September 3, 1986; Heller and Adler, "A Chemical Investigation," p. 98; Jumper, "Overview," and "Science"; and J. Jackson and R. Rogers, cited in C. Murphy, "Shreds of Evidence," *Harper's,* 263.1578 (November 1981): 42–65.
22. Jackson, "Blood and Possible Images."
23. G. R. Lavoie, *Unlocking the Secrets of the Shroud,* (Allen, Texas: Thomas More, 1998); G. R. Lavoie, B. B. Lavoie, V. J. Donovan, and S. J. Ballas, "Blood on the Shroud of Turin: Part I," *Shroud Spectrum International* 7 (June 1983): 15–19; and J. P. Jackson, "The Radiocarbon Date and How the Image Was Formed on the Shroud," *Shroud Spectrum International* 28/29 (September/December 1988): 2–12.
24. V. D. Miller and S. F. Pellicori, "Ultraviolet Fluorescence Photography of the Shroud of Turin," *Journal of Biological Photography* 49.3 (July 1981): 71–85; Adler, "Chemical Investigations" and "Lecture"; Heller, *Report,* Jumper et al., "Comprehensive Examination," and Gonella, "Scientific Investigation."
25. R. Bucklin, "The Shroud of Turin: Viewpoint of a Forensic Pathologist," *Shroud Spectrum International* 1.5 (December 1982): 3–10.
26. Personal communication with Don Lynn, March 21, 1984; and Heller, *Report.*

CHAPTER FOUR

1. J. Walsh, *The Shroud,* (New York: Random House, 1963), pp. 39, 40, 106, 129.
2. W. C. McCrone and C. Skirius, "Light Microscopical Study of the Turin 'Shroud,' I," *Microscope* 28 (1980): 105–113; W. C. McCrone, "Light Microscopical Study of the Turin 'Shroud,' II," *Microscope* 28 (1980): 115–128; and W. C. McCrone, "Light Microscopical Study of the Turin 'Shroud,' III," *Microscope* 29 (1981): 19-38.
3. L. A. Schwalbe and R. N. Rogers, "Physics and Chemistry of the Shroud of Turin," *Analytica Chimica Acta* 135 (1982): 3–49, 24; R. N. Rogers, "Chemical Considerations Concerning the Shroud of Turin," in *Proceedings of the United States Conference of Research on the Shroud of Turin* (Albuquerque, N.M.: Holy Shroud Guild, March 1977): 131–135; Alan D. Adler, personal communications on August 5 and September 3, 1986; Adler, "The Turin Shroud Lecture," Department of Chemistry, Queen Mary College, London, July 20, 1984; and A. D. Adler, "Chemical Investigations on the Shroud of Turin," in *Mystery of the Shroud of Turin Interdisciplinary Symposium,* video,

(Elizabethtown, PA: Elizabethtown College, February 15, 1986).

4. E. J. Jumper, A. D. Adler, J. P. Jackson, S. F. Pellicori, J. H. Heller, and J.R. Druzik, "A Comprehensive Examination of the Various Stains and Images on the Shroud of Turin," in *ACS Advances in Chemistry No. 205, Archaeological Chemistry III*, J. B. Lambert, ed. (American Chemical Society, 1984), pp. 447–476: 454–55; Schwalbe and Rogers, "Physics and Chemistry"; Rogers, "Chemical Considerations"; and Adler, personal communications.
5. S. F. Pellicori, "Spectral Properties of the Shroud of Turin," *Applied Optics* 19.12 (1980): 1913–1920; and V. D. Miller and S. F. Pellicori, "Ultraviolet Fluorescence Photography of the Shroud of Turin," *Journal of Biological Photography* 49.3 (July 1981): 71–85.
6. R. Gilbert, Jr., and M. M. Gilbert, "Ultraviolet-Visible Reflectance and Fluorescence Spectra of the Shroud of Turin," *Applied Optics* 19.12 (1980): 1930–1936, and Pellicori, "Spectral Properties," p. 1919.
7. R. A. Morris, L. A. Schwalbe, and J. R. London, "X-Ray Fluorescence Investigation of the Shroud of Turin," *X-Ray Spectrometry* 9.2 (1980): 40–47, 45; Schwalbe and Rogers, "Physics and Chemistry," pp. 16, 17, 31; Jumper et al., "Comprehensive Examination"; Adler, "Lecture" and personal communications.
8. Morris et al., "X-Ray Fluorescence"; Schwalbe and Rogers, "Physics and Chemistry," pp. 16, 17, 31; Heller, *Report on the Shroud of Turin* (New York: Houghton Mifflin, 1983); Adler, "Lecture" and "Chemical Investigations"; and Jumper et al.,"Comprehensive Examination."
9. E. J. Jumper, "An Overview of the Testing Performed by the Shroud of Turin Research Project with a Summary of Results," *IEEE 1982 Proceedings of the International Conference on Cybernetics and Society* (October 1982): 535–537; Adler, "Lecture" and "Chemical Investigations"; R. W. Mottern, R. J. London, and R. A. Morris, "Radiographic Examination of the Shroud of Turin: A Preliminary Report," *Materials Evaluation* 38.12 (1979): 39–44; and Schwalbe and Rogers, "Physics and Chemistry," pp. 18, 31.
10. Mottern et al., "X-Ray Fluorescence"; Jumper, "Overview"; Jumper, "Science, the Shroud and the Public," *Mystery*; Schwalbe and Rogers, "Physics and Chemistry"; Jumper et al., "Comprehensive Examination"; Gonella, "Scientific Investigation of the Shroud of Turin: Problems, Results and Methodological Lessons," in *Turin Shroud—Image of Christ?* (Hong Kong: Cosmos Printing Press Ltd., 1987), pp. 29–40; Adler, "Lecture," "Chemical Investigations," and personal communications.
11. Jumper, "Overview" and "Science"; Adler, "Lecture." "Chemical Investigations," and personal communications.
12. Heller, *Report;* Jumper et al., "Comprehensive Examination," p. 453; and J. S. Accetta and J. S. Baumgart, "Infrared Reflectance Spectroscopy and Thermographic Investigations of the Shroud of Turin," *Applied Optics* 19.12 (1980): 1921–29.
13. Heller, *Report*, pp. 194–95; Schwalbe and Rogers, "Physics and Chemistry," p. 13; Heller and Adler, "A Chemical Investigation," p. 93; Adler, "Chemical Investigations" and "Lecture"; and Jumper et al., "Comprehensive Examination," pp. 454–455.
14. Schwalbe and Rogers, "Physics and Chemistry," p. 21.
15. Pellicori, "Spectral Properties"; Gilbert, Jr., and Gilbert, "Ultraviolet-Visible Reflectance"; Jumper et al., "Comprehensive Examination"; Heller, *Report;* Schwalbe and Rogers, "Physics and Chemistry," p. 31; Pellicori, personal communication, February 25, 1987; and Adler, personal communications.
16. McCrone, "Microscopical Study," II and III.
17. Heller, *Report;* Schwalbe and Rogers, "Physics and Chemistry"; p. 13; J. H. Heller and A. D. Adler, "A Chemical Investigation of the Shroud of Turin," *Can. Soc. Forens. Sci. J.* 14.3 (1981): 81–103; and Adler, "Chemical Investigations."
18. The Biuret-Lowry and ultraviolet fluorescence testing was performed by Alan Adler, Vern Miller, and Sam Pellicori, and cited in Heller, *Report.*
19. A. D. Adler, "Updating Recent Studies on the Shroud of Turin," *Archaeological Chemistry: Organic, Inorganic, and Biochemical Analysis,* (1996), pp. 223–228.
20. Heller, Report, p. 183; Schwalbe and Rogers, "Physics and Chemistry," p. 13; Heller and Adler, "A Chemical Investigation"; Adler, "Chemical Investigations" and "Lecture"; and Jumper et al., "Comprehensive Examination," p. 454.

21. Heller, *Report*, p. 182; Schwalbe and Rogers, "Physics and Chemistry," p: 13; Heller and Adler, "A Chemical Investigation"; Adler, "Chemical Investigations" and "Lecture"; and Jumper et al., "Comprehensive Examination," p. 454.
22. Heller, *Report*, p. 197–198.
23. Solvents used by Heller and Adler to extract the fibril color were: methanol, ethanol, benzene, toluene, acetone, concentrated HCl, concentrated H_2SO_4, carbon tetrachloride, chloroform, pyridine, ethyl acetate, dimethylforamide, concentrated NH_4OH, 8N KOH, cyclohexane, ether, morpholine, dioxane, water, 30 percent H_2O_2, and hydrazine.
24. Jumper et al., "Comprehensive Examination," p. 454; Heller, *Report*; Heller and Adler, "A Chemical Investigation"; Adler, "Lecture," "Chemical Investigations," and personal communications.
25. Jumper et al., "Comprehensive Examination," p. 453.
26. Ibid., p. 454; Heller, *Report*; Heller and "A Chemical Investigation"; Adler, "Lecture," "Chemical Investigations," and personal communications.
27. Heller and Adler, "A Chemical Investigation," p. 93.
28. Jumper et al., "Comprehensive Examination," p. 455; Heller and Adler, "A Chemical Investigation"; Adler, "Chemical Investigations"; Heller, *Report*; Schwalbe and Rogers, "Physics and Chemistry," p. 15; and Morris et al., "X-Ray Fluorescence."
29. Heller, *Report*; Heller and Adler, "A Chemical Investigation"; Adler, "Chemical Investigations" and "Lecture"; and Jumper et al., "Comprehensive Examination," p. 465.
30. Heller and Adler, "A Chemical Investigation," p. 97; and Heller, *Report*, p. 194.
31. Jumper, "Science"; Heller and Adler, "A Chemical Investigation," p. 98; Jumper et al., "Comprehensive Examination," pp. 463, 465; Adler, "Chemical Investigations" and "Lecture"; Schwalbe and Rogers, "Physics and Chemistry," p. 15; and Heller, *Report*, p. 180.
32. Heller and Adler, "A Chemical Investigation," p. 98.
33. Ibid., p. 97; Jumper et al., "Comprehensive Examination," pp. 460, 468; Adler, "Chemical Investigations" and "Lecture."
34. Heller, *Report*, pp. 193–96; Riggi, as cited in Heller and Adler, "A Chemical Investigation"; Adler, "Chemical Investigations"; Schwalbe and Rogers, "Physics and Chemistry"; Jumper et al., "Comprehensive Examination," p. 465; McCrone and Skirius, "Microscopical Study," I pp. 110–111; McCrone, "Microscopical Study," II, pp. 115, 123, and III, p. 20.
35. Heller, *Report*, pp. 194–95; and Heller and Adler, "A Chemical Investigation," pp. 97, 100.
36. Schwalbe and Rogers, "Physics and Chemistry," p. 12; Heller, *Report*; Jumper et al., "Comprehensive Examination," p. 454; Jumper, "Science"; Adler, "Lecture" and "Chemical Investigations."
37. Heller, *Report*; Schwalbe and Rogers, "Physics and Chemistry," p. 11; Heller and Adler, "A Chemical Investigation," p. 86; Adler, "Chemical Investigations"; and G. Riggi, "The Dusts of the Holy Shroud of Turin: Progress Report on the Turin Section of STURP," New London, Conn., October 9, 1981.
38. L. Fossati, "Copies of the Holy Shroud," Parts 1, 2, and 3, *Shroud Spectrum International* 12 (September 1984): 7–23 and 13 (December 1984): 23–39; Adler, "Chemical Investigations" and "Lecture"; Heller, *Report*; and Schwalbe and Rogers, "Physics and Chemistry," p. 11. It can be documented that, at least, fifty paintings have been laid on the Shroud itself. Tiny fragments of paint also fell from the ceiling frescoes of the royal palace while the Shroud was being examined.
39. Jumper et al., "Comprehensive Examination," p. 455.
40. Heller and Adler, "A Chemical Investigation," p. 98.
41. Jumper et al., "Comprehensive Examination," p. 454; Schwalbe and Rogers, "Physics and Chemistry," p. 13; Heller, *Report*; Adler, "Lecture"; and Gonella, "Scientific Investigations."
42. McCrone and Skirius, "Microscopical Study," I, pp. 107–08; McCrone, "Microscopical Study," II, pp. 116–17; and Heller, *Report*.
43. Heller, *Report*, p. 177.
44. W. Spector, *Handbook of Biological Data*, p. 52, cited in Heller and Adler, "A Chemical Investigation," p. 100.
45. Jumper et al., "Comprehensive Examination";

Heller, *Report*; Schwalbe and Rogers, "Physics and Chemistry," pp. 15, 31; and Heller and Adler, "A Chemical Investigation," p. 86.

46. Heller and Adler, "A Chemical Investigation," p. 97; Adler, "Lecture"; and Jumper et al., "Comprehensive Examination," pp. 460, 468.
47. Walter C. McCrone, "Authenticity of Medieval Document Tested by Small Particle Analysis," *Analytical Chemistry 48* (1976), 676A–679A.
48. T. A. Cahill, R. N. Schwab, B. H. Kusko, R. A. Eldred, G. Moller, D. Dutschke, and D. L. Wick, "The Vinland Map, Revisited: New Compositional Evidence on Its Ink and Parchment," *Analytical Chemistry, 59* (1987), 829–833. It should be noted that even wormholes on the map came into play, as they were identically matched to wormholes in two documents that accompanied the map. This was also the case with water stains in the accompanying documents, which have, like the Vinland Map, been dated to ca. 1440. Some of this associated evidence was not even found until after the map was discovered, thus making a forger's effort all the more improbable.

CHAPTER FIVE

1. P. Vignon, *The Shroud of Christ* 1902 (New Hyde Park, NY: University Books, 1970).
2. E. J. Jumper, "Considerations of Molecular Diffusion and Radiation as an Image Formation Process on the Shroud," *Proceedings of the 1977 United States Conference of Research on the Shroud of Turin* (Albuquerque, N.M.: Holy Shroud Guild, March 1977), 182–89.
3. J. P. Jackson, E. J. Jumper, and W. R. Ercoline, "Correlation of Image Intensity on the Turin Shroud with the 3-D Structure of a Human Body Shape," *Applied Optics* 23.14 (July 1984): 244–70; and J. P. Jackson, E. J. Jumper, and W. R. Ercoline, "Three-Dimensional Characteristic of the Shroud Image," *IEEE 1982 Proceedings of the International Conference on Cybernetics and Society* (October 1982): 559–575.
4. Alan Adler, personal communication on February 2, 1988; L. A. Schwalbe and R. N. Rogers, "Physics and Chemistry of the Shroud of Turin," *Analytica Chimica Acta* 135 (1982): 3–49; and J. Nickell, "The Shroud of Turin—Unmasked," *The Humanist* (January/February 1978).
5. J. Rinaudo, "Protonic Model of Image Formation on the Shroud of Turin," *Third International Congress on the Shroud of Turin,* Turin, Italy, June 5–7, 1998, p.2.
6. Jackson et al., "Correlation"; and Jackson et al., "Three-Dimensional Characteristic."
7. J. H. Heller, *Report on the Shroud of Turin* (New York: Houghton Mifflin, 1983), p. 208.
8. Jackson et al., "Correlation"; E.J Jumper, A. D. Adler, J. P. Jackson, S. F. Pellicori, J. H. Heller, and J. R. Druzik, "A Comprehensive Examination of the Various Stains and Images on the Shroud of Turin," *ACS Advances in Chemistry No. 205 Archaeological Chemistry* (1984): 447–476.; S. F. Pellicori, "Spectral Properties of the Shroud of Turin," *Applied Optics* 19.12 (1980): 1913–1920; and E. Jumper, cited in C. Murphy, "Shreds of Evidence," *Harper's* 263.1578 (November 1981): 42–65.
9. Pellicori, "Spectral Properties"; S. F. Pellicori and M. S. Evans, "The Shroud of Turin Through the Microscope," *Archaeology* (January/February 1981): 32-43; S. F. Pellicori and R. A. Chandos, "Portable Unit Permits UV/vis Study of the Shroud," *Industrial Research and Development* (February 1981): 186–89; and V. D. Miller and S. F. Pellicori, "Ultraviolet Fluorescence Photography of the Shroud of Turin," *Journal of Biological Photography* 49.3 (July 1981): 71–85.
10. Schwalbe and Rogers, "Physics and Chemistry"; Jackson et al., "Correlation."
11. John Jackson, personal communication, February 1, 1988.
12. Jackson, personal communication, August 20, 1989; Jackson et al., "Correlation"; and Schwalbe and Rogers, "Physics and Chemistry."
13. Sister Damian of the Cross [E. L. Nitowski], *The Field and Laboratory Report of the Environmental Study of the Shroud in Jerusalem* (Salt Lake City: Carmelite Monastery, 1986).
14. S. Rodante, "The Imprints of the Shroud Do Not Derive Only from Radiations of Various Wavelengths," *Shroud Spectrum International* 7 (June 1983): 21–24; and S. Rodante, "Oily Migma and Shroud Marks. Exclusion of Apparent Death," in *La Sindone Scienz E Fede* (a cura di, Lamberto, Coppini e Francesco Cvazzuti, 1978).

15. G. R. Lavoie, B. B. Lavoie, V. J. Donovan, and J. S. Ballas, "Blood on the Shroud of Turin: Part I," *Shroud Spectrum International* 7 (June 1983): 15–19; and J. P. Jackson, "The Radiocarbon Date and How the Image Was Formed on the Shroud," *Shroud Spectrum International* 28/29 (September/December 1988): 2–12.
16. P. Barbet, *A Doctor at Calvary* (Garden City, N.Y.: Doubleday, 1963); R. K. Wilcox, *Shroud* (New York: Bantam Books, 1979); J. Volkringer, *The Holy Shroud; Science Confronts the Imprints.* Translated from the French, Procure du Carme de l'Action des Graces, Paris, 1942 (Australia: The Runciman Press: 1991).
17. A. A. Mills, "Image Formation on the Shroud of Turin," *Interdisciplinary Science Reviews* 20.4 (1995): 319–327.
18. C. Knight and R. Lomas, *Hiram Key: Pharaohs, Freemasons and the Discovery of the Secret Scrolls of Jesus* (London: Century, 1996).
19. L. A. Garza-Valdes, *The DNA of God?* (London: Hodder & Stoughton, 1998); L. A. Garza-Valdes et Faustino Cervantes-Ibarrola, "Biogenic Varnish and the Shroud of Turin." L'Identification Scientifique de l'homme du Linneul Jesus de Nazareth: *Actes Du Symposium Scientifique International,* Rome, 1993. Paris: Francois-Xavier Guibert, 1995; Dr. Garza-Valdes also stated this at a public conference on the Shroud of Turin held at the University of Southern Indiana on February 12, 1994; he has repeated this claim a number of times to other Shroud researchers. See also I. Wilson, *The Blood and the Shroud* (New York: The Free Press, 1998), p. 84.
20. Alan Adler, personal communications, 1996–1997; John Jackson, personal communications, 1996.
21. Personal communications with Adler, 1997.
22. Personal communications with Adler, 1997; personal communication with Jackson, 1996.
23. Garza-Valdes, *The DNA of God?*, pp. 57, 114–115.
24. J. Tyrer, 1981. "Looking at the Turin Shroud as a Textile," *Textile Horizons,* Manchester, U.K., (December 1981): 20–23, 21.
25. Nickell, "Unmasked"; J. Nickell, "The Shroud of Turin—Solved," *The Humanist* (November/December 1978); J. Nickell, "The Turin Shroud: Fake? Fact? Photograph?" *Popular Photography* 85.5 (November 1979): 97–99; J. Nickell, "The Shroud," *Christian Life* 41 (February 1980); J. Nickell, "The Shroud of Turin Is a Forgery," *Free Inquiry* 1.3 (Summer 1981): 28–30; and J. Nickell, *Inquest on the Shroud of Turin* (Prometheus Books, 1987).
26. Adler, "The Turin Shroud Lecture"; A. D. Adler, "Chemical Investigations on the Shroud of Turin," in *Mystery of the Shroud of Turin on Interdisciplinary Symposium,* video (Elizabethtown, PA: Elizabethtown College, February 15, 1986); Alan Adler, personal communications on August 5 and September 3, 1986; Schwalbe and Rogers, "Physics and Chemistry," p. 24; and R. N. Rogers, "Chemical Considerations Concerning the Shroud of Turin," in *Proceedings of the United States Conference of Research on the Shroud of Turin* (Albuquerque, N.M.: Holy Shroud Guild, March 1977), pp. 131–135.
27. Jackson et al., "Correlation"; and Jackson et al., "Three-Dimensional Characteristic."
28. Maloney, "Modern Archaeology, History and Scientific Research on the Shroud of Turin," *Mystery.*
29. McCrone, "Microscopical Study of the Turin Shroud," *Mystery;* and McCrone, personal communication, February 11, 1988.
30. Heller, *Report;* J. H. Heller and A. D. Adler, "A Chemical Investigation of the Shroud of Turin," *Can. Soc. Forens. Sci. J.* 14.3 (1981): 81–103; and Jumper et al., "Comprehensive Examination."
31. Adler, personal communication, February 24, 1988.
32. Heller, *Report;* Heller and Adler, "A Chemical Investigation"; Jumper et al., "Comprehensive Examination"; Adler, "Lecture"; and "Chemical Investigations"; and Schwalbe and Rogers, "Physics and Chemistry."
33. W. E. Morton and J.W.S. Hearle, *Physical Properties of Textile Fibers* (London: Chapman and Hall, 1955), cited in Jackson, "Foldmarks as a Historical Record of the Turin Shroud," *Shroud Spectrum International* 11 (June 1984): 6–29.
34. Jackson, "Foldmarks" and personal communications, spring 1988.
35. Jackson, "Foldmarks."
36. Jackson, personal communications, spring 1988.

37. I. Wilson, *The Shroud of Turin* (Garden City, NY: Doubleday, 1978); and Jackson, "Foldmarks."
38. Jackson, personal communications, spring 1988.
39. Schwalbe and Rogers, "Physics and Chemistry."
40. Don Lynn, personal communication, February 8, 1988; and Jackson et al., "Correlation."
41. *The Mysterious Man of the Shroud,* directed by Terry Landeau, CBS documentary, aired April 1, 1997; *Unsolved Mysteries,* Updates, John Cosgrove and Terry Dunn Meuer, exec. producers, aired September 1994.
42. *The Mysterious Man; Unsolved Mysteries.*
43. G. Ashe, "What Sort of Picture?" *Sindon* (1966): 15–19.
44. Schwalbe and Rogers, "Physics and Chemistry."
45. Jackson et al., "Correlation"; and Jackson et al., "Three-Dimensional Characteristic."
46. Jumper et al., "Comprehensive Examination"; Jackson et al., "Correlation"; Heller, *Report;* and Gonella, "Scientific Investigation of the Shroud of Turin: Problems, Results and Methodological Lessons," in *Turin Shroud—Image of Christ?* (Hong Kong: Cosmos Printing Press Ltd., 1987), pp. 29–40.
47. Jackson et al., "Correlation."
48. J. P. Jackson, "A Problem of Resolution Posed by the Existence of a Three-Dimensional Image on the Shroud of Turin," in *Proceedings,* p. 223; and Schwalbe and Rogers, "Physics and Chemistry."
49. Jackson, personal communication, February 1, 1988; see also W. R. Ercoline, R. C. Downs, Jr., and J.P. Jackson, "Examination of the Turin Shroud for Image Distortions," *IEEE 1982 Proceedings of the International Conference on Cybernetics and Society* (October 1982): 576–579.
50. Jumper, "Molecular Diffusion."
51. Schwalbe and Rogers, "Physics and Chemistry."
52. J. Walsh, *The Shroud* (New York: Random House, 1963).
53. Jackson et al., "Correlation"; and Jackson et al., "Three-Dimensional Characteristic."
54. Jackson et al., "Correlation."
55. Peter Schumacher, Panel Discussion, *Shroud of Turin Research Conferences,* Richmond, Va., June 18–20, 1999.
56. Adler, personal communication, February 1999; Gil Lavoie, personal communication, March 31, 1999.
57. P. Schumacher, Panel Discussion.
58. G. Carter, "Formation of the Image on the Shroud of Turin by X Rays: A New Hypothesis," *ACS Advances in Chemistry No. 205 Archaeological Chemistry III.* J. B. Lambert, ed. American Chemical Society, 1984, pp. 425–446, 430.
59. Schwalbe and Rogers, "Physics and Chemistry."
60. Jackson et al., "Correlation"; and Jackson et al., "Three-Dimensional Characteristic."
61. H. Kersten and E. Gruber, *The Jesus Conspiracy* (New York: Barnes and Noble Books, 1995).
62. *Leonardo's Shroud,* Arts & Entertainment Network. February 15, 1996.
63. Kersten and Gruber, *The Jesus Conspiracy,* p. 298.
64. L. Picknett and C. Prince, *Turin Shroud.* (New York: HarperCollins, 1994); N.P.L. Allen, "Verification of the Nature and Causes for the Photo-negative Images on the Shroud of Lirey—Chambéry—Turin" 1995; Allen, "How Leonardo did not fake the Shroud of Turin" 1995; Allen, "A Reappraisal of Late-Thirteenth-Century Responses to the Shroud of Lirey—Chambéry—Turin," 1994; Allen, "Is the Shroud of Turin the First Recorded Photograph?" 1993, http://www.petech.ac.zalshroud/nature.htm (October 1, 1997).
65. Allen, "Is the Shroud of Turin," p. 4.
66. Picknett and Prince, *Turin Shroud,* photo section and p. 167.
67. B. Schwortz, Lecture given in Turin, Italy, on June 3, 1998.
68. Allen, "How Leonardo did not," p. 4.
69. Allen, "How Leonardo did not," p. 6.
70. Allen, "How Leonardo did not," p. 7.
71. Allen, "Verification," pp. 5, 7.
72. Picknett and Prince, *Turin Shroud,* p. 168.
73. Picknett and Prince, *Turin Shroud,* pp. 162, 167–169; Allen, "How Leonardo did not," p. 4; "Verification," p. 7; and "Startling Findings on the Shroud of Turin," *Impetus* 13.
74. Allen, "Is the Shroud a Photo?" p. 3.
75. Picknett and Prince; *Turin Shroud,* photo section while the authors use ultraviolet lamps would have used the sun as his source of light,": Allen, "Verification," p. 7.

76. Allen, "How Leonardo did not," pp. 4, 7.
77. J. P. Jackson, "Blood and Possible Images of Blood on the Shroud," *Shroud Spectrum International,* 24 (September 1987): 3–11.
78. Allen, "How Leonardo did not," pp. 4, 5.
79. Allen, "How Leonardo did not," p. 4; Picknett and Prince, *Turin Shroud,* p. 163.
80. Allen, "Verification," pp. 7, 11; and "How Leonardo did not," p. 4.
81. Personal communications with Ernst Keller, Karl Zeiss Co., November, 1997.
82. Allen, "How Leonardo did not," p. 4.
83. Allen, "A Reappraisal."
84. Allen, "How Leonardo did not," p. 4.
85. I. Wilson and V. Miller, *The Mysterious Shroud* (Garden City, N.Y.: Doubleday, 1986), p. 78.
86. Personal communications with Adler, February 15, 2000.
87. Personal communications with STURP chemists Larry Schwalbe and Al Adler, February 14–15, 2000.
88. Personal communications with Adler, February 15, 2000.
89. Jackson et al., "Correlation"; and Jackson et al., "Three-Dimensional Characteristic."
90. Heller, *Report;* and personal communication with STURP scientist Sam Pellicori, December 16, 1987.
91. Statement released by STURP, October 1981.
92. Interview of Heller on KMOX Radio, St. Louis, Missouri, December 29, 1983.

CHAPTER SIX

1. T. W. Case, *The Shroud of Turin and the C-14 Dating Fiasco* (Cincinnati: White Horse Press, 1996), p. 93.
2. *Frontiers of Science,* May/June 1981, cited in F.C. Tribbe, *Portrait of Jesus?* (New York: Stein and Day, 1983), p. 154.
3. T. Grieder and A. B. Mendoza, "La Galgada—Peru Before Potter," *Archaeology* (March/April 1981).
4. G. Raes, "Examination of the 'Sindone'" in *Report of the Turin Commission on the Holy Shroud,* trans., ed. M. Jepps, (London: Screenpro Films, 1976), pp. 79–83.
5. R. Drews, *In Search of the Shroud of Turin* (Totowa, NJ: Rowman & Allanheld, 1984).
6. I. Wilson, *The Mysterious Shroud* (Garden City, N.Y.: Doubleday, 1986).
7. A. Braulik, "Altaegyptische Weberei," *Dinglers Polytechnisches Journal* 311 (1899): 13, fig. 2, cited in E. A. Wuenschel, "The Truth About the Holy Shroud," *American Ecclesiastical Review* (1953): 10.
8. A. Braulik, "Altaegyptische Weberei," p. 31, fig. 2, cited in Wuenschel, "Truth," p. 10.
9. A. Braulik, *Altaegyptische Gewebe* (Stuttgart: A. Bergstraesser, 1900), p. 15, fig. 22, cited in Wuenschel, "Truth," p. 10.
10. Braulik, *Altaegyptische Gewebe,* fig. 27, and "Altaegyptische Weberei," p. 13, Fig. 3, and p. 46, fig. 53, cited in Wuenschel, "Truth," p. 10.
11. L. Roth, *Ancient Egyptian and Greek Looms* (Bankfield Museum Notes, 2nd series, no. 2, 1913, pp. 24, 84–96, figs. 6, 7, 9, 10), cited in Wuenschel, "Truth," p. 10.
12. O. Petrosillo and E. Marinelli, *The Enigma of the Shroud: A Challenge to Science* (San Gwann, Malta: PEG, 1996), pp. 198–199.
13. I. Wilson, *The Shroud of Turin,* rev. ed. (Garden City, NY: Doubleday Image Books, 1979) and Wilson, *Mysterious Shroud.* See also W. Bulst, *The Shroud of Turin* (Milwaukee: Bruce, 1957), p. 29.
14. Petrosillo and Marinelli, *Enigma of the Shroud,* p. 198.
15. J. Tyrer, "Looking at the Turin Shroud as a Textile," *Shroud Spectrum International,* 6 (March 1983): 35–45, 38.
16. Ibid., p. 40.
17. F. Testore, *Le Sainte Suaire, Examine et prelevements effectues le 21 April 1988,* Symposium, Paris, France, 7–8 September 1989, p. 5.
18. According to Savio, in *Report.*
19. Raes, "Examination of the 'Sindone,'" and Curto, "The Turin Shroud: Archaeological Observations Concerning the Material and the Image," in *Report.*
20. Wilson, *Shroud of Turin.*
21. *Mishnah,* trans. H. Danby (Oxford: Oxford UP, 1954), Division I Zera'im, Tractate 4 Kila-im, 8:1 and 9:1; Division II Mo'ed, Tractate 7 Betzah, 1:10; Division IV Nizikin, Tractate 5 Makkoth, 3:8–9; Division VI Tohorôth, Tractate 5 Parah, 12:9, and Tractate 12 Uktzin, 2:6, cited in I. Wilson, *Shroud of Turin,* p. 303.
22. E. J. Jumper, A. D. Adler, J. P. Jackson, S. F. Pellicori, J. F. Heller, and J. R. Druzik, "A Comprehensive Examination of the Various Stains and Images on the Shroud of Turin," *Archaeological Chemistry III* 205 (1984): 447–476.

23. Jumper et al., "Comprehensive Examination."
24. British Society for the Turin Shroud, Newsletter 8, October 1984.
25. I. Wilson, *Shroud of Turin,* p. 48.
26. Tribbe, *Portrait?,* p. 84; T. Humber, *The Sacred Shroud* (New York: Pocket Books, 1974), p. 51; G. R. Habermas and K. E. Stevenson, *Verdict on the Shroud* (Ann Arbor, MI: Servant Books, 1981), pp. 36, 118–119; and P. Barbet, *A Doctor at Calvary* (Garden City, N.Y.: Doubleday Image Books, 1963), p. 48. See also Acts 16:22–23, 37–40.
27. Bulst, *Shroud of Turin,* p. 68; A. Pauly, G. Wissowa, and W. Kroll, eds., *Real-Encyclopedia der Klassischen Altertumwissenschaft* (Stuttgart, 1893), entries under "Hasta," "Lancea," and "Pilum," cited by Wilson, *Shroud of Turin,* p. 48.
28. W. Meacham, "The Authentication of the Turin Shroud: An Issue in Archaeological Epistemology," *Current Anthropology* 24.3 (June 1983): p. 290.
29. G. M. Ricci, "Historical, Medical and Physical Study of the Holy Shroud," in *Proceedings of the 1977 United States Conference of Research on the Shroud of Turin,* March 23–24, 1977 (Albuquerque, N.M.: Holy Shroud Guild, 1977), p. 58; R. K. Wilcox, *Shroud* (New York: Bantam Books, 1979), pp. 41, 180; and Meacham, "Authentication," p. 292.
30. V. Tzaferis, "Crucifixion—The Archaeological Evidence," *Biblical Archaeology Review* II (January/February 1985): pp. 44–53; V. Tzaferis, "Jewish Tomabs at and Near Giv'at ha-Mivtar, Jersusalem," *Israel Exploration Journal* 20.1/2 (1970): pp. 18–32; N. Haas, "Anthropological Observations on the Skeletal Remains from Giv'at ha-Mivtar," *Israel Exploration Journal* 20.1/2 (1970): pp. 38–59.
31. Jackson, Jumper, Mottern, and Stevenson, "The Three-Dimensional Image on Jesus' Burial Cloth," in *Proceedings.*
32. Z. H. Klawans and K. E. Bressef, ed. *Handbook of Ancient Greek and Roman Coins* (Racine, Wis.: Western Publishing Company, 1995), pp. 21, 24.
33. F. L. Filas, *The Dating of the Shroud of Turin from Coins of Pontius Pilate,* 2nd ed. (Youngstown, Ariz.: Cogan Productions, 1982).
34. Filas, *Dating of the Shroud.*
35. Rev. Adam J. Otterbein, CSSR., personal communication on September 23, 1986. Rev. Otterbein was the president of the Holy Shroud Guild, to whom Father Filas left his Pilate coin collection on his death in 1985.
36. Filas, *Dating of the Shroud;* Tribbe, *Portrait?;* and A. Brame, "The Dating of the Shroud of Turin: Two Rare, Previously Unrecognized, Lituus Dipleta Issued A.D. 24–25 by Valerius Gratus and A.D. 29-30 by Pontius Pilate," *The Augustan* 22.94 (1984): 66–78. Brame has claimed that the lituus could also have been used on a coin minted by Pilate's predecessor, Valerius Gratus, but this coin would have been minted in A.D. 24–25, which would still easily place it in the time of Christ.
37. Filas, *Dating of the Shroud.*
38. Ibid.
39. Ibid., pp. 11–12.
40. Ibid.
41. Ibid., p. 5.
42. Ibid., p. 7.
43. F.L. Filas, *Dating of the Shroud.*
44. Ibid., p. 7.
45. M. Gichon, "Excavations at En Boqeq," *Qadmoniot* 12 (1970): 138–41 (in Hebrew), cited in R. Hachlili, "Was the Coin-on-Eye Custom a Jewish Burial Practice in the Second Temple Period?" *Biblical Archaeologist* 46 (Summer 1983): 147–153; also discussed in W. Meacham, "On the Archaeological Evidence for a Coin-on-Eye Jewish Burial Custom in the First Century A.D." *Biblical Archaeologist* 49 (March 1986): 56–59.
46. Meacham, "Archaeological Evidence."
47. Ibid., p. 58.
48. R. Hachlili, "Ancient Burial Customs Preserved in Jericho Hills," *Biblical Archaeology Review* 5.4 (July 1979): 28–35, and Z. Greenhut, "Burial Cave of the Caiaphas Family," *Biblical Archaeology Review* (September/October 1992): 28–36, 76.
49. Hachlili, "Ancient Burial Customs," pp. 34–35.
50. Meacham, "Archaeological Evidence," p. 57.
51. Hachlili and Killebrew, "Coin-on-Eye Custom," p.150.
52. Meacham, "Archaeological Evidence," p. 57.
53. Ibid., p. 57.
54. M. Moroni, "Pontius Pilate's Coin on the Right Eye of the Man in the Holy Shroud, in the Light of the New Archaeological Findings," *History, Science, Theology and the*

Shroud, ed. A. Berard, (St. Louis: Richard Nieman, 1991), pp. 276–301, figs. 6, 7.

55. R. Reich, "Caiaphas' Name Inscribed on Bone Boxes," *Biblical Archaeology Review* (September/October, 1992): 39–44, 76; Greenhut, "Burial Cave of the Caiaphas Family."
56. Reich, "Caiaphas Name Inscribed," p. 38.
57. Reich, "Caiaphas Name Inscribed"; Greenhut,"Burial Cave of the Caiaphas Family."
58. Reich, "Caiaphas Name Inscribed, p. 41.
59. Josephus, *Antiquities* 18.35 and 18.95, as stated in Reich, "Caiaphas' Name Inscribed," pp. 41 and 76.
60. Meacham, "Archaeological Evidence," p. 58.
61. L. Y. Rahmani, "Jason's Tomb," *Israel Exploration Journal* 17 (1967): 61–113; E. L. Sukenik, "A Jewish Tomb North-West of Jerusalem," *Tarbiz* 1 (1930): pp. 122–124 (in Hebrew); and A. Kloner, "The Necropolis of Jerusalem in the Second Temple Period" (doctoral thesis, Hebrew University of Jerusalem, 1980), all cited in Hachlili and Killebrew, "Coin-on-Eye Custom," pp. 151–52.
62. A. Kloner, "The Necropolis of Jerusalem," cited in Hachlili and Killebrew, "Coin-on-Eye Custom," p. 152.
63. Hachili and Killebrew, "Coin-on-Eye Custom," p. 152 (oral communication with Y. Meshorer). The locations of the coins that this and the previous endnote refer to were not provided by Hachlili and Killebrew.
64. Hachlili and Killebrew, "Coin-on-Eye Custom," p. 152.
65. Greenhut, "Burial Cave of Caiaphas Family," p. 36.
66. Sister Damian of the Cross, OCD (Dr. Eugenia L. Nitowski), *The Field and Laboratory Report of the Environmental Study of the Shroud in Jerusalem,* (Salt Lake City: Carmelite Monastery, 1986).
67. Sister Damian of the Cross, *Field and Laboratory Report;* J. Kohlbeck and E. Nitowski, "New Evidence May Explain Image on Shroud of Turin," *Biblical Archaeology Review,* 12.4 (July/August 1986).
68. Sister Damian of the Cross, *Field and Laboratory Report;* Kohlbeck and Nitowski, "New Evidence."
69. Kohlbeck and Nitowski, "New Evidence."
70. R. Levi-Setti, G. Crow, Y. L. Wang, "Progress in High Resolution Scanning Ion Microscopy and Secondary Ion Mass Spectrometry Imaging Microanalysis," *Scanning Electron Microscopy* 2 (1985): 535–52.
71. Kohlbeck and Nitowski, "New Evidence."
72. P. Maloney, "Modern Archaeology, History and Scientific Research on the Shroud of Turin," in *The Mystery of the Shroud of Turin: An Interdisciplinary Symposium,* video, (Elizabethtown, Penn.: Elizabethtown College, February 15, 1986), citation of letter from Dr. M. Frei.
73. M. Frei, "Nine Years of Palinological Studies on the Shroud," *Shroud Spectrum International,* 3 (June 1982): 3–7.
74. Ibid.
75. Ibid.
76. W. Bulst, "The Pollen Grains on the Shroud of Turin," *Shroud Spectrum International* 10 (March 1984): 20–28.
77. Ibid.
78. This entire argument can be found in the article by Bulst, "The Pollen Grains on the Shroud of Turin."
79. A. Danin, A. Whanger, U. Baruch, M. Whanger, *Flora of the Shroud of Turin* (St. Louis: Missouri Botanical Garden Press, 1999) pp. 14–15, 24.
80. M. Whanger and A. Whanger, *The Shroud of Turin: An Adventure of Discovery* (Franklin, Tenn.: Providence House Publishers, 1998), p. 78.
81. Whanger and Whanger, *The Shroud: An Adventure,* p. 80; *CBS Evening News,* interview of Dr. Danin, April 12, 1997.
82. Whanger and Whanger, *The Shroud; An Adventure,* p. 79.
83. *CBS Evening News,* interview of Dr. Danin, April 12, 1997; A. Danin, lecture at the Missouri Botanical Garden, St. Louis, MO, June 6, 1997; A. Whanger, "Flowers on the Shroud: Current Research," *CSST News* 1.1 (November 1997). Drs. Danin, Whanger, and Baruch are also able to conclude on the basis of extensive floral images and/or pollens on the Shroud of four narrowly distributed flowers that they could only derive from the vicinity of Jerusalem. A. Danin et al., *Flora of the Shroud,* p. 18.
84. *CBS Evening News,* interview of Dr. Danin; A. Danin, lecture at the Missouri Botanical Garden; Whanger and Whanger, *The Shroud: An Adventure,* p. 78.
85. Danin et al., *Flora of the Shroud,* p. 22.
86. Whanger and Whanger, *The Shroud: An Adventure,* pp. 74–75, 80.

87. Danin, lecture at the *Third International Congress.* Danin et al., *Flora of the Shroud,* p. 16.
88. Wilcox, *Shroud,* p. 133; see also T. D. Stewart, cited in Wilcox, who finds that the man is Caucasian.
89. Bulst, *Shroud of Turin,* p. 104.
90. H. Daniel-Rops, *Daily Life in Palestine at the Time of Christ* (London: Weidenfeld, 1962); H. Daniel-Rops, *Daily Life in the Time of Jesus* (New York: Hawthorn Books, 1962); H. Gressman, "Festschrifte for K. Budde," Appendix to *Zeitschrift für die alttestamentliche Wissenschaft* 34 (1920): 60–68, cited in Bulst, *Shroud of Turin;* and Wilson, *Shroud of Turin,* p. 47. Also according to several Orthodox Jewish rabbis and scholars, as cited in Habermas and Stevenson, *Verdict on the Shroud,* p. 36.
91. Gressman, "Festschrifte for K. Budde," cited in Bulst, *Shroud of Turin;* Daniel-Rops, *Palestine;* Daniel-Rops, *Time of Jesus;* and Wilson, *Shroud of Turin,* p. 47.
92. *Mishnah* (Shab. 23:5); John 11:44, 20:7; see also *The Jewish Encyclopedia,* vol. 3, pp. 434–436, 1925 edition, as cited in J. Iannone, *The Mystery of the Shroud of Turin* (New York: Alba House, 1998), p. 42.
93. J. P. Jackson et al., "Three-Dimensional Image," in *Proceedings,* p. 91; and J. P. Jackson, in *Silent Witness.* Scientific evidence for the chin band may be indicated in the vertical directionality study. J. Jackson, "The Vertical Alignment of the Frontal Image," *Shroud Spectrum International,* 32/33 (Sept/Dec 1989): 3–26, 12. Scientists are not certain what caused the impressions along the sides of the face, nor what looks like an unbound pigtail at the back of the head. Some think these impressions might be because, as with any ancient linen, the lots of the thread used in making the cloth vary. But such variations occur throughout the cloth, not just in these areas. In other places on the cloth where the lots of the weave vary, the image does not seem affected at all. That the image would be noticeably affected only in these particular locations, but nowhere else, would certainly be an amazing coincidence.
94. V. Barclay, *Catholic Herald,* March 18, 1960, cited in Wilcox, *Shroud,* and Tribbe, *Portrait?;* E. Wilson, *The Scrolls from the Dead Sea* (London: W.H. Allen, 1955), p. 60; and R. P. de Vaux, "Fouille au Khirbet Qumran," *Revue Biblique* 60 (1953): 102.
95. Maloney, "Modern Archaeology," *Mystery.*
96. Ian Dickinson, "Preliminary Details of New Evidence for the Authenticity of the Shroud: Measurement by the Cubit," *Shroud News* 58 (April 1990): 4-7.
97. S. Gansfield, *Code of Jewish Law,* rev. ed., trans. H. E. Goldin (New York: Hebrew Publishing, 1927), Ch. CXCVII, No. 9, pp. 99–100.
98. E. A. Wuenschel, "The Shroud of Turin and the Burial of Christ/II—John's Account of the Burial," *Catholic Biblical Quarterly* 8 (1946): 161–66.
99. Meacham, "Authentication"; and Bulst, *Shroud of Turin,* pp. 90–91.
100. K. Little, "The Holy Shroud of Turin and the Miracle of the Resurrection," *Christian Order* (April 1994): 218–31.
101. V. Tzaferis, "Crucifixion—The Archaeological Evidence," *Biblical Archaeology Review* 11 (January/February 1985): 44-53.
102. Tzaferis, "Crucifixion"; and V. Tzaferis, "Jewish Tombs at or near Giv'at ha-Mivtar, Jerusalem," *Israel Exploration Journal* 20/1.2 (1970): 18–32.
103. John 18:30–31.
104. Don Lynn, Image Processing Specialist and STURP scientist, personal communication on June 3, 1984.
105. Tzaferis, "Crucifixion."
106. E. M. Meyers, J. F. Strange, and C. L. Meyers, *Excavations at Ancient Meiron, Upper Galilee, Israel* (Durham, NC: American Schools of Oriental Research and Duke University, 1981), according to Dr. Michael Fuller, chairman, Sociology Department, St. Louis Community College at Florissant Valley.
107. S. Kraus, *Talmudische Archaologie* (Leipzig: 1910–1911), cited in Meacham, "Authentication."
108. J.A.T. Robinson, "The Shroud of Turin and The Grave-Clothes of the Gospels," in *Proceedings;* and A. Feuillet, "The Identification and the Disposition of the Funerary Linens of Jesus' Burial According to the Data of the Fourth Gospel," *Shroud Spectrum International* 4 (September 1982) 13–23, reprint from *La Sindone E La Scienze,* II Congresso Internationale di Sindonologia, Turin, Italy, 1978.
109. *Temple Scroll* 64:10–13; Philo, *De Spec. Leg.* 3.28 S. 151–52, *De Post.* 61, *De Som.* 2.31

S. 213. See also the statement by Josephus: "The Jews are so careful about funeral rites that even those who are crucified because they were found guilty are taken down and buried before sunset." J.W. 4.5.2. S. 317. All above according to R. E. Brown, "The Burial of Jesus (Mark 15:42–47)," *Catholic Biblical Quarterly* 50 (1988).

110. Daniel-Rops, *Palestine;* Daniel-Rops, *Time of Jesus;* Wilson, *Shroud of Turin,* pp. 56–57; Barbet, *Calvary,* pp. 163–165; Tribbe, *Portrait?* pp. 89–92; Robinson, "Grave-Clothes," pp. 24-25; P. Vignon, "The Problem of the Holy Shroud," trans. E.A. Wuenschel, *Scientific American* 93.163 (1937): 162–64; Humber, *Sacred Shroud,* p. 66; E. A. Wuenschel, *Self-Portrait of Christ* (Esopus, N.Y.: Holy Shroud Guild, 1957), p. 48; and J. Marino, *The Burial of Jesus and the Shroud of Turin* (Anaheim: Jeff Richards, 1999). (The amount of spices mentioned in John would be about 50–75 lbs. in our weight today.)

111. Daniel-Rops, Palestine; Daniel-Rops, *Time of Jesus;* Wilson, *Shroud of Turin,* pp. 56–57; Barbet, *Calvary,* pp. 163–165; Tribbe, *Portrait?* pp. 89–92; Robinson, "Grave-Clothes," pp. 24–25; P. Vignon, "The Problem of the Holy Shroud," Humber, *Sacred Shroud,* p. 66; Wuenschel, *Self-Portrait,* p. 48; and Marino, *The Burial of Jesus.*

112. Wilson, *The Blood and the Shroud* (New York: The Free Press, 1998), p. 55; see also I. Wilson, *Mysterious Shroud,* p. 46.

113. B. Lavoie, G. Lavoie, D. Klutstein, and J. Regan, "In Accordance with Jewish Burial Custom, the Body of Jesus Was Not Washed," *Shroud Spectrum International* 3 (June 1982): 8–17.

114. Lavoie et al., "In Accordance with Jewish Burial Custom," p. 15.

115. *Mishnah,* Division II Mo'ed, *Shabbath* 23:5, p. 120. See also Robinson, "Grave-Clothes," p. 27; and Tribbe, *Portrait?* p. 89.

116. Personal communication from Robinson to STURP scientist Don Lynn in 1979.

117. J. A. T. Robinson, "Grave-Clothes," p. 23.

118. J. A. T. Robinson, "Grave-Clothes," p. 23.

CHAPTER SEVEN

1. D. Crispino, "1204: Deadlock or Springboard?" *Shroud Spectrum International,* 4 (September 1982): 24–30, P. Dembowski, "Sindone in the Old French Chronicle of Robert de Clari," *Shroud Spectrum International* 2 (March 1982): 13–27 I. Wilson, *The Shroud of Turin,* (Garden City, NY: Doubleday, 1979), p. 169; I. Wilson, *The Mysterious Shroud,* (Garden City, NY: Doubleday, 1986), p. 104; M. Green, "Enshrouded in Silence: In Search of the First Millennium of the Holy Shroud," *Ampleforth Journal* 74 (Autumn 1969): 321–345; See also R. Andes, as stated in F. C. Tribbe, *Portrait of Jesus?* (New York: Stein and Day Publishers, 1983), p. 56.
2. Greek text in A. Heisenberg, *Nicholas Mesarites-Die Palas-revolution des Johannes Comnenos* (Wurzburg, 1907), p. 30, according to I. Wilson, *The Shroud of Turin,* pp. 167–68.
3. *Story of the Image of Edessa,* Appendix C, in Wilson, *The Shroud of Turin; Evagrius' Ecclesiastical History,* From 431 to 594 A. D. (London: Samuel Bagster and Sons: 1846); *St. John of Damascus on the Divine Images,* trans. by D. Anderson, (Crestwood, N.Y.: St. Vladimir's Seminary Press, 1980); *Acts of the Holy Apostle Thaddaeus,* trans. in A. Roberts and J. Donaldson, eds., *The Ante-Nicene Fathers,* Vol. 8 (New York: Charles Scribner's Sons, 1899; rpt. Grand Rapids, Mich.: Eerdmans, 1951); *The Doctrine of Addai,* as stated in J. B. Segal, *Edessa The Blessed City* (Oxford, 1970).
4. Wilson, *The Mysterious Shroud,* p. 110.
5. P. Vignon, *The Shroud of Christ* (Westminster, 1902; trans. New Hyde Park, N.Y.: University Books, 1970); E. A. Wuenschel, *Self-Portrait of Christ* (Esopus, N.Y.: Holy Shroud Guild, 1957); Green, "Enshrouded in Silence."
6. A. Whanger and M. Whanger, "Polarized image overlay technique: a new image comparison method and its applications," *Applied Optics* 24.6 (March 15, 1985): 766–72; Wilson, *The Mysterious Shroud,* color plates 23–27; Tribbe, *Portrait of Jesus?*
7. M. Whanger and A. Whanger, *The Shroud of Turin: An Adventure of Discovery* (Franklin, Tenn.: Providence House Publishers, 1998); Alan Whanger, personal communications, February 10 and April 16, 2000. See also Tribbe, *Portrait of Jesus?,* p. 241.
8. I am indebted to Dr. Alan Adler, who first pointed out this line of reasoning to me.
9. D. Mercieri, "Ancient coin portrays shroud-like Jesus," Image (Turin Shroud Center of Colorado Newsletter) 3.1 (Spring 1995): 6–7.

10. W. Bulst, "The Pollen Grains on the Shroud of Turin," *Shroud Spectrum International* 10 (March 1984): 20–28; M. Frei, "Nine Years of Palinological Studies on the Shroud," *Shroud Spectrum International* 3 (June 1982): 3–7.
11. M. Symeon, "De Const. Porph. et Romano Lecapeno," sec. 50, p. 491 of Ms. and 748 of *Corpus scriptorum historiae byzantinae* (Bonn, 1878) as stated in Wilson, *The Shroud of Turin,* p. 116.
12. Wilson, *The Shroud of Turin,* p. 115.
13. This term is first used by Evagrius in his *Ecclesiastical History* written in the sixth century and continues to be used to describe the Image of Edessa and Mandylion until its disappearance in the thirteenth century.
14. Wilson, *The Shroud of Turin,* p. 112.
15. J. Heller, KMOX radio interview, St. Louis, Mo., on December 29, 1983.
16. Wilson, *The Shroud of Turin;* Tribbe, *Portrait of Jesus?;* R. Drews, *In Search of the Shroud of Turin* (Totowa, NJ: Rowman & Allanheld, 1984).
17. *Story of the Image of Edessa,* appendix C, para. 15, in Wilson, *The Shroud of Turin,* p. 280.
18. A. Grabar, "La Sainte Face de Laon et le Mandylion dans l'art orthodoxe," *Seminarium Kondakovianum* (Prague, 1935), p. 16, as stated in Wilson, *The Shroud of Turin,* p. 114.
19. Wilson, *The Shroud of Turin,* pp. 114–115.
20. *Vita Alexius,* Monday, August 5, Ulr.III, Massiman, 176 T. according to I. Wilson in "The Shroud and the Mandylion: A reply to Professor Averil Cameron," *Turin Shroud—Image of Christ?* (Hong Kong: Cosmos Printing Press Ltd., 1987), pp. 19–28; See also W. Bulst, *The Shroud of Turin* (Milwaukee: Bruce, 1957), pp. 42, 125.
21. Matthew 27:59; Mark 15:46; Luke 23:53.
22. E. Von Dobschutz, *Christusbilder* (Leipzig 1899), Beilage III, pp.· 130–135, as stated in R. Drews, *In Search of the Shroud,* pp. 39, 46–48; See also Wilson in "The Shroud and the Mandylion," pp. 23–24.
23. Von Dobschutz, *Christusbilder,* Document 30b (Kap. V), p. 189, according to Drews, *In Search of the Shroud;* John of Damascus *De.Fid.Orth.* IV, 16, in Migne, *Patrologia Graeca* 94, 1173 according to Wilson in "The Shroud and the Mandylion."
24. E. Von Dobschutz, *Christusbilder,* Document 71 (Kap V), p. 217, according to Wilson, "The Shroud and the Mandylion," and Drews, *In Search of the Shroud.*
25. Wilson, *The Shroud of Turin,* pp. 119–120.
26. *Acta Thaddaei 3* (from *Acta apostolorum apocrypha,* ed. R.A. Lipsius (Leipzig, 1891), I, p. 274; trans. in A. Roberts and J. Donaldson, eds., *The Ante-Nicene Fathers* (Grand Rapids, MI.: Eerdmans, 1951), Vol. VIII, pp. 558–59, as stated in I. Wilson, "The Shroud and the Mandylion"; see also Wilson, *The Shroud of Turin,* pp. 120, 307.
27. "Liturgical Tractate" according to E. Von Dobschiitz, *Christusbilder,* Beilage II, C pp. 110–114 as stated in Wilson "The Shroud and the Mandylion," pp. 21, 27; "Monthly Lection," as stated in Drews, *In Search of the Shroud of Turin,* pp. 39–40, 116.
28. Paul Maloney states 88 pollen grains were counted from a 2 cm^2 location from the dorsal sidestrip and 163 on a tape from the same size area on the left arm, but that about 300 were counted from a comparable size area near the face. "Is the Shroud of Turin Really Medieval?" *Newsletter of the Association of Scientists and Scholars International for the Shroud of Turin,* Ltd. (ASSIST). 1 (1): 5-7.
29. J. P., Jackson, "Foldmarks as a Historical Record of the Turin Shroud," *Shroud Spectrum International* 11 (June 1984): 6–29.
30. Wilson, *The Mysterious Shroud,* p. 120.
31. *Story of the Image of Edessa; Evagrius' Ecclesiastical History; St. John of Damascus on the Divine Images; Acts of the Holy Apostle Thaddaeus; The Ante-Nicene Fathers; The Doctrine of Addai,* as stated in Segal, *Edessa The Blessed City.*
32. The principal Syriac texts are: "The Doctrine of Addai," published by Dr. W. Cureton in *Ancient Syriac Documents Relative to the Earliest Establishment of Christianity in Edessa,* 1864, from two manuscripts of the fifth and sixth centuries from the Nitrian collection; *The Doctrine of Addai the Apostle,* translated by G. Phillips and Wright, 1876, from a manuscript then in the Imperial Library of St. Petersburg, according to I. Wilson, *The Shroud of Turin.*
33. Wilson, *The Shroud of Turin,* p. 128.
34. *Story of the Image of Edessa,* Appendix C, in I. Wilson, *The Shroud of Turin.*
35. "The Teaching of Thaddaeus the Apostle," trans., in A. Roberts and J. Donalson, eds., *The Ante-Nicene Fathers,* Vol. 8, p. 665, as stated in Wilson, *The Shroud of Turin,* p. 131.
36. J. Wilkinson, *Egeria's Travels,* rev. ed, (Jerusalem: Ariel Publishing House, Warminster, England: Aris & Phillips, 1981).

37. S. Runciman, "Some Remarks on the Image of Edessa," *Cambridge Historical Journal* III (1929–31).
38. *Evagrius' Ecclesiastical History,* from A.D. 431 to 594.
39. Evagrius, "Ecclesiastical History," original text in Migne, *Patrologia graeca,* 86.2: 2748–2749, translation from Bohn's Ecclesiastical Library (1854), as stated in Wilson, *The Shroud of Turin,* p. 137.
40. Green, "Enshrouded in Silence"; Wilson, *The Shroud of Turin;* Wilson, *The Mysterious Shroud;* Humber, *The Sacred Shroud* (New York: Pocket Books, 1974); R. Wilcox, *Shroud* (New York: Bantam Books, 1979); Tribbe, *Portrait?*
41. As stated in Wilson, *The Shroud of Turin;* D. Scavone, "The History of the Shroud to the 14th Century," *History, Science, Theology and the Shroud,* ed. A. Berard, (St. Louis: Richard Nieman, 1991), pp. 171–204; Tribbe, *Portrait?*
42. The one account that mentions any image of Jesus prior to the sixth century is found in the "Doctrine of Addai" written in about A.D. 400. This account describes the image as an ordinary painting. It does not even describe it as having been painted on cloth. Unlike the Image of Edessa, no miracles are associated with this painting, nor is it described as "not made by human hands" or as an image of Christ miraculously imprinted onto cloth.
43. E. J. Jumper, A. D. Adler, J. P. Jackson, S. F. Pellicori, J. H. Heller, and J. R. Druzik, "A Comprehensive Examination of the Various Stains and Images on the Shroud of Turin," *ACS Advances in Chemistry No. 205 Archaeological Chemistry III,* ed. J. B. Lambert, American Chemical Society (1984), pp. 447–76, p. 453.
44. Jumper et al., "Comprehensive Examination," p. 453.
45. Attorney Jack Markwardt points out that if the Shroud is folded once widthwise and once lengthwise (as you would a towel, for example), that several small circular burn marks on the dorsal side line up over each other and match. These burn marks are separate from the burn and scorch marks left by the fire of 1532. He speculates that the small round burn marks resulted from pitch-soaked firebrand being administered to the cloth folded in this manner while the Edessans were trying to start the fire that burned Chosroes's timber mound in 544. One could speculate that small burning wood, embers, or sparks could have caused them, as well. Markwardt further speculates the Shroud, at that point, could have been placed in its long-standing doubled in four fold configuration within a frame to conceal the damning evidence of their treatment of the revered cloth.
46. W. Meacham, "The Authentication of the Turin Shroud: An Issue in Archaeological Epistemology," *Current Anthropology* 24.3 (June 1983), pp. 283–311.
47. W. H. Carroll, "The Dispersion of the Apostles: Jude and the Shroud," *Faith and Reason* (Fall 1981): 235–243.
48. J. B. Chabot, "Anonymi auctoris Chronicon ad annum Christi 1234 pertinens," *Carpus scriptorum christianorom orientalium,* 81–82, Scr. Syri 36–37, 1953, quoted and translated in Segal, *Edessa The Blessed City.*
49. Translation from Green, "Enshrouded in Silence," p. 333.
50. Vatican Library Coder No. 5696, fol. 35, published in P. Savio *Ricerche storiche sulla Santa Sindone* (Turin, 1957), footnote 31, p. 340; translation by Green, as stated in Wilson, *The Shroud of Turin,* pp. 158, 312.
51. Ordericus Vitalis, *Historia ecclesiastica,* part III, bk. IX, 8, "De Gestis Bolduini Edessae principatum obtinet," according to Wilson, *The Shroud of Turin,* pp. 158–312.
52. Wilson, *The Shroud of Turin,* Chapter XVIII.
53. Ibid.
54. Ibid.
55. J. Jackson and R. Jackson, "New Evidence that the Shroud of Turin Pre-Dates the Radiocarbon Date by Centuries," *Third International Congress on the Shroud of Turin, Turin,* Italy, June 5–7, 1998.
56. Ibid.
57. H. Evans and W. Wixon, eds., *The Glory of Byzantium: Art and Culture of the Middle Byzantine Era,* A.D. 843–1261 (New York: The Metropolitan Museum of Art, 1997); Jackson and Jackson, "New Evidence."
58. Jackson and Jackson, "New Evidence."
59. Bulst, *Das Turiner grabtuch und das Christusbild.*
60. Translated by and reported in the *British Society for the Turin Shroud Newsletter* 18 (January 1988): pp. 7–8.
61. Ibid.
62. Translation from P. Johnstone, *The Byzantine Tradition in Church Embroidery,* (London, 1967), p. 54, as stated in Wilson, *The Shroud of Turin,* pp. 157–312.

63. Wilson, *The Shroud of Turin,* pp. 145–146 and 156–157.
64. *St. John of Damascus on the Divine Images,* trans. by D. Anderson, in A. Roberts and J. Donaldson, eds., *The Ante-Nicene Fathers;* Theodore of Studium, in Migne, *Patrologia graeca* 177.64, p. 1288, as stated in Wilson, *Shroud of Turin,* p. 148 Segal, *Edessa The Blessed City,* p. 215; also see S. Runciman, "Some Remarks on the Image of Edessa," *Cambridge Historical Journal* III.3 (1931): 238–52.
65. This theory was originated by Ian Wilson more than twenty years ago and is presented in great detail in his book *The Shroud of Turin.* See in particular chapter XIX.
66. Tribbe, *Portrait?* p. 57.
67. Wilson, *The Shroud of Turin,* Chapter XIX.
68. de Puy, *History of the Military Order of the Templars* (Paris, 1713), as cited in Tribbe, *Portrait?* p. 57.
69. Wilson, *The Shroud of Turin,* p. 183.
70. Ibid.
71. All such descriptions here and elsewhere of the Templar image or idol can be found in Wilson, *The Shroud of Turin,* Chapter XIX.
72. *Chronicles of St. Denis,* art. III, quoted in de Puy, *Histoire de l' Ordre Militaire des Templiers* (1713), p. 25, as stated in Wilson, *The Shroud of Turin,* p. 189.
73. Wilson, *The Shroud of Turin.*
74. See Crispino, "1204: Deadlock or Springboard?"
75. P. Savio, *Ricerche storiche sulla Santa Sindone* (Turin, 1957), according to Prof. D. Scavone, personal communication, August 18, 1987. A summarized version of Savio's theory appears in Tribbe, *Portrait?*
76. As stated in Wilson, *The Mysterious Shroud,* and in N. Currer-Briggs, *The Holy Grail and the Shroud of Christ* (Middlesex: England, ARA Publications, 1984).
77. Wilson, *The Mysterious Shroud.*
78. As stated in Currer-Briggs, *The Holy Grail and the Shroud of Christ.*
79. J. Walsh, *The Shroud* (New York and Toronto: Random House, 1963), pp. 44–45; Vignon, *The Shroud of Christ,* pp. 56–57; Tribbe, *Portrait?* pp. 62–63; J. Iannone, *The Mystery of the Shroud of Turin* (New York: Alba House, 1998); and W. K. Muller, as stated in the *British Society for the Turin Shroud Newsletter* 13 (April 1986).
80. Wuenschel, *Self-Portrait of Christ;* L. Fossati, "A Critical Study of the Lirey Documents," *Shroud Spectrum International* 41 (December 1992), p. 4.
81. Translation by Herbert Thurston found in Appendix B in Wilson, *The Shroud of Turin.*
82. After this account, the Shroud was thereafter referred to by most people not as the true Shroud of Christ, but as a painting or even a likeness or representation, as did the Avignon Pope Clement VII (although its exhibitors and subsequent popes certainly did not). The Avignon Pope not only allowed the exhibit to continue, but imposed perpetual silence upon Bishop d'Arcis about this matter. For an additional account on the background and motives attributed to the Avignon Pope and to the de Charny family, see Wilson, *The Shroud of Turin,* Chapter XX, particularly p. 208.
83. Scavone's views can be found in his review of a work by Robert Babinet published in the *British Society for the Turin Shroud Newsletter,* 46 (November/December 1997): 36–37.
84. The discovery of the Chevalier manipulation was made by Hilda Leynen of Antwerp and the total explanation is contained in D. Crispino, "Literary Legerdemain," *Shroud Spectrum International,* "Spicilegium" (1996): 63–66.
85. Crispino, "Literary Legerdemain," p. 64.
86. L. Fossati, "The Lirey Controversy," *Shroud Spectrum International* 8, (September 1983): 24–34, Crispino, "Literary Legerdemain," p. 65.
87. Crispino, "Literary Legerdemain," p. 66.
88. Fossati, "The Lirey Controversy," pp. 28–30.
89. Bishop de Poitiers would subsequently also allow his niece to marry Geoffrey II de Charny, who arranged for the Shroud to be exhibited in 1389. It is questionable that such permission would have been granted if scandal had followed the de Charny family name.
90. *Promptuarium Sacrarum Antiquitatum Tricassinae Dioecesis,* 1610, as stated in E. A. Wuenschel, "The Holy Shroud of Turin: Eloquent Record of the Passion," *The Ecclesiastical Review,* 93 (November 1935), p. 444; Humber, *The Sacred Shroud.*
91. Frei, "Nine Years of Palinological Studies"; Bulst, "The Pollen Grains."
92. S. Shafersman, in letter to Walter McCrone in the *Microscope,* 30 (1982): 344–352.

93. P. Maloney, "Modern Archaeology, History and Scientific Research on the Shroud of Turin," in *Mystery of the Shroud of Turin: An Interdisciplinary Symposium.* video, Elizabethtown, PA: Elizabethtown College, (February 15, 1986).

94. A. Danin, A. Whanger, U. Baruch, M. Whanger, *Flora of the Shroud of Turin* (St. Louis: Missouri Botanical Garden Press, 1999) p. 24; Maloney, "Modern Archaeolog."

CHAPTER EIGHT

1. W. Meacham, "Radiocarbon Measurements and the Age of the Turin Shroud; Possibilities and Uncertainties," in *Turin Shroud—Image of Christ?* (Hong Kong: Cosmos Printing Press Ltd., 1987), pp. 41–56; O. Petrosillo, and E. Marinelli, *The Enigma of the Shroud,* translated from Italian (San Gwann, Malta: PEG 1996); I. Wilson, "Is This the News We Have Been Waiting For?" *British Society for the Turin Shroud Newsletter* 14, September 1986: 3–4.; D. Sox, *The Shroud Unmasked* (Basingstroke, Hampshire: Lamp, 1988); N. Rufford, "Vatican Steels Itself for 'Fake' Result." London *Times,* August 7, 1988, pp. 1–3; D. Nelson, R. Morlan, J. Vogel, J. Southen, and D. Having, "New Dates on Northern Yukon Artifacts: Holocene Not Upper Pleistocene," *Science* 232 (1986): 749–751.

2. Meacham, "Radiocarbon Measurements"; Rufford, "Vatican Steels Itself"; Nelson et al., "New Dates on Northern Yukon Artifacts."

3. R. Stuckenrath, Jr., "On the Care and Feeding of Radiocarbon Dates," *Archaeology* 18 (1965): 277–281.

4. A. Goude, *Environmental Change* (Oxford: Clarendon, 1977), p. 10, as cited in Meacham, "Radiocarbon Measurements," p. 43.

5. F. E. Zeuner, *Dating the Past.* (Darien, CT: Hafner, 1970), pp. 341–346, as cited in Meacham, "Radiocarbon Measurements," p. 43.

6. P. Jennings, "Still Shrouded in Mystery," *30 Days in the Church and in the World* 1.7 (1988): 70–71.

7. T. Phillips, "Shroud Irradiated with Neutrons?" *Nature* 337 (1989): 594.

8. Ibid.

9. Jennings, "Still Shrouded," p. 71.

10. Ibid.

11. R. Hedges, "Hedges Replies," *Nature* 337 (1989): 594.

12. B. Kelly, "Turin Shroud,": *New Scientist* Vol. 119, September 1988; Statement confirmed by Dr. Robert Otlet of the Harwell Laboratory and by Prof. Edward Hall of the Oxford Laboratory according to I. Wilson in the *British Society for the Turin Shroud Newsletter* 20 (October 1988): p. 14.

13. J. Rinaudo, "Protonic Model of Image Formation on the Shroud of Turin," *Third International Congress on the Shroud of Turin,* Turin, Italy, June 5–7, 1998.

14. Rinaudo, "Protonic Model of Image Formation," pp. 5–6; J. Rinaudo, "A Sign for Our Time," *Shroud Sources Newsletter,* May/June 1996, pp. 2–4, J. Rinaudo, in *BSTS Newsletter,* No 38, August/September 1994, pp. 13–16.

15. This is a well-known scientific technique. For examples, see Rinaudo, "Protonic Model," p. 6; Rinaudo, "A Sign"; S. Pellicori and M. Evans, "The Shroud of Turin Through the Microscope," *Archaeology,* 34: January/February 1981, 34–43, S. F. Pellicori, "Spectral Properties of the Shroud of Turin," *Applied Optics* 19: (June 15, 1980): 1913–1920; G. G. Gray, "Determination and Significance of Activation Energy in Permanence Tests," in *Preservation of Paper and Textiles of Historic and Artistic Value,* Advances in chemistry series 164 (Washington, D.C.: American Chemical Society, 1977) as cited in Pellicori, "Spectral Properties"; S. F. Pellicori and R. A. Chandos, "Portable Unit Permits UV/vis Study of 'Shroud,'" *Industrial Research & Development* 23 (February 1981), 23: 186–189.

16. Rinaudo, "Protonic Model," p. 4 and "A Sign."

17. Rinaudo, "Protonic Model," p. 6, "A Sign," and in *BSTS Newsletter* No. 39.

18. M. Moroni, F. Barbesino, and M. Bettinelli, "Verification of an Hypothesis of Radiocarbon Rejuvenation," *Third International Congress on the Shroud of Turin,* Turin, Italy, June 5–7, 1998. See also M. Moroni, F. Barbesino, and M. Bettinelli, "Possible Rejuvenation Modalities of the Radiocarbon Age of the Shroud of Turin," *Shroud of Turin International Research Conference,* Richmond, VA, June 18–20, 1999.

19. Moroni et al., "Possible Rejuvenation," p. 5.

20. Dr. Rinaudo, personal communication, April 29, 1999.
21. A. Adler, "Updating Recent Sudies on the Shroud of Turin," *Archaeological Chemistry: Organic, Inorganic, and Biochemical Analyses,* Mary Virginia Orna, ed. American Chemical Society (1996): 223–228.
22. Adler, "Updating Recent Studies," p. 225.
23. Alan Adler, personal communication, December 1997.
24. A. D. Adler, A. Whanger, and M. Whanger, "Concerning the Side Strip on the Shroud of Turin," http://www.shroud.com/adler2.htm, p. 5, (October 27, 1997).
25. John Jackson, personal communications, 1989–1993.
26. A video of this refolding was supplied to me by Dr. Alan Whanger, Duke University; he participated along with scientists John Jackson and Kevin Moran, and Shroud expert Rev. Albert (Kim) Dreisbach.
27. J. Chickos and J. Uang, "Chemical Modifications of Cellulose. The Possible Effects of Chemical Cleaning on Fatty Acids Incorporated in Old Textiles," Submitted for publication.
28. J. H. Heller and A. D. Adler, "A Chemical Investigation of the Shroud of Turin," *Can. Soc. Forens. Sci. J.* 14.3 (1981): 81–103; J. H. Heller, *Report on the Shroud of Turin* (New York: Houghton Mifflin, (1983); Dr. Alan Adler, personal communication, January 1998.
29. J. Heller and A. Adler, as stated in T. W. Case, *The Shroud of Turin and the C-14 Dating Fiasco* (Cincinnati: White Horse Press, 1996), p. 76; W. Meacham in "Turin Shroud Dated to A.D. 200–1000," press release, October 14, 1988; Several STURP scientists have also confirmed this finding in informal communications to myself and other researchers.
30. Heller and Adler, as stated in Case, *The Shroud and the C-14 Fiasco.* According to archaeologist W. Meacham, STURP scientists also informed him of this, as stated in "Turin Shroud Dated to 200–1000 A.D." press release; STURP scientist Thomas D'Muhula also stated this at a lecture on May 11, 1995, according to J. Kerlin at http://childrensermons.com/shroud/present.l.htm. Several STURP scientists have also confirmed these datings to me and other researchers after this information was first published.
31. T. D'Muhula, lecture, May 11, 1995.
32. Vernon Miller, STURP photographer, personal communications, 1989; See also J. Marino, "The Shroud of Turin and the Carbon 14 Controversy," *Fidelity,* (February 1989): 35–47.
33. Dr. Alan Adler, personal communications, June 1998 and February 1999; see also Adler et al., "Concerning the Side Strip."
34. Miller, personal communications, 1989. See also Marino, "The Shroud of Turin."
35. Meacham, "Radiocarbon Measurements," p. 49.
36. R. Morgan. "World Reaction to Carbon Dating a Farce," *Shroud News* 49 (October 1988): 3–18.
37. "Rogue Fibers Found in the Shroud," *Textile Horizons* (December 1988): 13.
38. G. Raes, "Examination of the 'Sindone,'" *Report of the Turin Commission on the Holy Shroud,* trans. ed. M. Jeeps, London: Screenpro Films, 1976) pp. 79–83.
39. Adler, "Updating Recent Studies," p. 225.
40. Adler, personal communications, 1989–1997.
41. W. Meacham, "Comments on the British Museum's Involvement in Carbon Dating the Turin Shroud," unpublished manuscript, 1988; See also, Marino, "The Shroud of Turin."
42. Adler et al., "Concerning the Side Strip," p. 5.
43. Ibid.
44. Rinaudo, "Protonic Model," pp. 4–5.
45. K. Clark, "Carbon Tests Prove Shroud Is Not Burial Cloth of Jesus," *Chicago Tribune,* October 14, 1988.
46. "Archaeological sherd dating: comparison of the dates with radiocarbon by Beta counting and accelerator techniques." By Johnson, Stipp, Tamers, Bonani, Suter, and Wolfli, paper presented at the International Radiocarbon Conference at Trondheim, Norway, 1985, as stated in W. Meacham, "Turin Shroud Carbon Dating," unpublished manuscript, 1988.
47. Meacham, "Radiocarbon Measurements."
48. R. Burleigh, M. Leese, and M. Tite, "An Intercomparison of Some AMS and Small Gas Counter Laboratories," *Radiocarbon,* Vol. 28, No. 2A (1986): 571–577, H. E. Gove, *Relic, Icon or Hoax?* (Bristol and Philadelphia: Institute of Physics Publishing, 1996), p. 77; G. Harbottle and W. Heino, "Carbon Dating the Shroud of Turin: A Test of

Recent Improvements in the Technique," *ACS Advances in Chemistry No. 220 Archaeological Chemistry IV,* ed. R. Allen. (American Chemical Society, 1989), pp. 313–320.
49. Burleigh et al., "An Intercomparison," p. 577.
50. Press Release, Atlanta International Center for Continuing Study of the Shroud of Turin, October 13, 1988; See also Marino, "The Shroud of Turin," p. 39.
51. Gove, *Relic?* p. 18.
52. Gove, *Relic?* p. 18.; Sox, *The Shroud Unmasked,* p. 84.
53. Harbottle and Heino, "Carbon Dating the Shroud"; Gove, *Relic?* p. 150; Sox, *The Shroud Unmasked,* p. 93; D. Sox, *The Image on the Shroud* (London: Mandala Books) (Unwin Paperbacks, 1981), p. 161.
54. Formal Proposal for Performing Scientific Research on the Shroud of Turin. Submitted by the Shroud of Turin Research Project, Inc., August 15, 1984.
55. R. Dinegar and L. Schwalbe, "Isotope Measurements and Provenance Studies of the Shroud of Turin," *Archaeological Chemistry IV;* Ralph O., Allen, ed., *Advances in Chemistry Series 220* (American Chemical Society: Washington, D.C., 1989), pp. 409–417.
56. Burleigh et al., "An Intercomparison," p. 571.
57. See comments of Garman Harbottle, Brookhaven National Laboratory, quoted in Marino, "The Shroud of Turin," p. 42.
58. J. Raloff, "Controversy Builds as Shroud Tests Near," *Science News* Vol. 133, April 16, 1988.
59. Paul Maloney, personal communications, January 1998.
60. Ibid.
61. Dinegar and Schwalbe, "Isotope Measurements and Provenance Studies," p. 412.
62. Sox, *The Shroud Unmasked,* p. 96; Petrosillo and Marinelli, *The Enigma of the Shroud,* p. 28.
63. H. E. Gove, "Turin Workshop on Radiocarbon Dating the Turin Shroud"; *Nuclear Instruments and Methods in Physics Research,* B29 (1987) 193–195, 194. See also Sox, *The Shroud Unmasked,* pp. 108–09; Petrosillo and Marinelli, *The Enigma of the Shroud,* pp. 34–35; and Marino, "The Shroud of Turin," p. 41.
64. Dinegar and Schwalbe, "Isotope Measurements and Provenance Studies," p. 413.
65. Sox, *The Shroud Unmasked,* p. 119
66. Gove, *Relic?* p. 237.
67. Michael Tite, letter to *Nature,* Vol. 332, April 7, 1988.
68. Personal communication to Paul Maloney from Garman Harbottle, April 22, 1988.
69. Petrosillo and Marinelli, *The Enigma of the Shroud,* p. 60.
70. Sox, *The Shroud Unmasked,* p. 133.
71. Ibid.
72. Petrosillo and Marinelli, *The Enigma of the Shroud,* p. 61.
73. Gove, *Relic?* pp. 155–56.
74. Petrosillo and Marinelli, *The Enigma of the Shroud,* pp. 65–67.
75. Sox, *The Shroud Unmasked,* pp. 136–37; Marino, "The Shroud of Turin," p. 38.
76. Interview of Dr. Michael Tite, at Paris Symposium on Shroud of Turin, September 1989, conducted by E. Marinelli and O. Petrosillo, in *Shroud News* 81 (February 1994).
77. Dr. Tite interview, Paris Symposium, September 8, 1989.
78. A. Morgan, "Turin Shroud 'May be Fake,'" *The Times,* July 25, 1988, p. 3.
79. I. Wilson, "Recent Publications," *British Society for the Turin Shroud Newsletter* 20. (October 1988), p. 19.
80. As stated in J. Dart, "Labs Find Turin Shroud Dates to Middle Ages, Scientist Hints," *Los Angeles Times,* September 21, 1988.
81. Wilson, "Recent Publications."
82. Sox, *The Shroud Unmasked,* pp. 143–147.
83. According to Luigi Gonella, as stated in Petrosillo and Marinelli, *The Enigma of the Shroud,* p. 121.
84. A. Coghlan, "Unexpected Errors Affect Dating Techniques," *New Scientist,* (September 30, 1989): 26.
85. Ibid.
86. Ibid.
87. Struckenrath, Jr., "On the Care and Feeding of Radiocarbon Dates," p. 281.
88. As quoted in P. Jennings, "Art Historian Not Convinced the Shroud Is a Fake," *Our Sunday Visitor,* (October 23, 1988), p. 24.
89. Press release, "Environmental Study of the Shroud of Jerusalem," October 15, 1988.
90. Dr. Warren Grundfest, lecture at the *Dallas Shroud Meeting,* Dallas, Texas, November 6–8, 1998; W. S. Grundfest, "Imaging Spectroscopy for the Non-Destructive Evaluation of Items of Historical Interest: Applications to the Shroud of Turin," *Shroud of Turin Conference,* Richmond, VA, June 18–20, 1999.
91. Grundfest Dallas lecture; Grundfest, "Imaging Spectroscopy."

92. Phillips, "Shroud Irradiated," p. 594; Hedges, "Hedges Replies," p. 594.
93. R. A. Morris, L. A. Schwalbe, and J. R. London, "X-Ray Fluorescence Investigation of the Shroud of Turin, *X-Ray Spectrometry* 9.2 (1980): 40–47; J. H. Heller and A. D. Adler, "A Chemical Investigation of the Shroud of Turin," *Can. Soc. Forens. Sci. J.* 14.3 (1981): 81–103.
94. Phillips, "Shroud Irradiated," p. 594. See also Hedges, "Hedges Replies," p. 594.
95. L. A. Schwalbe and R. N. Rogers, "Physics and Chemistry of the Shroud of Turin," *Analytica Chimica Acta 135* (1982): 3-49, 47.
96. Phillips, "Shroud Irradiated," p. 594.
97. Gove, *Relic?* p. 154.
98. Schwalbe and Rogers, "Physics and Chemistry," p. 44.
99. The Shroud's owners also must consider that, if this abundance of blood was shed by Jesus Christ, the time and purpose for which it was shed, along with the unique event that caused the neutron flux, could be much better illustrated if a small fragment of it was tested for the worldwide public.
100. The amount of additional C-14 could also be subtracted from the 1988 ratios for comparative purposes and to ascertain if any of the other contaminants discussed in this chapter remained on the 1988 samples at the time of dating.
101. Gove, *Relic?* p. 154.
102. Small samples could be removed from non-image areas of the Shroud and carbon dated after the spectra-imaging technology has been applied to it and the least intrusive locations with the least amount of contaminants have been identified. If the Shroud was irradiated with a neutron flux, while all parts of it would have been affected, some would have been more affected. The most noticeable evidence for a neutron flux, and the most noticeable difference with any nonbody image parts of the cloth (except the blood) would most likely come from a sample taken from the body image. If the Shroud received a dose of radiation from the body, it would probably be most evident on a sample taken from the body image. More neutron flux, with a resulting effect on the cloth's radiocarbon age, probably would have occurred on the body image than on any other part of the cloth itself. If dating from the body image was to be considered, only the smallest piece possible from the least intrusive area of the body image should be removed. Care should be taken to remove a sample with as little three-dimensional or other information as possible; nor should it contain any blood or scourge marks. Since the left thigh is a large area that is well encoded, whose elevated nature runs its entire length and contains body image between the scourge marks, the tiniest sample necessary to date should possibly be taken from there. Alternatively, for similar reasons, a sample from the left buttock should possibly be removed from the dorsal area for this purpose.

CHAPTER NINE

1. D. H. Sox, *The Shroud Unmasked* (Basingstoke, Hampshire: The Lamp Press, 1988), p. 95.
2. E. T. Hall, "The Turin Shroud: An Editorial Postscript," *Archaeometry* 31.1 (1989): 92–95, See H. E. Gove, *Relic, Icon or Hoax?* (Bristol and Philadelphia: Institute of Physics Publishing, 1996), p. 104 wherein it is also reported that Harbottle of the Brookhaven Laboratory stated that Gove was "famous for grabbing the ball and running with it."
3. Gove, *Relic?* p. 106.
4. Ibid., pp. 6, 7.
5. Ibid., p. 8.
6. Ibid., p. 48.
7. Ibid., p. 57.
8. Ibid., p. 87.
9. Ibid., p. 206.
10. Ibid., pp. 96, 97.
11. Ibid., p. 98.
12. Ibid.
13. Ibid., p. 99.
14. Ibid., p. 100.
15. Ibid., p. 101.
16. Ibid.
17. Ibid.
18. Ibid., pp. 100, 102.
19. Ibid., p. 93.
20. Ibid., p. 155.
21. Ibid., p. 93.
22. Ibid., p. 112.
23. Ibid., p. 113.
24. Ibid.
25. Ibid.

26. Ibid., p. 164.
27. Meacham, "Radiocarbon Measurement and the Age of the Turin Shroud: Possibilities and Uncertainties," in *Turin Shroud: Image of Christ* (Hong Kong: Cosmos Printing Press Ltd., 1987), 41–56.
28. Ibid., p. 48.
29. Ibid., p. 47.
30. Jennings, "Still Shrouded in Mystery," *30 Days in the Church and in the World* I.7 (1988): 70–71; Hedges, "Hedges Replies," p. 594; J. Cornwell, "Science and the Shroud," *The Tablet*, (January 14 1989): pp. 36–38.
31. Hedges, "Hedges Replies," p. 594.
32. Hedges, "Hedges Replies," p. 594; Cornwell, "Science and the Shroud," p. 37.
33. Meacham, "C-14 Dating of the Shroud," treasureseeker@hotmail.com, February 16, 1998.
34. Gove, *Relic?*, p. 155.
35. Ibid., p. 165.
36. Ibid., pp. 191, 192.
37. Ibid., pp. 192, 193.
38. Ibid., p. 192.
39. Ibid., p. 214.
40. Ibid., p. 216.
41. Organization known as ASSIST (Association of Scientists and Scholars International for the Shroud of Turin) and the BSTS (British Society for the Turin Shroud), along with various Italian scientists had also submitted proposals and questions.
42. R. Dinegar, and L. Schwalbe, "Isotopic Measurements and Provenance Studies of the Shroud of Turin," *Archaeological Chemistry* IV; R. O. Allen, ed. (American Chemical Society: Washington, D.C.) 1989.
43. M. DeNiro, L. Sternbert, B. Marino, and J. Druzik, *Geochim. Cosmochim. Acta*, as cited in Dinegar and Schwalbe, "Isotopic Measurements and Provenance Studies," p. 116.
44. Dinegar and Schwalbe, "Isotopic Measurements and Provenance Studies," p. 116.
45. I. Wilson, *The Blood and the Shroud* (New York: The Free Press, 1998), p. 183.
46. O. Petrosillo and E. Marinelli, *The Enigma of the Shroud*, (San Gwann, Malta: PEG, 1996), p. 42.
47. Ibid., p. 99.
48. *Il Giornale*, May 12, 1989, as quoted in Petrosillo and Marinelli, *The Enigma*, pp. 119–120.
49. Petrosillo and Marinelli, *The Enigma*, p. 99.
50. Ibid., pp. 99, 100.
51. Gove, *Relic?*, p. 6.
52. Ibid., p. 72.
53. Sox, *The Shroud Unmasked*, p. 69.
54. Gove, *Relic?*, p. 51.
55. Ibid.
56. Ibid., p. 51, 52.
57. Ibid., p. 89. See also pp. 84, 93.
58. Dinegar and Schwalbe, "Isotope Measurements and Provenence Studies," p. 3.
59. Gove, *Relic?* p. 53.
60. Gove, *Relic?* p. 264. See also Sox, *The Shroud Unmasked*, p. 151.
61. Gove, *Relic?* p. 265.

CHAPTER TEN

1. L. A. Schwalbe and R. N. Rogers, "Physics and Chemistry of the Shroud of Turin," *Analytica Chemica Acta* 135 (1982): 3–49; R. H. Dinegar, "The 1978 Scientific Study of the Shroud of Turin," *Shroud Spectrum International* 4 (September 1982): 3–12; E. J. Jumper, A. D. Adler, J. P. Jackson, S. F. Pellicori, J. H. Heller, and J. R. Druzik, "A Comprehensive Examination of the Various Stains and Images on the Shroud of Turin," *ACS Advances in Chemistry No. 205 Archaeological Chemistry III*, J. B. Lambert, ed. (American Chemical Society, 1984): pp. 447–76.
2. Schwalbe and Rogers, "Physics and Chemistry"; Dinegar, "The 1978 Scientific Study."
3. Schwalbe and Rogers, "Physics and Chemistry"; Dinegar, "The 1978 Scientific Study"; Jumper et al., "A Comprehensive Examination."
4. Jumper et al., "A Comprehensive Examination"; J. H. Heller and A. D. Adler, "A Chemical Investigation of the Shroud of Turin," *Can. Soc. Forens. Sci. J.* 14:3 (1982): 81–103.
5. R. Gilbert, Jr., and M. M. Gilbert, "Ultraviolet-Visible Reflectance and Fluorescence Spectra of the Shroud of Turin," *Applied Optics* 19:12 (June 1980): 1930–1936; Schwalbe and Rogers, "Physics and Chemistry."
6. Gilbert, Jr., and Gilbert, "Ultraviolet-Visible Reflectance"; Schwalbe and Rogers, "Physics and Chemistry."
7. Schwalbe and Rogers, "Physics and Chemistry"; Dinegar, "The 1978 Scientific Study."
8. Jumper et al., "A Comprehensive Examination."

9. S. Pellicori and M. Evans, "The Shroud of Turin Through the Microscope," *Archaeology* (1981): 34: 34–43.
10. Jumper et al., "A Comprehensive Examination," p. 456.
11. Pellicori and Evans, "The Shroud Through the Microscope"; S. F. Pellicori, "Spectral Properties of the Shroud of Turin," *Applied Optics* (June 15 1980): 1913–1920, G. G. Gray, "Determination and Significance of Activation Energy in Permanence Tests," in *Preservation of Paper and Textiles of Historic and Artistic Value,* Advances in Chemistry series 164 (Washington, DC: American Chemical Society, 1977), as cited in Pellicori, "Spectral Properties"; S. Pellicori and R.A. Chandos, "Portable Unit Permits UV/vis Study of 'Shroud,' " Industrial Research & Development, February 1981, 23: 186–189, 187; J. Rinaudo, "Protonic Model of Image Formation on the Shroud of Turin," *Third International Congress on the Shroud of Turin,* Turin, Italy, June 5–7, 1998; J. Rinaudo, "A Sign for Our Time," *Shroud Sources Newsletter,* May/June 1996, pp. 2-4; J. Jackson, E. Arthurs, L. Schwalbe, R. Sega, D. Windisch, W. Long, E. Stappaerts, "Infrared Laser Heating for Studies of Cellulose Degradation," *Applied Optics,* 15 Sept., 1988, 27:3937-3943.
12. S. Pellicori, "Spectral Properties"; J. Rinaudo, "Protonic Model," p. 6, and "A. Sign"; Jumper et al., "A Comprehensive Examination."
13. The following is a list of those who have specifically stated that some form of radiation caused or was a principal cause of the images on the Shroud: Dr. John Jackson, Dr. Jean-Baptiste Rinaudo, Dr. Kitty Little, Dr. Art Lind, Dr. Giles Carter, Dr. John Heller, Dr. Gil LaVoie, Dr. August Accetta, Dr. Luigi Gonella, Dr. Sebastiano Rodante, Dr. Eberhard Lindner, Dr. Alan and Mary Whanger, Dr. Thaddeus Trenn, Dr. Edward Hall, Peter Schumacher, developer of the VP-8 Image Analyzer, Kevin Moran, Image Processing Specialist—Kodak, Barrie Schwortz, STURP photographer, scientist Peter Carr, Oswald Scheuerman, Ray Rogers, writer Goeffrey Ashe, authors Kenneth Stevenson and Gary Habermas, writer Joe Marino and this author. This list does not include the names of many more scientists and experts who privately think the same thing but, lacking absolute proof, have never stated so publicly. Even those (such as Prof. Nicholas Allen, Lynn Picknett and Clive Prince) who advocate forgery theories, thinking that radiant energy is a principal cause of the body images on the Shroud.
14. L. Gonella, "Scientific Investigation of the Shroud of Turin: Problems, Results and Methodological Lessons." in *Turin Shroud—Image of Christ?* (Hong Kong: Cosmos Printing Press Ltd. 1987), pp. 29–40, 31.
15. As quoted in W. McDonald, "Science and the Shroud," *The World and I,* (Oct. 1986): 420–428, 426.
16. Gonella, "Scientific Investigation," p. 31.
17. J. P Jackson, E. J. Jumper, and W. R. Ercoline, "Correlation of Image Intensity on the Turin Shroud with the 3-D Structure of a Human Body Shape," *Applied Optics* 23.14 (July 1984): 2244-70.
18. G. F. Carter, "Formation of the Image on the Shroud of Turin by X Rays: A New Hypothesis," *ACS Advances in Chemistry No. 205 Archaeological Chemistry* III, J.B. Lambert, ed. American Chemical Society, 1984, pp. 425–446; G. Carter Interview, May 23, 1999, http://earthfiles.com/earth025.html.
19. Carter, "Formation of the Image," p. 435.
20. See for example, Dr. Alan Whanger, surgeon, in M. and A. Whanger, *The Shroud of Turin: An Adventure of Discovery* (Providence House Publishers, Franklin, Tenn., 1998), pp. 111–115; A. D. Accetta, "Experiments with Radiation as an Image Formation Mechanism," *Shroud of Turin International Research Conference,* Richmond, Va., June 18–20, 1999; A. D. Accetta, Lecture at the Dallas Shroud Meeting, Dallas, Tex., November 6–8, 1998; J. P. Jackson, "Is the Image on the Shroud Due to a Process Heretofore Unknown to Modern Science?" *Shroud Spectrum International* 34 (March 1990): 3-29, 18; J. P. Jackson, "An Unconventional Hypothesis to Explain All Image Characteristics Found on the Shroud Image," *History, Science, Theology and the Shroud,* A. Berard, ed. (St. Louis: Richard Nieman, 1991), pp. 325–344, 333–335.
21. Carter, "Formation of the Image," p. 431.
22. Dr. John Jackson, personal communications, 1991–1994; M. and A. Whanger, *The Shroud of Turin,* pp. 116–117; Carter, "Formation of the Image," p. 433; Accetta, "Experiments

with Radiation" and Dallas lecture.

23. Accetta, "Experiments with Radiation" and Dallas lecture; Jackson, "An Unconventional Hypothesis," pp. 333–335; Jackson, personal communications, 1991–1994.
24. Carter, "Formation of the Image," pp. 433–434; M. and A. Whanger, *The Shroud of Turin*, pp. 117–118; Dr. Accetta, personal communication, August 9, 1999.
25. Carter, "Formation of the Image," p. 433.
26. M. and A. Whanger, *The Shroud of Turin*, p. 118.
27. K. Little, "The Holy Shroud and the Miracle of the Resurrection," *Christian Order* (April 1994): 218–231.
28. A. D. Accetta, "Experiments with Radiation" and Dallas lecture. While the source of radiation is only part of the explanation of the cause of the image, Dr. August Accetta injected himself with technetium-99, which allowed his external body to emit low-energy radiation that could be captured using a gamma camera and a vertical collimator, and was able to reproduce Shroud image characteristics such as soft tissue and skeletal information, noncontact imaging, and vertical alignment, along with borderless density shading.
29. Jackson, "Is the Image," p. 5.
30. Ibid.
31. Rinaudo, "Protonic Model," "A Sign," and in *British Society for the Turin Shroud Newsletter*, No. 38, August/September 1994, pp. 13–16.
32. Rinaudo, "Protonic Model" and "A Sign."
33. Ibid.
34. Ibid.
35. Rinaudo, "Protonic Model"; Jumper et al., "A Comprehensive Examination."
36. Rinaudo, "Protonic Model" and "A Sign."
37. Rinaudo, "Protonic Model," p. 4.
38. Jackson, "Is the Image?" p. 5.
39. G. R. LaVoie, B. B. LaVoie, and A. D. Adler, "Blood on the Shroud of Turin: Part III," *Shroud Spectrum International* 20 (Sept. 1986): 3-6; G. R. LaVoie. *Unlocking the Secrets of the Shroud*, (Allen, Tex.: Thomas More, 1998) pp. 101–111; Jackson, "Is the Image?" pp. 7–8, 15–16, 20–21; Jackson, "An Unconventional Hypothesis," p. 337.
40. Jackson, "Is the Image?" pp. 11–14, 22.
41. Jackson, "Is the Image?" p.14, (see also pp. 7–8, 15, 20–23); Jackson, "An Unconventional Hypothesis," p. 341 (see also pp. 332, 340–343).
42. Rinaudo, "Protonic Model," p. 4.
43. Carter, "Formation of the Image," p. 437.
44. Jackson, "Is the Image?" p. 9.
45. R. Gilbert, Jr., and M. Gilbert, "Ultraviolet-Visible Reflectance," p. 1935.
46. K. Little, "The Formation of the Shroud's Body Image," *British Society for the Turin Shroud Newsletter*, No. 46, November/December 1997, pp. 19-26, 20.
47. Little, "The Holy Shroud," p. 225. Only a small fraction of the body needs to have disintegrated to produce the Shroud's images and change its radiocarbon date. Hence, the explosion, if any, would have been small and consistent with Matthew 27:51–53 and 28:2.
48. McDonald, "Science and the Shroud," p. 426.
49. Little, "The Formation," pp. 20–22, and "The Holy Shroud," p. 226.
50. Little, "The Formation," p. 20, and "The Holy Shroud," p. 225.
51. Little, "The Formation," p 25.
52. According to Dr. Little, if a nuclear disintegration of the body occurred the nucleus would split into protons, alpha particles and neutrons, with gamma radiation being emitted and released. Little, "The Formation," pp. 22–23, and "The Holy Shroud," p. 225. By analogy, this would also be the most likely radiation emitted if the body dematerialized into these same or similar particles and components. However, X rays or ultraviolet light, which are next to gamma rays in the electromagnetic spectrum, could also be candidates.
53. Even if the cloth had been tucked under the legs, it would not maintain any pressure against the leg, which would be necessary for complete and intricate encoding of the scourge marks by direct contact since the cloth would drape or fall away from the leg by gravity. The dorsal images clearly indicate that significant parts of the upraised left leg were never in contact with the cloth initially, or even when other parts of the leg were subsequently encoded. The same reasoning can be applied to the back of the right leg.
54. Since blood marks and body images are found together where initial direct contact on the dorsal side would not have existed (such as the small of the back, lower back,

sole of the right foot, and across most of the width of the back of the head) and since these blood marks do not appear to have been broken or to have pulled away from the cloth, the blood marks could not have been caused solely by direct contact. These facts are consistent with the blood marks and body images at these locations and along the entire dorsal side, having been encoded by the means described in the Historically Consistent Method.

55. John 20: 1, 14-17; Matt. 28: 1, 7–9; Luke 24: 1; Mark 16: 2, 9–12.
56. John 20: 14–17; Luke 24: 13; Mark 16: 12–13.
57. Jackson, "An Unconventional Hypothesis," p. 343.
58. T. Trenn, "X-File on the Shroud," interview by Linda Moulton Howe, in *British Society for the Turin Shroud Newsletter*, No. 49, June/July 1999, pp. 16–22. Interview originally published on the website Earthfiles (http://earthfiles.com).
59. Little, "The Formation," p. 24.
60. Ibid.
61. Ibid.
62. Little, "The Formation," p. 26. Only a tiny fraction of the atoms are needed to disintegrate to encode the Shroud's images. Possible explanations for the dematerialization of the remaining atoms of the body are discussed below.

One possible explanation as to what happened to the man in the Shroud under the Historically Consistent or a related method, or to the historical Jesus Christ, was first introduced in a highly respected scientific journal in 1935 by Albert Einstein and Nathan Rosen, "The Particle Problem in the General Theory of Relativity," *Physical Review* 48 (1935): 73–77. They first devised the concept of a shortcut in space time-travel based on Einstein's theory of general relativity that allows a person or object to pass through a bridge or "wormhole" in space and time. According to modern physicists, mathematical theories of space-time travel are not only possible under Einstein's theory of general relativity, but these wormholes are completely consistent with tested theories of gravity and would allow travel between two points in different universes or two points within the same universe. This form of travel could circumvent the speed of light barrier and may even permit travel to past or future times. The famous British physicist Steven Hawking has published and lectured on wormholes, and his bestselling book, *A Brief History of Time*, devotes whole chapters to this subject. At this time the science of wormholes is not only mature, but in the words of physicist Matt Visser in *Lortenzian Wormholes: From Einstein to Hawking*, ". . . the theoretical analysis of Lortenzian wormholes is 'merely' an extension of *known physics*—no new physical principles of fundamentally new physical theories are involved." (Woodbury, N.Y.: American Institute of Physics, 1996) p. 369. NASA also has an interest in wormholes for space travel as evidenced by its hosting a workshop at Caltech's Jet Propulsion Laboratory entitled, "Advanced Quantum/Relativity Theory Propulsion Workshop" in May, 1994, where wormholes were a major topic of discussion.

A key element of this theory is that, as matter passes through the wormhole, the entrance and exit mouths of the hole gain and lose mass. An object could even traverse the wormhole as energy and return to its former mass upon its exit. If the unexplained disappearance of the man in the Shroud, or the historical Jesus Christ, was connected or related to this theory, the entrance mouth to the wormhole would be the point of the body's departure. The Shroud itself would have been right at the mouth of the entrance and may have received some of the increase in mass in the form of the basic building blocks of matter—protons, neutrons, electrons and alpha particles. Space-time travel could even be said to be a possible means for Jesus to have traveled between heaven and earth.

Another expanation of what occurs during the Historically Consistent Method is that the body turned from mass to energy, during the image-encoding phase. With the body turning to energy, the cloth would begin to fall by gravity through the body region; however, as it fell it would interrupt the almost completed process and receive small particles such as protons, neutrons, electrons, and alpha particles from the body that had not yet completely changed to energy. All of the subsequent steps and encoding features found in the first historically consistent method would be found in this similar, historically consistent process. An imperceptibly small amount of protons and alpha particles would be sufficient to encode the superficial body images on the Shroud. All of the nonimage features also produced by the first method, such as the cloth strengthening features and the formation of additional C-14 in the Shroud, would occur with this related or similar method.

If the energy then turned back to mass within a short period of time, no scientific principles would be violated nor would explosions occur. This conversion back to mass would also be consistent with the historical accounts of what Jesus' body could do following his resurrection.

63. For all such cross-section information in this paragraph, the neutrons were at 14 MeV. (Incidentally, carbon, oxygen and nitrogen, the principal components of linen, give off only insignificant amounts of gamma rays, protons, and alpha particles, and they do not give off any deuterium when hit by neutrons. Since these components are uniformly distributed throughout the cloth, these insignificant amounts cannot affect the contrast between the Shroud's superficial body image and the rest of the cloth. When neutrons hit nitrogen and C-13, they turn to C-14.)
64. For all such cross-section information in this paragraph, the neutrons were at thermal energy.
65. F. C. Tribbe, "Enigmas of the Shroud of Turin," *Sindon,* No. 33 (December 1984) Turin, Italy.
66. John 20: 14–19, 26; and 21: 4; Luke 24: 15–31; Mark 16: 12.

CHAPTER ELEVEN

1. These historical accounts purport to record what eyewitnesses to these events were preaching and testifying to, or were written by the eyewitnesses themselves. John 19:35; Luke 1: 1–3; John 20: 30, 31; Acts 10: 39–42; Acts 2: 22–24; Acts 1: 1–3; 2 Peter 1: 1–16. See also the statements of the apostle John attributed to him by Papias, bishop of Heirapolis (A.D. 130) and student of the apostle, concerning the book of Mark, found in Eusebius, *Ecclesiastical History* III.39: "The Elder [apostle John] used to say this also: 'Mark, having been the interpreter of Peter, wrote down accurately all that he [Peter] mentioned, whether saying the doings of Christ, not, however, in order. For he was neither a hearer nor a companion of the Lord; but afterwards, as I said, he accompanied Peter, who adapted his teachings as necessity required, not as though he were making a compilation of the sayings of the Lord. So then Mark made no mistake, writing down in this way some of the things as he [Peter] mentioned them; for he paid attention to this one thing, not to omit anything that he had heard, not to include any false statement among them.'" Papias also commented about the Gospel of Matthew: "Matthew recorded the oracles in the Hebrew [i.e., Aramaic] tongue."
2. J. W. Montgomery, *History and Christianity* (Downers Grove, IL: Inter-Varsity Press, 1971), p. 29.
3. F. F. Bruce, *The Books and the Parchments,* rev. ed., (London: Pickering and Inglis Ltd., 1963), p. 178.
4. 1 Corinthians 15:3–8; Matthew 28:9, 10, 28:16–20; Mark 16:9, 16:14–20; Luke 24:13–52; John 20:14–29; 21:1–25; Acts 1:1–9, 2:22–24, 32, 3:15, 4:33, 5:30–32, 13:30, 31.
5. 1 Corinthians 15:3–15 is recognized by all historians and scholars as having been written no later than the A.D. mid-50s. Furthermore, many scholars think that the historical accounts contained in the Acts of the Apostles and the Synoptic Gospels were written prior to or no later than the A.D. mid to late 60s. Part of their reasons is that the book of Acts closes with Paul awaiting trial under house arrest in A.D. 59, after having been in custody for two years in Rome, yet we are not told of the outcome. These historical sources make no mention at all of the very critical facts of Paul's death, Peter's death, or the complete destruction of Jerusalem. Most historians agree that these events occurred in the mid to late 60s.

The destruction of Jerusalem, the centuries-old capital and center of Jewish culture, would have been as traumatic to Jews as a civil war. Several of the books of the New Testament center on Paul's travels around the ancient world. Peter was the leader of the apostles and the rock upon which Jesus built his church. Similarly, the death of James, the brother of Jesus himself, is not mentioned. James did not believe in the divinity of Jesus (John 7:5; cf. Mark 3:21), but after he saw the resurrected Christ (I Corinthians 15:7), became a prominent leader in the Jerusalem Church. Josephus (Ant. 20.9.1) and Hegesippus (in Eusebius, *Hist. Eccl.* 2:23) both tell of the martyrdom of James in Jerusalem ca. A.D. 62 or 66. The absence of the deaths of all three of these important leaders is all the more glaring when we read of the deaths of other lesser-known individuals and martyrs, such as Stephen, in ca. A.D. 36 and James, the brother of John, in ca. A.D. 41–44, clearly mentioned in Acts 7:54–8:1 and 12:1, 2. J. Finegan, *The Archaeology of the New Testament* (London: Croon Held Ltd; Boulder, Col.: Westview Press, Inc., 1981).

6. Present-day archaeology also confirms the Old Testament accounts back until about the time of Solomon's reign. However, in 1995, Egyptologist David Rohl asserted in a landmark book that there is actually a wealth of known archaeological and historical evidence to corroborate Old Testament accounts from the time of Solomon all the way back to the times of Joseph and Jacob described in Genesis. This discovery was made after Rohl realized that the traditional chronology of certain Egyptian rulers and dynasties, established by eighteenth-century scholars, was incorrect.

   The periods or dates that we assign to Old Testament times and events actually are based upon the Egyptian dates or chronologies. People in those times did not assign dates as a certain year B.C. or A.D., as in our present system, but designated dates as occurring in a numbered year of a named ruler; that is, in the tenth year of Solomon's reign, or the fifteenth year of Ramses' reign. If the Egyptian chronology is inaccurate, the calendar years assigned to Old Testament events, will be correspondingly incorrect. Rohl asserts that with a corrected Egyptian chronology, a wealth of already discovered archaeological and historical evidence now corresponds with and corroborates the people, events and histories described over the course of many centuries in the Old Testament. D. Rohl, *Pharaohs and Kings* (New York: Crown Publishers, Inc., 1995).
7. Matthew 17:12, 2, 5–7; Mark 9:2–4; Luke 9:28, 29, 34, 35. In two other eyewitness accounts in Acts, Jesus appears to Paul and his companioins in a light from heaven. Although Jesus temporarily blinded Paul, the light did not harm his body or clothing, or those of his companions (Acts 9:3-8; 26: 13-16).
8. W. Meacham, "The Authentication of the Turin Shroud: An Issue in Archaeological Epistemology," *Current Anthropology*, 24.3, (June 1983), p. 294.
9. K. Stevenson and G. Habermas, *Verdict on the Shroud*, (Ann Arbor, MI: Servant Books, 1981), p. 151.
10. STURP scientist Alan Adler has made this vivid statement to the author in several personal communications, as well as to others.

APPENDIX A

1. G. D. Painter, introduction to the new edition, *The Vinland Map and the Tarter Relation* (New Haven and London: Yale University Press, 1995).
2. W. C. McCrone, *Chemical Analytical Study of the Vinland Map. Report to Yale University Library* (New Haven, 1974), p. 4.
3. W. C. McCrone, "Authenticity of Medieval Document Tested by Small Particle Analysis," *Analytical Chemistry* 48 (1976): 676A–679A.
4. Painter, introduction to the new edition, *The Vinland Map.*
5. McCrone, *Chemical Analytical Study of the Vinland Map*, table 2, as cited in Painter, introduction to the new edition, *The Vinland Map.*
6. Painter, introduction to the new edition, *The Vinland Map*, p. x.
7. Ibid.
8. T. A. Cahill et al., *Further Elemental Analyses of the Vinland Map, Tarter Relation, and the Speculum Historiale. Report to Yale University Beineke Library*, 1985, as cited in Painter.
9. Painter, introduction to the new edition, *The Vinland Map*, p. xi.
10. Ibid.

APPENDIX B

1. Indeed, this is how it is translated in numerous versions of the Bible, specifically the Catholic Family Edition, Douay-Confraternity, Westminster, Phillips Modern English, Monsignor Knox Edition, and Amplified Bible. See also the translation of Rev. A. Feuillet, *Shroud Spectrum International* 4 (Sept. 1982) p. 15; and the translations of M. Levesque and Bailey's dictionary found in P. Barbet, *A Doctor at Calvary* (Garden City, NY: Doubleday Image Books, 1963), p. 168. Additional references

appear in E.A. Wuenschel, "The Shroud of Turin and the Burial of Christ" (parts I and II), *Catholic Biblical Quarterly* 7 (1945) and 8 (1946), where Wuenschel asserts that "to prepare the dead for burial [is] unquestionably the meaning in John's context." Also see T. Hunter, *The Sacred Shroud* (New York: Pocket Books, 1974), p. 66; and P. Vignon, *The Shroud of Christ* (New Hyde Park, N.Y.: University Books, 1970).

APPENDIX C

1. A. Feuillet, "The Identification and the Disposition of the Funerary Linens of Jesus' Burial According to the Fourth Gospel," *Shroud Spectrum International* 4 (September 1982): 13–23. See also W. Bulst, *The Shroud of Turin* (Milwaukee: Bruce, 1957), p. 85.
2. The passage is omitted in only one Greek manuscript.
3. M. Green, "Enshrouded in Silence: In Search of the First Millennium of the Holy Shroud," *Ampleforth Journal* 74 (Autumn 1969): 321–345.
4. Bulst, *Shroud of Turin*, p. 91; P. Barbet, *A Doctor at Calvary* (Garden City, N.Y.: Doubleday Image Book, 1963), p. 167; A. Feuillet, "The Identification and the Disposition."

APPENDIX D

1. W. Meacham, "The Authentication of the Turin Shroud: An Issue in Archaeological Epistemology," *Current Anthropology* 24.3 (June 1983): 287.
2. M. Green, "Veronica and Her Veil: The Growth of a Christian Legend," *The Tablet* 31 (December 1966): 1470–1471.
3. Macarius of Magnesia, *Apocritus*, ed. C. Blondel (published posthumously by P. Foucart, Paris, 1876), p. 1, trans. T. W. Crafer (London: SPCK, 1919). See also T. W. Crafer, "Macarius Magnesius: A Neglected Apologist," *Journal of Theological Studies* 8 (1906–07): 401–423, 546–571, as stated in I. Wilson, *The Shroud of Turin*, rev. ed. (Garden City, N.Y.: Doubleday Image Books, 1979), pp. 110, 136.
4. Green, "Veronica and Her Veil."
5. Wilson, *Shroud of Turin*; M. Green, "Veronica and Her Veil."
6. Peter Mallius, canon at St. Peter's, according to Green, "Veronica and Her Veil," and Wilson, *Shroud of Turin*.
7. Wilson, *Shroud of Turin*.
8. Ibid.
9. Ibid.
10. Green, "Veronica and Her Veil."
11. Green, "Veronica and Her Veil"; Wilson, *Shroud of Turin*; and F. C. Tribbe, *Portrait of Jesus?* (New York: Stein and Day, 1983).
12. Green, "Veronica and Her Veil"; Wilson, *Shroud of Turin*; Tribbe, *Portrait of Jesus?*; R. W. Hynek, *The True Likeness* (London: Sheed and Ward, 1951); S. Runciman, "Some Remarks on the Image of Edessa," *Cambridge Historical Journal* III.3 (1931): 238–252.

APPENDIX E

1. L. A. Garza-Valdes, *The DNA of God?* (London: Hodder & Stoughton, 1998); *The Mysterious Man of the Shroud*, directed by Terry Landau, CBS documentary aired on April 1, 1997; L. A. Garza-Valdes and F. Cervantes-Ibarrola; "Biogenic Varnish and the Shroud of Turin," *L'Identification Scientifique de l'homme du Linneul Jesus de Nazareth: Actes Du Symposium Scientifique International*, Rome, 1993 (Paris: Francois-Xavier Guibert, 1995). Dr. Garza also stated this at a public conference on the Shroud of Turin held at the University of Southern Indiana on February 12, 1994, and has repeated this claim a number of times to other Shroud researchers.
2. Dr. Alan Adler, personal communications in 1996–1997
3. Ibid.
4. Ibid.
5. J. Jackson, "A Scientific Evaluation of the Shroud's Radiocarbon Date," *Shroud of Turin Conference*, Richmond, VA, June 18–20, 1999; B. Walsh, "The 1988 Radiocarbon Dating Reconsidered," *Shroud of Turin Conference*, Richmond, VA, June 18–20, 1999.
6. J. Jackson, "A Scientific Evaluation."
7. H. E. Gove, S. J. Mattingly, A. R. David, and L. A. Garza-Valdes, "A Problematic Source of Organic Contamination of Linen," *Nuclear Instrument and Methods in Physics Research B* 123 (1997): 504–507.
8. Gove et al., "A Problematic Source," p. 505.
9. *The Mysterious Man of the Shroud*.
10. Ibid.
11. Ibid.
12. Adler, personal communication, December 20, 1997.
13. Garza-Valdes, *The DNA of God?* pp. 41, 42, 45.

14. Ibid.

APPENDIX F

1. A. Ivanov and D. Kouznetsov, "Biophysical correction to the old textile radiocarbon dating results," *L'Identification Scientifique de l'homme du Linneul Jesus de Nazareth:* Actes Du Symposium Scientifique International, Rome, 1993 (Paris: Francois-Xavier Guibert, 1995), pp. 229–236; D. A. Kouznetsov, A. A. Ivanov, and P. R. Veletsky, "A re-evaluation of the Radiocarbon Date of the Shroud of Turin Based on Biofractionation of Carbon Isotopes and a Fire-Simulating Model," *Archaeological Chemistry: Organic, Inorganic, and Biochemical Analyses,* Mary Virginia Orno, ed., (American Chemical Society, 1996): 229–247; D. A. Kouznetsov, A. A. Ivanov, and P. R. Veletsky, "Effects of fires and biofractionation of carbon isotopes on results of radiocarbon dating of old textiles: the Shroud of Turin," *Journal of Archaeological Science* 23 (1996): 109–121.
2. A. J. T. Jull, D. J. Donahue, P. E. Damon, "Factors That Affect the Apparent Radiocarbon Age of Textiles," *Archaeological Chemistry: Organic, Inorganic, and Biochemical Analyses,* Mary Virginia Orno, ed., (American Chemical Society, 1996): 248–253; A. J. T. Jull, D. J. Donahue, P. E. Damon, "Factors Affecting the Apparent Radiocarbon Age of Textiles: A Comment on 'Effects of Fires and Biofractionation of Carbon Isotopes on Results of Radiocarbon Dating of Old Textiles: The Shroud of Turin,' by D. A. Kouznetsov et al.," *Journal of Archaeological Science* 23 (1996): 157–160.
3. G. Salet, "To put an end to Ivanov and Kouznetsov's theories," *Revue Internationale Du Linneul de Turin* 3, (Hiver, 1996–1997).
4. J. Jackson to the editor, *British Society for the Turin Shroud, Newsletter* 43 (June/July 1996): 39–40; D. Kouznetsov, Letter to the Editor, *British Society for the Turin Shroud, Newsletter* 42 (January 1996): 32–36.
5. J. Jackson and K. Propp, "On the evidence that the radiocarbon date of the Turin Shroud was significantly affected by the fire of 1532," *"non fait de main d'homme"* Actes du III Symposium Scientifique International, Nice, 1997 (Paris: Marie-Alix Doutrebente, 1998), pp. 61–82.
6. Dr. Alan Adler, personal communication, February 23, 1999.

APPENDIX H

1. A. D. Adler, "Updating Recent Studies on the Shroud of Turin," *Archaeological Chemistry: Organic, Inorganic, and Biochemical Analyses,* Mary Virginia Orno, ed., (American Chemical Society, 1996): 223–228; A. D. Adler, "The Shroud of Turin—Blood Tests," interview by Linda Moulton Howe, May 23, 1999, http://earthfiles.com/earth026.html; A. D. Adler, A. Whanger, M. Whanger, "Concerning the Side Strip on the Shroud of Turin," http://www.shroud.com/adler2.htm; A. Danin, A. Whanger, U. Barach, M. Whanger, *Flora of the Shroud of Turin* (St. Louis: Missouri Botanical Graden Press, 1999) pp. 23–24; M. Guscin, *The Oviedo Cloth* (Cambridge: The Lutterworth Press, 1998) p. 30.
2. Guscin, *The Oviedo Cloth,* p. 56; Adler, "The Shroud of Turin—Blood Tests"; Danin et al., *Flora of the Shroud,* pp. 23–24.
3. Danin et al., *Flora of the Shroud,* pp. 11–15, 23, 24, 37.
4. Guscin, *The Oviedo Cloth,* p. 22.
5. I. Wilson, "Controversy Over Oviedo Cloth Radiocarbon Dating," *British Society for the Turin Shroud Newsletter* 50 (November 1999): 12. See also Guscin, *The Oviedo Cloth,* pp. 76–84.
6. Guscin, *The Oviedo Cloth,* p. 77.

ACKNOWLEDGMENTS

I couldn't begin to state all the areas or subjects in which Joe Marino provided information and resources throughout this entire project. His acquisition, organization, and information on the Shroud is unsurpassed. I easily visited him hundreds of times, and on just about every trip went back with copies of materials for use in the book. Our conversations on the phone probably exceeded a thousand, and whenever I inevitably inquired about an item of information, no matter how small or obscure, not only had he heard of it, he was well aware of its contents and would then invariably say, "I think I know where I have that," and sure enough he would have it.

Even more impressive than the breadth of his material and his knowledge of it was his willingness to share it with everyone. He has provided this information to people around the world through his continuous correspondence, updates, and responses. He has built or substantially enhanced entire libraries on this subject by his willingness to copy any and everything in his library for the use of others. His dedication to and knowledge of this subject is unsurpassed, and whatever contributions are made by this book, my good friend Joe deserves a substantial amount of the credit.

Words cannot describe my indebtedness to Art Lind, a McDonnell Douglas Fellow and experimental physicist who studied chemical structures of polymers over a long career with McDonnell Douglas Aerospace and the Boeing Company. His insights and explanations as to what may have been occurring in the experiments of Mario Maroni and Jean-Baptiste Rinaudo, and to the Shroud during its image-encoding event and the years following, were invaluable. He codeveloped the Historically Consistent Method and provided countless hours of expertise and explanations on a range of scientific matters that I could never have afforded or acquired elsewhere.

The breadth of his knowledge and understanding in the scientific areas relating to the Shroud was greater than any one scientist I ever came across. This is all the more remarkable in light of the fact his exposure to the Shroud has only been over the last several years. His explanations were also the most concise and understandable that I ever received. He has two other qualities that sometimes can be rare for scientists or any other people—he has an abundance of common sense, and is one of the nicest guys you'll ever meet. Of course, he never thought of charging me for anything even though he is now a scientific consultant. Yet, far more substantively than financially, I owe him an enormous debt—that I could never repay. We all owe him an enormous debt.

I would like to thank Marla Watson and Martha Baker, who each rewrote about half the book. They had a very difficult task of rewriting and editing a very complicated and technical subject written by a very redundant first-time author. Marla's expertise was in the more technical scientific and medical chapters. She organized and rewrote this technical information in a superb fashion, while enhancing its substantive points and arguments. Martha rewrote the first as well as the last several chapters. She taught me to put more of myself into the book and made the writing much smoother. If there's one mistake I made in writing the book, it was not listening to everything these ladies suggested—instead of accepting only most of their suggestions. Their rewriting and editing abilities, along with their general advice about writing and publishing, were always impressive. Whatever overused adjectives, redundant passages, and complicated tedious evidence is found in this book is the fault of the author or the material.

Debbie Bauer is a lawyer who has either been a neighbor or worked in the same office throughout most of this project. She has helped me in countless ways during this time. She has edited several chapters, typed parts of proposals, fixed the computer software and hardware, expanded and watched over the office while I was absent for months at a time, handled many of my legal cases in addition to her own, and did a great job at all of it. She displayed enormous patience and flexibility allowing me to meet every real and perceived deadline throughout. She was far more patient and helpful than I can say, or ever deserved. She easily has the finest combination of toughness and sweetness, and is more fun than any attorney I know.

Esther Noble has been my office landlady and friend throughout this project. She probably doesn't charge anyone in the building the going market price for their offices. She's established a scholarship for students where she used to teach. She has also loaned money to me and many others on generous terms to help us through tight business situations. Only because of her actions was I able to devote most of my time to writing this manuscript while carrying on a law practice to support myself and family. All the while she has continued to accumulate very sizable assets through her financial wizardry. She won't like me telling you, but she is eighty-seven years young and is the most unforgettable person I ever met. She is the ideal capitalist that I would like to become.

Dick and Sandy Nieman have made many sizable donations to advance scientific knowledge and research on the Shroud, even though some of the recipients did not attempt to complete their stated or required tasks. Yet Dick and Sandy continued to quietly support research in a number of areas and sponsor conferences and publications on the Shroud. Some of the most important evidence on the subject of the Shroud has resulted from their long-time support and interest. Their support is the rarest and most difficult to find, and is appreciated almost as much as their long-time friendship.

John Schulte has reproduced hundreds of slides on my behalf throughout the entire project. This has entailed countless trips, purchases, and retakes, which he admirably performed. John, Dick, Joe, Chuck, and myself and others have borrowed and loaned each other numerous articles, photos, tapes, etc., and spent hours and hours reviewing ideas and points. His wonderful wife, Mary, has not only endured this, but has always welcomed and encouraged us. John has given many presentations on the Shroud and would do just about anything to advance the understanding of the subject.

Chuck Hampton has a long-time interest in the Shroud and a degree in chemistry. We conducted numerous experiments in his basement, which taught us a lot and helped lead to bigger projects by far-more-qualified scientists. Chuck was the initial sounding board for many scientific ideas. Our many conversations on scientific and religious concepts helped develop and advance the progress of image-forming models and methods. He and his wife Sandy's friendship have always been a joy. Sandy endured all our experiments and fed me many suppers; her folks even put up and endured other Shroud researchers for a week at their house.

Ed Cherbonnier, former head of the Department of Religion at Trinity College in Hartford, Connecticut, has reviewed, corrected, and edited several chapters. He discovered and sent me several important sources, donated his own money for research, and spent many hours behind the scenes trying to raise additional sums. His class and diplomacy in this regard was sometimes hampered by my lack of such comparable qualities. We have maintained a mutual interest, friendship, and correspondence for a number of years despite his living modestly half the time overseas. His knowledge, grace, and kindness have always been admirable and helpful.

I would like to thank fellow attorneys Steve Andreyuk and Mike Brazil, who gave advice on and edited the last chapter. Their merger into our office has not only lightened the financial overhead, but their advice and handling of some of my legal cases has brought me added income and valuable time to complete the manuscript. In addition, I greatly appreciated their flexibility and tolerance in allowing our secretary to type various drafts in order to meet publishing deadlines.

I would especially like to thank our legal secretary, Patti Wolfe, for single-handedly taking care of four lawyers and large parts of the manuscript. We have given her a job that two people should be sharing, yet she has maintained her grace, met every deadline and assignment, and done it all efficiently. Worse yet, she has had to endure my lack of patience with computer hardware and software that does not always perform as represented, and whose programs, formatting, and/or memories always find a way to be incompatible, regardless of what you try. Patty's biblical knowledge has also proven to be very helpful on a number of occasions. She and her husband, Richard, have always given me encouragement and support in a lot of ways.

I would also especially like to thank Carol Esmar, who voluntarily typed several drafts of several chapters. She typed some of the most difficult parts entailing early drafts of chapters or major revisions. She and her husband, Jack, are long-time friends whose interest and help have always been forthcoming. Carol, who has worked for several years at McDonnell Douglas and the Boeing Company, first put me in touch with the scientist who led me to Art Lind.

I want to sincerely thank Jim and Delores Truetken, David and Joyce Meyer,

Charles and Carolyn Parlato, Jack Markwardt, and Ben Wiech for their donations, which helped complete and advance important scientific research that has been included in this book. My thanks are also extended to Dick Fennell, who helped this project in a variety of ways, and to Don Voss, for introducing me to Jim and Dick. I want to thank Beth DeShelter for thinking of the title of the book. Debbie Deadrick spent more time typing this manuscript than anyone and typed the hardest parts—the very earliest drafts. Her unsurpassed ability and enthusiasm were extremely valuable in those first several years. I am also indebted to Roger Carlson, who spent many hours of production time on the book. I greatly appreciated the excellent quality and extensive efforts of Maria Avila Ottinger and Joe Bonefeste, whose translations of foreign medical and scientific articles into English were extremely halpful. Thanks also to Jim Chickos and Jack Uang for their experiments, and to Jim for spending many hours discussing chemical reactions with me.

I want to thank my friend and fellow author Howard Rosenthal, whose interest and advice throughout the entire project has been particularly helpful in several areas. My friends Bob Goby and Mike Noltkamper have also given me valuable advice about corpses, rigor mortis, and methods that attempt to produce images with such characteristics. I would also like to thank the following friends who spent a number of hours reading or typing chapters and offering advice: my brother Steve, Wanda Snedeker, Carla Fletcher, Christine Barks, Vicki Knepper, Carol Williams, Tom Kitta, Lesa Gaebeline, Laura Brown, Debbie Hickerson, Nancy Muchnick, Charlie Bowden, and Fred and Nancy Moore. I'm grateful for all the quarterly reports, income tax filings, and free advice given by Joan Willenberg and Dick Schmidt through the years, and to Mart and Sue Markwell for the same things and for being there in the good times and bad times. Many thanks to Ron Buretta and Bruce Purdom for their valuable legal investigative work, and to the whole gang in the 3550 building for all their years of interest and patience.

I am very grateful to all the scientists, doctors, and so many other professionals who allowed me to interview them for sometimes hours at a time, and for countless times over two decades. Their dedication, brilliance, and patience has been a source of inspiration from the beginning. I can't begin to name or express my gratitude to all of the Shroud researchers and enthusiasts with whom I've developed long and lasting friendships over the course of twenty years, many of which are cited in the book. All have advanced the research on the Shroud by sharing, evaluating, and debating the evidence, whether they published it or not. We have all been a part of a larger ongoing effort to ascertain the truth about a subject and its initiating events or cause, which could be the most pivotal and relevant issue of this or any age.

I would like to note the typical dedication of one Shroud researcher, Dorothy Crispino. She spent years acquiring many important articles in other languages and/or from other eras, which still have great relevance today, and then translated and published them in English for the first time in her journal, *Shroud Spectrum International.* She did this in addition to acquiring, writing, and publishing the latest scholarship and research on a wealth of other Shroud subjects. She lived quite modestly and lost significant sums of money for many years while doing this, so that all this very important research could become known to a new generation and large segment of researchers. Almost all Shroud researchers and authors have used these articles for multiple purposes for many years. Because of the efforts of this energetic octogenarian, a much larger world public now benefits from this extensive research.

I could not close without thanking my publisher, George de Kay, for the graciousness and kindness that he has always shown me. I want to particularly thank Rik Schell, the book's editor, whose knowledge, insight, and questions very definitely shaped and improved this book.

Most of all, I would like to thank my wife, Mary Rose, who has endured a number of concerns and worries over the last eight years because of this project. During the project's very worst times she dug in and supported me and the book more than ever. In addition, she typed healthy portions of the manuscript and helped edit the beginning and ending chapters. Moreover, she was the principal rewriter for the most important sentences in the book—the last two sentences.

PHOTO CREDITS

Fig. 1. Courtesy Brooks Institute
Fig. 2. Copyright 1978, Vernon Miller
Fig. 3. Copyright 1978, Vernon Miller
Fig. 4. Copyright 1978, Vernon Miller
Fig. 5. Copyright 1978, Vernon Miller
Fig. 6. Copyright, 1978, Vernon Miller
Fig. 7. Courtesy Dr. Sebastiano Rodante
Fig. 8. Copyright 1978, Vernon Miller
Fig. 9. Courtesy Dr. Sebastiano Rodante
Fig. 10. Copyright 1978, Vernon Miller
Fig. 11. Courtesy Ian Wilson
Fig. 12. Tipolitografia F. lli Scaravaglio e C.
Fig. 13. Vernon Miller
Fig. 14. Copyright 1978, Vernon Miller
Fig. 15. Copyright 1978, Vernon Miller
Fig. 16. Copyright 1978, Vernon Miller
Fig. 17. Courtesy Pauline Books & Media
Fig. 18. Copyright 1978, Vernon Miller
Fig. 19. Dr. Robert Bruce-Chwatt
Fig. 20. Dr. Robert Bruce-Chwatt
Fig. 21. Courtesy Dr. Sebastiano Rodante
Fig. 22. Courtesy Dr. Sebastiano Rodante
Fig. 23. Wellcome Institute for the History of Medicine
Fig. 24. Courtesy Holy Shroud Guild
Fig. 25. Copyright 1978, Vernon Miller
Fig. 26. Copyright 1978, Vernon Miller
Fig. 27. Copyright 1978-81, Ernest Brooks II, Vernon Miller, and Barrie Schwortz
Fig. 28. Courtesy Jean Lorre, Jet Propulsion Laboratory
Fig. 29. Copyright 1978, Vernon Miller
Fig. 30. Courtesy Holy Shroud Guild
Fig. 31. Vernon Miller
Fig. 32. Vernon Miller
Fig. 33. Vernon Miller
Fig. 34. Courtesy Dr. Alan Adler
Fig. 35. Courtesy Dr. Alan Adler
Fig. 36. Courtesy Dr. Gil Lavoie and Shroud Spectrum International
Fig. 37. Courtesy Dr. Gil Lavoie and Shroud Spectrum International
Fig. 38. Courtesy Dr. Gil Lavoie and Shroud Spectrum International
Fig. 39. Courtesy Dr. John Jackson and Optical Society of America
Fig. 40. Courtesy Dr. John Jackson and Optical Society of America
Fig. 41. Courtesy Dr. John Jackson and Optical Society of America
Fig. 42. Courtesy Dr. John Jackson and Optical Society of America
Fig. 43. Courtesy Dr. Eugenia Nitowski
Fig. 44. Courtesy Dr. Eugenia Nitowski
Fig. 45. Courtesy Dr. Eugenia Nitowski
Fig. 46. Courtesy Dr. Sebastiano Rodante and Shroud Spectrum International
Fig. 47. Courtesy Dr. John Jackson and Optical Society of America
Fig. 48. Courtesy Dr. John Jackson and Optical Society of America
Fig. 49. Courtesy Dr. Walter McCrone
Fig. 50. Courtesy Dr. John Jackson and Optical Society of America
Fig. 51. Courtesy Dr. John Jackson and Optical Society of America
Fig. 52. Courtesy Dr. John Jackson and Optical Society of America
Fig. 53. Courtesy Dr. John Jackson and Optical Society of America
Fig. 54. Courtesy Dr. John Jackson and Optical Society of America
Fig. 55. Courtesy Dr. John Jackson and Optical Society of America
Fig. 56. Courtesy Dr. John Jackson and Optical Society of America
Fig. 57. Elmar R. Gruber, Bretten
Fig. 58. Elmar R. Gruber, Bretten
Fig. 59. Copyright © 1994 by Lynn Picknett & Clive Prince. Reprinted by permission of HarperCollins Publishers, Inc.
Fig. 60. Courtesy Kevin Moran
Fig. 61. Nicholas Allen
Fig. 62. Courtesy Kevin Moran
Fig. 63. Nicholas Allen
Fig. 64. Courtesy Kevin Moran
Fig. 65. Nicholas Allen
Fig. 66. Nicholas Allen
Fig. 67. Gernsheim Collection of the University of Texas at Austin, reproduced by Remi Van Halst.

Fig. 68. P. Vigorureux
Fig. 69. Photo Researchers, Inc.
Fig. 70. Courtesy Dr. John Jackson and Optical Society of America
Fig. 71. Courtesy Dr. John Jackson and Optical Society of America
Fig. 72. Courtesy Dr. John Jackson and Optical Society of America
Fig. 73. Courtesy Dr. John Jackson and Optical Society of America
Fig. 74. Vernon Miller
Fig. 75. Courtesy Pauline Books & Media
Fig. 76. Fr. Francis Filas
Fig. 77. Fr. Francis Filas
Fig. 78. Fr. Francis Filas
Fig. 79. Dr. Max Frei and Shroud Spectrum International
Fig. 80. Dr. Max Frei and Shroud Spectrum International
Fig. 81. Dr. Max Frei and Shroud Spectrum International
Fig. 82. Dr. Max Frei and Shroud Spectrum International
Fig. 83. Dr. Max Frei and Shroud Spectrum International
Fig. 84. Dr. Max Frei and Shroud Spectrum International
Fig. 85. Courtesy Dr. Avinoam Danin and Dr. Alan Whanger
Fig. 86. Israel Exploration Journal
Fig. 87. British Museum
Fig. 88. Werner Bultst, S.J.
Fig. 89. British Museum
Fig. 90. Courtesy Shroud Spectrum International
Fig. 91. Art Resource
Fig. 92. Courtesy Ian Wilson and Frank Tribbe
Fig. 93. Art Resource
Fig. 94. Art Resource
Fig. 95. Art Resource
Fig. 96. Art Resource
Fig. 97. Art Resource
Fig. 98. Courtesy the Ampleforth Journal
Fig. 99. Courtesy the Ampleforth Journal
Fig. 100. Courtesy the Ampleforth Journal
Fig. 101. John Gitchell
Fig. 102. Art Resource
Fig. 104. Lennox Manton
Fig. 105. Art Resource
Fig. 107. Courtesy Shroud Spectrum International
Fig. 108. John Gitchell
Fig. 110. Courtesy Dr. Alan Whanger and the Holy Shroud Guild
Fig. 111. Courtesy Dr. Alan Whanger and the Holy Shroud Guild
Fig. 112. Courtesy Dr. Alan Whanger
Fig. 113. Courtesy Dr. Alan Whanger
Fig. 114. Courtesy Dr. Alan Whanger
Fig. 115. Courtesy Dr. Alan Whanger
Fig. 116. Courtesy Ian Wilson
Fig. 117. Copyright 1978, Ernest Brooks, II
Fig. 118. Grabar, la Sainte Face de Laon
Fig. 119. Grabar, la Sainte Face de Laon
Fig. 120. Courtesy Holy Shroud Guild
Fig. 121. Vernon Miller
Fig. 122. Pray Mss. Fol. 27 v., Budapest
Fig. 123. Ian Wilson
Fig. 124. Courtesy Ian Wilson
Fig. 125. Courtesy Dan Scavone
Fig. 126. Courtesy Ian Wilson
Fig. 127. British Museum
Fig. 128. Courtesy Br. Bruno Bonnet-Eymard
Fig. 130. Bibliotheque Nationale
Fig. 131. Courtesy Ian Wilson
Fig. 132. Courtesy Dr. Art Lind
Fig. 133. Aurelio Ghio
Fig. 134. Vernon Miller
Fig. 135. Giuseppe Enrie and Aldo Guerreschi
Fig. 136. Courtesy Dr. James Chikos
Fig. 137. Courtesy Dr. Gil Lavoie and Shroud Spectrum International
Fig. 138. Courtesy Dr. Gil Lavoie and Shroud Spectrum International
Fig. 139. Courtesy Dr. Gil Lavoie and Shroud Spectrum International

INDEX